Pollen:
Development and Physiology

Pollen:
Development and Physiology

Edited by

J. HESLOP-HARRISON

LONDON BUTTERWORTHS

THE BUTTERWORTH GROUP

ENGLAND
Butterworth & Co (Publishers) Ltd.
London: 88 Kingsway WC2B 6AB

AUSTRALIA
Butterworth & Co (Australia) Ltd.
Sydney: 20 Loftus Street
Melbourne: 343 Little Collins Street
Brisbane: 240 Queen Street

CANADA
Butterworth & Co (Canada) Ltd.
Toronto: 14 Curity Avenue, 374

NEW ZEALAND
Butterworth & Co (New Zealand) Ltd.
Wellington: 49/51 Ballance Street
Auckland: 35 High Street

SOUTH AFRICA
Butterworth & Co (South Africa) (Pty) Ltd.
Durban: 33/35 Beach Grove

First published 1971

© The several contributors named in the List of Contents, 1971

ISBN 0 408 70149 8

Filmset and printed Offset Litho by
Cox & Wyman Ltd.
London, Fakenham and Reading

Foreword

If we can talk of a primogeniture in sporology, that right must be bestowed upon pollen physiology, the oldest branch not only of sporology but of plant physiology as a whole. It was the intelligent precursors of plant physiologists of today who in their native country Mesopotamia made the startling discovery that the output of dates increased if flowering branches of what we call male trees were placed among fertile, date-producing specimens.

Now, several thousand years after the Mesopotamian palynological prophase, modern palynologists gratefully realise that they are in a very enviable scientific position: their objects of study, pollen grains and spores, all constitute, so to say, an *alter ego* of the plant that produced them, a sort of a world, a *palynogea*, of its own. They can be amassed in great quantities; their shapes are usually very regular; they are not too large and difficult to handle; and like sea-urchin eggs they can easily be experimented with.

The reviews and abstracts published in this book constitute an up-to-date synopsis of new acquisitions and ideas within the fertile field of pollen physiology, concerning the living pollen grains, their growth and essential biological functions. Nevertheless a tremendous amount of work still remains to be done. In this connection Linnaeus's well-known words, 'Natura in minimis maxima'—'Nature is greatest in its very smallest details'—will no doubt appear to be particularly true. Difficulties will, however, tend to increase *ad infinitum* when, leaving relatively large items behind us, we proceed to successively smaller subjects. Will there ever be a real end, a static skeleton of thoroughly unmasked details? Or will, paradoxically yet perhaps even more likely, the sense of Shelley's words, 'Naught may endure but mutability', remain the ultimate fate or principle of Science?

August 1970

G. ERDTMAN,
Palynological Laboratory,
Solna,
Sweden

Preface

In the late summer of 1969, three meetings on various aspects of the physiology, biochemistry, structure and genetics of pollen were held in the State of Washington, U.S.A., two in Pullman, and one in Seattle. The occasion brought together biologists engaged in pollen research from many parts of the world, and offered the opportunity of reviewing the field in an unusually comprehensive manner. This volume is based largely upon papers presented at these meetings; but it is not simply a verbatim record. The authors of the principal papers have extended their texts to provide reviews covering several major areas of pollen research, and additional material has been included to provide some cover of certain topics not touched upon in the meetings themselves. The product, although by no means a compendium of current work on pollen, is thus more in the nature of a textbook than a volume of proceedings. The research reports given at the meetings have not been neglected, however, and extended abstracts of these are included in the text.

The two meetings held at Pullman were the Divisional Symposium of the Pacific Division of the American Association for the Advancement of Science (August 20) and an International Conference on Pollen and Pollen Physiology (August 21–22). The meeting at Seattle was the Symposium on Microsporogenesis and the Fine-structure of Pollen (August 25), which formed part of the work of the XIth International Botanical Congress.

The participants in the meetings at Pullman were warmly appreciative of the contribution of Professor Adolph Hecht of the Department of Botany, Washington State University. As Chairman of the A.A.A.S. Pacific Division Meeting Committee, Dr Hecht arranged the Divisional Symposium, and he was also responsible for the local organisation of the International Conference on Pollen and Pollen Physiology. This Conference was the successor to that held in 1964 at the University of Nijmegen, The Netherlands, planned by Professor H. F. Linskens and his colleagues. The Pullman Conference was under the general auspices of the A.A.A.S., and I know that I speak for all those who took part in the meeting in thanking the Association for its invaluable support.

J. Heslop-Harrison

Royal Botanic Gardens,
Kew,
Richmond, Surrey

Contents

SECTION THREE

POLLEN AND POLLEN TUBE METABOLISM

SECTION FIVE

INCOMPATIBILITY

Review

Abstracts

Nucleus and Cytoplasm in Microsporogenesis

Physiology and Biochemistry of Meiosis in the Anther

Jörg J. Sauter, *Institute of Forest Botany, University of Freiburg, West Germany*

INTRODUCTION

Meiosis is one of the most striking cellular activities. It consists of a series of finely co-ordinated physiological, biochemical, cytological and morphological processes which accompany the principal event: the reduction of the diploid parental chromosome set to the haploid set. Meiosis in the anther of Angiosperms provides an excellent system for studying the events associated with the different meiotic stages, because the meiocytes are easily accessible and are available in large numbers in a fairly well-synchronised state. Recent reviews dealing with aspects of meiosis in the anther include those of Linskens [32], Taylor [62] and Vasil [66]. The present article is restricted mainly to an evaluation of the more recent findings on the physiology and biochemistry of meiosis in anthers, with special regard to DNA, RNA and protein synthesis and the relation of those activities to different meiotic developmental stages. Consideration of other typical meiotic events, such as chromosome pairing, formation of chiasmata, exchange of genetic material and structural organisation of meiotic chromosomes, has been excluded, and the role of the tapetum in microsporogenesis is not touched upon, since aspects are dealt with in the article by Echlin elsewhere in this volume (p.41).

Considering the fact that during meiosis the DNA in the meiocytes is engaged in two different groups of activities, namely in the replication, recombination and distribution of the genetic information to the four haploid microspores, and in the control of this process itself by ordered release of information, our interest may be focused mainly on the following problems:

(*a*) How is meiosis initiated in the anther?

3

(*b*) What are the principal biochemical events occurring during the different meiotic stages?

(*c*) Does the meiotic chromosome condensation render the whole cell inactive with respect to RNA and protein synthesis, or is there an ordered release of information controlling the meiotic processes itself?

(*d*) How is the energy supply, which depends on the maintenance of the activity of different enzyme systems, maintained during those stages in which the DNA-dependent RNA and protein synthesis seem to be drastically reduced?

The results so far obtained with the aid of biochemical, cytochemical and autoradiographic methods are still somewhat contradictory, and in evaluating different reports it is essential to pay special attention to experimental conditions.

THE INDUCTION OF MEIOSIS

There is very little direct information on the induction of meiosis in anthers. During the premeiotic stage the sporogenous cells divide by asynchronous mitoses. When a certain number of pollen mother cells (PMCs) is formed, the somatic mitoses stop, and the premeiotic interphase begins, followed by a remarkable synchronisation of all further meiotic events. This fact is well known, but it is unknown how these early meiotic events are initiated in the anther.

A striking phenomenon is the accumulation of RNA and proteins in the premeiotic stage, which has been demonstrated autoradiographically, biochemically, cytochemically and also electron microscopically by several authors, to be mentioned later. Foster and Stern [14, 15], Stern [55] and Hotta and Stern [20] found transient peaks in concentration of presumed nucleic acid precursors in *Lilium* and *Trillium*, particularly deoxyribosides, which appeared to precede the premeiotic DNA synthesis as well as the microspore mitosis. Nevertheless, it does not appear to be reasonable to assume that such changes in concentration of nucleosides are involved in the *induction* of meiosis. From the results of Stone, Miller and Prescott [56] concerning mitosis in *Tetrahymena* we must rather assume that changes in the nucleoside pool size do not initiate meiosis, but that the initiation of meiosis does involve changes in the nucleosidic pool [13]. Addition of nucleic acids to the nutrient media was reported as having a promoting effect on the progress of meiosis, but failed to induce meiotic processes in excised anthers in the experiments of Vasil [64, 65] and Pereira and Linskens [44].

Furthermore, the free or protein-bound sulphhydryl groups which are considered to play an important role in the metabolism of the dividing cell have been shown to undergo remarkable variations in concentration during meiosis [34, 54]. Pereira and Linskens [44], however, investigating the significance of sulphhydryl compounds for induction and progress of meiosis, concluded that the concentration of sulphhydryl groups is not the inducing factor. But it should be kept in mind that all these studies were impaired by difficulties in culturing premeiotic anthers.

Since meiosis is not initiated in anthers when excised during premeiotic stages, one has to assume that the stimulus for the induction of meiosis in the PMCs originates probably in some other part of the plant, and is transmitted then to the anther. This view is supported by the earlier experiments of Gregory [17], who cultured small branches of tomato plants bearing flower buds at different stages of development: branches bearing premeiotic anthers showed normal pollen development only if they were grafted on to healthy flowering stocks. Similarly, Clutter and Sussex [5] have concluded from their experiments on the culture of excised fern leaves that 'meiosis requires a stimulus which arises in the vegetative shoot of the plant'. It is difficult, however, to imagine how such a meiosis-inducing substance passes through the prospective tapetal cells, triggering the meiosis only in the prospective PMCs. In this connection, the recent findings of Kanatani *et al.* [28] in the starfish *Asturias amurensis* may be of some interest. Here it was found that a neural substance, a polypeptide, acts on the ovary, inducing the production of a second substance, which was shown to be a meiosis-inducing factor, *sensu strictu*. This meiosis-inducing substance could be isolated and identified: it proved to be heat-stable, and not to be a peptide, but most probably 1-methyladenine.

On the other hand, it should not be overlooked that in some micro-organisms—for instance, baker's yeast—meiosis is directly triggered by changes in environmental conditions, such as the disappearance of glucose from the growth medium. Among others, Croes [6, 7] showed that this event initiates sequences of new synthetic and enzymatic activities, not to be discussed here, which lead to meiosis. On the basis of these findings, it may be asked whether the tapetum, which surrounds the PMCs in the anthers and which is differentiated just at the beginning of meiosis, could bring about such a glucose-starvation or other starvation effect in the PMCs, inducing in a similar way the early meiotic events. In this context, it is noteworthy that Taylor [62] emphasises that DNA synthesis starts at the top of the anthers, the place where such starvation effect would be expected to appear at first.

It is obviously much too early to speculate on whether an inducing substance or inducing conditions are the factors triggering meiosis in the anther. It may be hoped, however, that progress in experimental technique, especially in culturing PMCs from the G_1-stage throughout the meiotic prophase, will cast some light on these very fundamental but still unsolved problems of anther development.

SOME PRINCIPAL PHYSIOLOGICAL AND BIOCHEMICAL EVENTS DURING MEIOSIS IN THE ANTHER

Once the meiosis is triggered, ordered sequences of various synthetic activities occur in the PMCs during meiotic development which now seem to be controlled by ordered release of genetic information in the meiocytes itself rather than by environmental conditions. This is demonstrated by the illuminating studies of Hotta, Stern and co-workers on meiocytes cultured *in vitro* [22, 25, 26, 27, 43]. Some of these synthetic activities, DNA, RNA and protein synthesis, may be considered below.

DNA synthesis

In PMCs, synthesis of DNA is usually found to be completed before the onset of the meiotic prophase, or at least by the end of the leptotene. This was shown by the early microphotometric measurements of Feulgen-stained nuclei in *Zea mays, Lilium, Tradescantia* and *Paeonia* [30, 57, 58]. Similar results were obtained in the autoradiographic studies of Plaut [45], Taylor and co-workers [41, 59, 60, 61, 63] and De [12] in several monocotyledonous plants. In none of these studies was there any indication of synthesis of DNA at stages later than early leptotene, although we must assume that the tracers would have moved freely into the meiocytes at least until pachytene. This was confirmed by Heslop-Harrison and Mackenzie [19], who used an autoradiographic method suitable for locating soluble tracers.

Besides this nuclear labelling, the occurrence of a cytoplasmic labelling was described in *Agapanthus* meiocytes by Lima-de-Faria [31] after the uptake of ^{3}H-thymidine. This cytoplasmic labelling was shown to be due, at least in part, to the synthesis of DNA in cytoplasmic organelles.

Furthermore, there are biochemical data on the changes in DNA content of PMCs during meiosis which do not agree at all with either the cytophotometric or the autoradiographic results. Sparrow *et al.* [53] described a continuous increase of DNA content in PMCs during the meiotic prophase in *Trillium*. Linskens and Schrauwen [35] recently reported a similar increase in DNA content of PMCs during meiosis. They found, for instance, a more than threefold increase in DNA content from leptotene until pachytene in the PMCs of one anther. Actually, we have no pertinent explanation for these observations. It cannot be decided yet whether the disagreement of the biochemical and the autoradiographic results is caused by changes in permeability of the PMCs for the labelled DNA precursors, as was argued by Linskens and Schrauwen [35], or whether it could be simply explained by variation in the number of PMCs squeezed out from a certain number of anthers at different meiotic stages.

Much more convincing are the results of Hotta, Ito and Stern [24] in *Lilium* and *Trillium*, which indicate that a very small amount of DNA remained undetected in the autoradiographic studies. These authors obtained an incorporation of ^{32}P- and ^{3}H-thymidine into DNA of PMCs during zygotene and pachytene. The amount, however, was only 0·3% of the DNA replicated during preleptotene. Initially one may have thought that this very small amount of DNA synthesis in zygotene and pachytene could be due to contamination with tapetal nuclei or with PMCs or earlier stages than leptotene, but inhibition experiments, as well as the characterisation of this DNA, clearly revealed that this delayed synthesis is an essential feature in meiosis. Inhibiting the delayed DNA synthesis with deoxyadenosine, the same authors found either zygotene arrest or fragmentation of chromosomes, depending on the time at which inhibition was effected [27].

The interpretation of these observations is still somewhat problematic. The simplest interpretation would be that crossing over occurs during the meiotic prophase and that DNA synthesis is a consequence of a break-repair mechanism. But the zygotene–pachytene DNA was found to have a lower average molecular weight and a somewhat altered base ratio. Furthermore, it seemed to be double-stranded and to be present in somatic nuclei. For these reasons the authors favour the interpretation that at least part of the DNA synthesised during the meiotic prophase is a chromosomal component which was not replicated during the premeiotic S-period. They speculate that 'this component is *functionally* an axial element, probably needed during the time of chromosome pairing'. This view could be supported by the recent finding of Hotta, Parchman and Stern [25], who showed that synthesis of DNA during the zygotene–pachytene stage requires simultaneous synthesis of certain nuclear proteins, 30% proving to be associated physically with the DNA. Selective inhibition of protein synthesis at the end of zygotene resulted in failure of chiasma formation [43].

Intimately related to these problems of meiotic chromosome organisation are the results on histone synthesis, which was found to occur during, as well as after, the DNA-synthesis period [3, 11, 12, 52], and the findings of Littau and co-workers [37, 39] on the role of histones in maintenance of chromatin structure, which can only be mentioned here.

In an electron microscopical study on chromatin organisation during and after synapsis in cultured microsporocytes of *Lilium*, Sen [51] recently reported that inhibition of DNA synthesis by mitomycin C at leptotene–zygotene prevented formation of synaptinemal complexes.

From all these results we may draw the conclusion that besides the main DNA synthesis period in preleptotene there is a slight synthesis of DNA and proteins during zygotene and pachytene which is obviously essential to chromosome pairing and chiasma formation. Whether the delayed DNA synthesis is caused by crossing-over processes or is involved in structural organisation of the synaptinemal complex has, however, to be elucidated by further investigations.

RNA synthesis

Remarkable changes in RNA synthesis and in RNA content were observed to occur in the PMCs during meiosis. The results so far obtained with the aid of cytochemical, autoradiographic, biochemical and electron microscopical methods are of special interest, as they can give information on (*a*) the physiological activities during meiotic stages, (*b*) whether and to what extent the condensing chromatin is able to be transcribed, and (*c*) whether the messengers coding for functions in the meiosis are present before or are only made during meiosis. However, it should be emphasised that these results do not yet agree completely.

The autoradiographic studies performed by Taylor [61] on *Lilium*, Albertini [1] on *Rhoeo discolor*, Das [8] on *Zea mays* and Sauter on *Paeonia* [47, 48] all revealed a very prominent RNA synthesis in the premeiotic stage, a remarkable and continuous drop of this synthesis during the meiotic prophase, and an arrest of any isotope incorporation into RNA from metaphase I until the end of meiosis. Whether the failure of RNA labelling from metaphase I onwards must be attributed to the sealing of PMCs with callose walls which may prevent the permeation of tracers into the cells in advanced meiotic stages, or whether there is only a very slight RNA synthesis probably of messenger RNA which may be extracted during the autoradiographic procedure, is still open to conjecture. It is of interest, however, that the results obtained in spermatogenesis of animals where meiocytes are not sealed with callose walls showed the same incorporation pattern [4, 10, 18, 40]: RNA synthesis was drastically reduced during the meiotic prophase, and stopped completely at least in metaphase and anaphase stages. This would favour the view that 'mitotic as well as meiotic chromosome condensation renders the whole cell inactive with respect to RNA synthesis' [9].

Furthermore, the autoradiographic studies showed that synthesis of ribosomal RNA becomes inhibited very early in the meiotic prophase [1, 8]. This is very consistent with the cytochemical results on RNA content of PMCs; Genevès [16], Moss and Heslop-Harrison [42] and Sauter and Marquardt [49, 50] found RNA content to decrease continuously from preleptotene until the tetrad stage. In a subtle ultrachemical investigation of RNA content of meiocytes in *Lilium* and *Trillium* which was paralleled by an observation of the ribosome population by electron microscopy, Mackenzie, Heslop-Harrison and Dickinson [38] similarly revealed that the meiocyte RNA content drastically decreases from zygotene to pachytene and still more during diplotene and diakinesis. In diplotene and diakinesis hardly any ribosomes were detected in cytoplasm by electron microscopy. Thus, the PMCs seem to become ribosome-depleted during these stages. In metaphase, however, the same authors observed a sharp increase in RNA and ribosome content, which contrasts with the autoradiographic as well as with the cytochemical results. Following the variations in nucleic acid content of a PMC fraction, Linskens and Schrauwen [35] recently found a similar increase in RNA content during metaphase in *Lilium*. The same authors [36] reported also a considerable rise in polysomes during metaphase. Mackenzie *et al.* [38] are of the opinion that the increase of ribosomes during metaphase may be due to a *de novo* synthesis at the expense of the pool of breakdown products released by degeneration of ribosomes originally present. Nevertheless, it may be questioned how metaphase chromosomes should be able to synthesise RNA.

In contrast to the autoradiographic results, Hotta and Stern [21–23] found transient peaks of RNA synthesis during the meiotic prophase, especially in pachytene–diplotene, and again in interkinesis and the tetrad stage, in *Trillium* meiocytes. This indicates that the autoradiographic studies may not have detected all of the

RNA synthesised. By the use of actinomycin, chloramphenicol, pulse labelling and nucleotide analyses of RNA digests, they could show that at least part of the RNA synthesised corresponds to mRNA. The main conclusion drawn from these studies is that meiosis obviously is accompanied by a set of phase-specific gene transcriptions. Although few genes are presumed to be active during distinct meiotic stages, the action of such genes seemed to be essential for normal meiotic development. This view is strongly supported by further experiments of Hotta and Stern [22] and those of Kemp [29], in which the effect of inhibiting RNA synthesis was investigated.

Protein synthesis

The results on protein synthesis during meiotic development in PMCs are relatively scanty. Incorporation studies came from Taylor [60, 61], who investigated the incorporation of ^{32}P- and ^{14}C-glycine into proteins of *Tulbaghia* and *Lilium* meiocytes. De [12] studied the synthesis especially of histones in *Tradescantia*. Albertini [2] and Sauter [46, 48] followed the incorporation of ^{3}H-leucine in *Rhoeo* and in *Paeonia*, respectively.

The incorporation pattern observed in these different plants revealed much the same situation: protein synthesis is most prominent in the premeiotic stage, decreases considerably when leptotene is reached, and remains quite low throughout the remainder of the meiotic prophase. As no additional incorporation occurred during spindle formation, Taylor [62] emphasises that spindle proteins must be synthesised before the spindle makes its appearance. From metaphase until the tetrad stage, protein synthesis was found to continue, in contrast to RNA synthesis, but it proved to be at its very minimum [2, 48, 61]. Similar results were obtained in spermatogenesis of animals [10, 40]. Since protein synthesis was found to occur in the absence of any RNA synthesis, it was assumed that it may be due to long-lived messenger RNA; to mitochondrial protein synthesis, which would be much less affected by meiotic chromosome condensation; or even to a supply from the tapetum [2, 40, 48]. However, further investigations are needed for the elucidation of all these problems.

The well-known results of Hotta and Stern [22] obtained by measuring the uptake of labelled amino acids in cultured meiocytes of *Trillium* contrast somewhat with the autoradiographic findings. These authors demonstrated the occurrence of definite peaks for incorporation in leptotene–zygotene, in late pachytene and in early diakinesis which were not observed in the autoradiographic studies. It may be asked whether these peaks in synthetic activity could be related to synthesis of soluble proteins, probably enzymes, which may have been extracted during autoradiographic procedure. Furthermore, by following protein synthesis when RNA synthesis was inhibited, Hotta and Stern [22] have accumulated convincing evidence that at least part of these proteins were coded for by newly synthesised messengers. The conclusion drawn from these studies, as mentioned above, is that the sequence of morphological changes associated with chromosome contraction and movement during meiosis is paralleled by stage-specific gene transcriptions.

Linskens [33], investigating the variations in protein and enzyme pattern during meiosis, confirmed this view. He showed that every meiotic stage is characterised by a specific protein pattern. Based on his observations, Linskens [33] similarly concluded that meiosis is accompanied by a series of stage-specific gene transcriptions. We should, however, have further information on the extent to which such protein synthesis may also be due to mitochondrial protein synthesis and to long-lived messenger RNA as well as to stage-specific transcription of nuclear DNA. In this connection, it may be added that mitochondrial enzymes, like succinic dehydrogenase and cytochrome oxidase, were found to remain very active throughout the meiosis in our own cytochemical studies (Sauter, unpublished).

Regarding the significance of protein synthesis during meiosis, some results obtained in the meiotic prophase are of special interest. The proteins synthesised concurrently with the replication of DNA in preleptotene were shown by De [12] to be mostly histones which are bound to DNA. Using biochemical methods, Sheridan and Stern [52] found a special 'meiotic histone' in meiocytes of lily and tulip which probably was synthesised during the same period and which persisted through meiosis and pollen development, suggesting again that it may have been intimately bound to the DNA synthesised. The autoradiographic studies have consistently revealed a labelling of chromatin in zygotene and pachytene [2, 12, 61], which indicates that proteins synthesised during this period are involved in structural organisation of chromosomes. Hotta, Parchman and Stern [25] accumulated evidence for the view that synthesis of certain nuclear proteins during zygotene and pachytene is intimately related to chromosome pairing and chiasma formation. As mentioned above, the inhibition of this protein synthesis with cycloheximide prevented DNA synthesis and caused failure of chiasma formation [43]. On the other hand, Sen [51] in his electron microscopical study of meiocytes in *Lilium* clearly showed that inhibition of DNA synthesis in leptotene and zygotene prevented the formation of synaptinemal complexes, whereas inhibition of protein synthesis did not interfere with its formation. Thus, one could conclude on a preliminary basis that prophase DNA synthesis is essential to pairing of homologous chromosomes, whereas protein synthesis is involved particularly in structural organisation of chromosomes. Whether this view would be pertinent, however, remains to be proved in further experiments.

REFERENCES

1 ALBERTINI, L., 'Étude autoradiographique des synthèses d'acide ribonucléique (RNA) au cours de la microsporogenèse chez le *Rhoeo discolor* Hance', *C.r. hebd. Séanc. Acad. Sci., Paris*, **260**, 651–3 (1965)

2 ALBERTINI, L., 'Étude autoradiographique des synthèses de protéines au cours de la microsporogenèse chez le *Rhoeo discolor* Hance', *C.r. hebd. Séanc. Acad. Sci., Paris*, **264**, 2773–6 (1967)

3 BOGDANOV, Y. F., LIAPUNOVA, N. A., SHERUDILO, A. I. and ANTROPOVA, E. N., 'Uncoupling of DNA and histone synthesis prior to prophase I of meiosis in the cricket *Grillus (Achaeta) domesticus* L.', *Expl Cell Res.*, **52**, 59–70 (1968)

4 BRASIELLO, A. R., 'Autoradiographic study of ribonucleic acid synthesis during spermatogenesis of *Asellus aquaticus* (crist. *isopoda*)', *Expl Cell Res.*, **53**, 252–60 (1968)

5 CLUTTER, M. E. and SUSSEX, I. M., 'Meiosis and sporogenesis in excised fern leaves grown in sterile culture', *Bot. Gaz.*, **126**, 72–8 (1965)

6 CROES, A. F., 'Induction of meiosis in yeast. I. Timing of cytological and biochemical events', *Planta*, **76**, 209–26 (1967 a)

7 CROES, A. F., 'Induction of meiosis in yeast. II. Metabolic factors leading to meiosis', *Planta*, **76**, 227–37 (1967 b)

8 DAS, N. K., 'Inactivation of the nucleolar apparatus during meiotic prophase in corn anthers', *Expl Cell Res.*, **40**, 360–4 (1965)

9 DAS, N. K., SIEGEL, E. P. and ALFERT, M., 'On the origin of labeled RNA in the cytoplasm of mitotic root tip cells of *Vicia faba*', *Expl Cell Res.*, **40**, 178–81 (1965 a)

10 DAS, N. K., SIEGEL, E. P. and ALFERT, M., 'Synthetic activities during spermatogenesis in the locust', *J. Cell Biol.*, **25**, 2, 387–95 (1965 b)

11 DAS, N. K. and ALFERT, M., 'Cytochemical studies on the concurrent synthesis of DNA and histone in primary spermacytes of *Urechis caupo*', *Expl Cell Res.*, **49**, 51–8 (1968)

12 DE, D. N., 'Autoradiographic studies of the nucleoprotein metabolism during the division cycle', *Nucleus, Calcutta*, **4**, 1–24 (1961)

13 DUSPIVA, F., 'Enzymatische Aspekte der Mitose'. In: *Problems of Reduplication in Biology* (Ed. P. SITTE), 120–36, Springer-Verlag, Berlin (1966)

14 FOSTER, T. S. and STERN, H., 'The accumulation of soluble deoxyribosidic compounds in relation to nuclear division in anthers of *Lilium longiflorum*', *J. biophys. biochem. Cytol.*, **5**, 178–92 (1959)

15 FOSTER, T. S. and STERN, H., 'Soluble deoxyribosidic compound in relation to duplication of deoxyribonucleic acid', *Science, N.Y.*, **128**, 653–4 (1958)

16 GENEVÈS, L., 'Évolution des infrastructures du cytoplasme dans le tissu sporifère des anthères de *Ribes rubrum* L. (Grossulariaceae)', *C.r. hebd. Séanc. Acad. Sci., Paris*, **262**, 72–5 (1966)

17 GREGORY, W. C., 'Experimental studies on the cultivation of excised anthers in nutrient solution', *Am. J. Bot.*, **27**, 687–92 (1940)

18 HENDERSON, S. A., 'RNA synthesis during male meiosis and spermiogenesis', *Chromosoma*, **15**, 345–66 (1964)

19 HESLOP-HARRISON, J. and MACKENZIE, A., 'Autoradiography of soluble (2-^{14}C) thymidine derivatives during meiosis and microsporogenesis in *Lilium* anthers', *J. Cell Sci.*, **2**, 387–400 (1967)

20 HOTTA, Y. and STERN, H., 'Transient phosphorylation of deoxyribosides and regulation of deoxyribonucleic acid synthesis', *J. biophys. biochem. Cytol.*, **11**, 311–19 (1961)

21 HOTTA, Y. and STERN, H., 'Inhibitors of protein synthesis during meiosis and its bearing on intracellular regulation', *J. Cell Biol.*, **16**, 259–79 (1963 a)

22 HOTTA, Y. and STERN, H., 'Synthesis of messenger-like ribonucleic acid and protein during meiosis in isolated cells of *Trillium erectum*', *J. Cell Biol.*, **19**, 45–58 (1963 b)

23 HOTTA, Y. and STERN, H., 'Polymerase and Kinase activities in relation to RNA synthesis during meiosis', *Protoplasmatologia*, **60**, 218–32 (1965)

24 HOTTA, Y., ITO, M. and STERN, H., 'Synthesis of DNA during meiosis', *Proc. natn. Acad. Sci. U.S.A.*, **56**, 1184–91 (1966)

25 HOTTA, Y., PARCHMAN, L. G. and STERN, H., 'Protein synthesis during meiosis', *Proc. natn. Acad. Sci. U.S.A.*, **60**, 575–82 (1968)

26 ITO, M. and STERN, H., 'Studies of meiosis *in vitro*. I. *In vitro* culture of meiotic cells', *Devl Biol.*, **16**, 36–53 (1967)

27 ITO, M., HOTTA, Y. and STERN, H., 'Studies of meiosis *in vitro*. II. Effect of inhibiting DNA synthesis during meiotic prophase on chromosome structure and behavior', *Devl Biol.*, **16**, 54–77 (1967)

28 KANATANI, H., SHIRAI, H., NAKANISHI, K. and KUROKAWA, T., 'Isolation and identification of meiosis inducing substance in starfish *Asterias amurensis*', *Nature, Lond.*, **221**, 273–4 (1969)

29 KEMP, C. L., 'The effects of inhibitors of RNA and protein synthesis on cytological development during meiosis', *Chromosoma*, **15**, 652–65 (1964)

30 KNAUER, J., 'Karyologische und absorptionsphotometrische Untersuchungen an Zellkernen verschiedener Gewebe in Antheren von *Paeonia tenuifolia*', Dissert. Univ. Freiburg (1960)

31 LIMA-DE-FARIA, A., 'Labeling of the cytoplasm and the meiotic chromosomes of *Agapanthus* with ^{3}H-thymidine', *Hereditas*, **53**, 1–11 (1965)

32 LINSKENS, H. F., 'Pollen physiology', *A. Rev. Pl. Physiol.*, **15**, 255–70 (1964)

33 LINSKENS, H. F., 'Die Änderung des Protein- und Enzym-Musters während der Pollenmeiose und Pollenentwicklung', *Planta*, **69**, 79–91 (1966)

34 LINSKENS, H. F. and SCHRAUWEN, J. A. M., 'Änderungen des Gehaltes an Sulfhydryl-Gruppen während der Pollenmeiose und Pollenentwicklung', *Biol. Pl., Praha*, **5**, 239–48 (1963)

35 LINSKENS, H. F. and SCHRAUWEN, J. A. M., 'Quantitative nucleic acid determinations in the microspore and tapetum fractions of lily anthers', *Proc. K. ned. Akad. Wet., Sect. C.*, **71**, 267–79 (1968 a)

36 LINSKENS, H. F. and SCHRAUWEN, J. A. M., 'Änderung des Ribosomen-Musters während der Meiosis', *Naturwissenschaften*, **2**, 91 (1968 b)

37 LITTAU, V. C., BURDICK, C. J., ALLFREY, V. G. and MIRSKY, A. E., 'The role of histones in the maintenance of chromatin structure', *Proc. natn. Acad. Sci. U.S.A.*, **54**, 1204–12 (1965)

38 MACKENZIE, A., HESLOP-HARRISON, J. and DICKINSON, H. G., 'Elimination of ribosomes during meiotic prophase', *Nature, Lond.*, **215**, 997–9 (1967)

39 MIRSKY, A. E., BURDICK, C. J., DAVIDSON, E. H. and LITTAU, V. C., 'The role of lysine-

rich histone in the maintenance of chromatin structure in metaphase chromosomes', *Proc. natn. Acad. Sci. U.S.A.*, **61**, 592–7 (1968)

40 MONESI, V., 'Synthetic activities during spermatogenesis in the mouse (RNA and protein)', *Expl Cell Res.*, **39**, 197–224 (1965)

41 MOSES, M. J. and TAYLOR, J. H., 'Desoxypentose nucleic acid synthesis during microsporogenesis in *Tradescantia*', *Expl Cell Res.*, **9**, 474–88 (1955)

42 MOSS, G. I. and HESLOP-HARRISON, J., 'A cytochemical study of DNA, RNA, and protein in the developing maize anther. II. Observations', *Ann. Bot.*, **31**, 555–72 (1967)

43 PARCHMAN, L. G. and STERN, H., 'The inhibition of protein synthesis in meiotic cells and its effect on chromosome behavior', *Chromosoma*, **26**, 298–311 (1969)

44 PEREIRA, R. A. S. and LINSKENS, H. F., 'The influence of glutathione and glutathione antagonists on meiosis in excised anthers of *Lilium henryi*', *Acta bot. neerl.*, **12**, 302–14 (1963)

45 PLAUT, W. S., 'DNA synthesis in the microsporocytes of *Lilium henryi*', *Hereditas*, **39**, 438–44 (1953)

46 SAUTER, J. J., 'Histoautoradiographische Untersuchung der Proteinsynthese während der Meiosis bei *Paeonia tenuifolia* L.', *Naturwissenschaften*, **55**, 187 (1968 a)

47 SAUTER, J. J., 'Histoaudiographische Untersuchungen zur Ribonucleinsäure-Synthese während der Meiosis bei *Paeonia tenuifolia* L.', *Naturwissenschaften*, **55**, 236 (1968 b)

48 SAUTER, J. J., 'Autoradiographische Untersuchungen zur RNS- und Proteinsynthese in Pollenmutterzellen, jungen Pollen und Tapetumzellen während der Mikrosporogenese von *Paeonia tenuifolia* L.', *Z. Pfl. Physiol.*, **61**, 1–19 (1969)

49 SAUTER, J. J. and MARQUARDT, H., 'Nucleohistone und Ribonukleinsäure-Synthese während der Pollenentwicklung', *Naturwissenschaften*, **54**, 546 (1967 a)

50 SAUTER, J. J. and MARQUARDT, H., 'Die Rolle des Nucleohistons bei der RNS- und Proteinsynthese während der Mikrosporogenese von *Paeonia tenuifolia* L.', *Z. Pfl. Physiol.*, **58**, 126–37 (1967 b)

51 SEN, S. K., 'Chromatin-organisation during and after synapsis in cultured microsporocytes of *Lilium* in presence of mitomycin C and cycloheximide', *Expl Cell Res.*, **55**, 123–7 (1969)

52 SHERIDAN, W. F. and STERN, H., 'Histones of meiosis', *Expl Cell Res.*, **45**, 323–35 (1967)

53 SPARROW, A. H., MOSES, M. J. and STEELE, R., 'A cytological and cytochemical approach to an understanding of radiation damage in dividing cells', *Br. J. Radiol.*, **25**, 182–8 (1952)

54 STERN, H., 'Variations in sulfhydril concentration during microsporocyte meiosis in the anthers of *Lilium* and *Trillium*', *J. biophys. biochem. Cytol.*, **4**, 157–61 (1958)

55 STERN, H., 'Aspects of deoxyriboside metabolism in relation to the mitotic cycle', *Ann. N.Y. Acad. Sci.*, **90**, 440–54 (1960)

56 STONE, G. E., MILLER, O. L. JR and PRESCOTT, D. M., 'H³-Thymidine derivative pools in relation to macronuclear DNA synthesis in *Tetrahymena pyriformis*', *J. Cell Biol.*, **25**, 171–7 (1965)

57 SWIFT, H., 'The constancy of DNA in plant nuclei', *Proc. natn. Acad. Sci. U.S.A.*, **36**, 643–54 (1950)

58 SWIFT, H., 'Quantitative aspects of nuclear nucleoproteins', *Int. Rev. Cytol.*, **2**, 1–76 (1953)

59 TAYLOR, J. H., 'Autoradiographic detection of incorporation of ³²P into chromosomes during meiosis and mitosis', *Expl Cell Res.* **4**, 164–73 (1953)

60 TAYLOR, J. H., 'Incorporation of phosphorus-32 into nucleic acids and proteins during microgametogenesis of *Tulbaghia*', *Am. J. Bot.*, **45**, 123–31 (1958)

61 TAYLOR, J. H., 'Autoradiographic studies of nucleic acids and proteins during meiosis in *Lilium longiflorum*', *Am. J. Bot.*, **46**, 477–84 (1959)

62 TAYLOR, J. H., 'Meiosis'. In: *Handbuch d. Pfl. Physiol.*, Vol. 18 (Ed. W. RUHLAND), 344–67, Springer-Verlag, Berlin (1967)

63 TAYLOR, J. H. and MCMASTER, R. D., 'Autoradiographic and microphotometric studies of desoxyribose nucleic acid during microgametogenesis in *Lilium longiflorum*', *Chromosoma*, **6**, 489–521 (1954)

64 VASIL, I. K., 'Cultivation of excised anthers *in vitro*—effect of nucleic acids', *J. exp. Bot.*, **10**, 399–408 (1959)

65 VASIL, I. K., 'Effect of nucleic acids on cell division in excised anthers', *Mem. Indian bot. Soc.*, **3**, 94–8 (1960)

66 VASIL, I. K., 'Physiology and cytology of anther development', *Biol. Rev.* **42**, 327–73 (1967)

The Cytoplasm and its Organelles During Meiosis

J. Heslop-Harrison, *Institute of Plant Development, University of Wisconsin, Madison, Wisconsin, U.S.A.*

INTRODUCTION

The haploid nuclei produced by meiosis in the anther normally express that part of the total genome appropriate to the male gametophyte, sporophytic functions remaining latent until after gamete formation and fusion. This must mean that a state of repression is imposed upon a large part of the genome, either in consequence of some intranuclear event accompanying meiosis, or because of the nature of the environment experienced by the haploid nuclei at the conclusion of the divisions. The evidence of apospory and apogamy, particularly from the lower archegoniates, adequately demonstrates that the alternation of generations is not obligatorily linked with meiosis and syngamy, so we are left with the conclusion that the sporophyte–gametophyte transition, although ultimately an expression of genetic potential, must depend in the short term upon extranuclear factors. In the anther, these factors must act through the agency of the cytoplasm inherited by the daughter cells of meiosis. This cytoplasm is derived from the quadripartition of that of the mother cell, which at one time provided the metabolic arena for the diploid parent nucleus. It may be argued, then, that there must be some rather far-reaching cytoplasmic reorganisation during the meiotic period, to prevent the carry-over of extranuclear diplophase information and to permit the institution of a new environment favourable to the expression of gametophyte functions.

It has been known for half a century or more that quite conspicuous changes do indeed proceed in the cytoplasm and organelles of the angiosperm microsporocyte concurrently with meiosis. Cytologists using light microscopy were aware that the cytoplasm underwent quite profound changes in stainability, and that the

16

organelles varied in prominence in a manner suggesting that they passed through some kind of degenerative episode during the meiotic prophase. The most notable early observations were those of Guilliermond's school [1, 2], especially those contained in the extensive studies of Py [3]. In retrospect, the results obtained in the 1920s and 1930s seem quite remarkable in view of the severe handicap of uncertain identification: at least four classes of cytoplasmic inclusion—plastids, mitochondria, spherosomes and dictyosomes—are present in the mother cells and spores, all of a size near the resolution limit of the light microscope. A major preoccupation of Guilliermond was indeed to make the distinction between the different classes of cytoplasmic inclusion, particularly between mitochondria and plastids, and to show that whatever the metamorphoses of each they were independent as organelle classes. Guilliermond was, furthermore, convinced that the 'chondriome' was composed of self-propagating elements and that continuity was preserved through the reduction division. This view stood in contrast to Wagner's [4], which was that new generations of chondriosomes arose *de novo* from hyaline cytoplasm during mother cell maturation, later to differentiate into the organelles visible in the mature pollen.

Electron microscopy has now led to a better characterisation of cytoplasmic constituents, and the continuity of plastids and mitochondria through somatic cell lineages is not in doubt. In the last few years biochemical and fine-structural information about the cytoplasmic changes associated with microsporogenesis has been accumulating, and it is now possible to interpret some of the earlier light-microscopic observations in a more precise manner. It is the purpose of this chapter to provide a brief review of the new information, particularly as it bears upon the continuity of organelles through microsporogenesis and changes in the cytoplasm perhaps connected with the sporophyte–gametophyte transition.

THE RIBOSOME POPULATION OF THE MEIOCYTE

A loss in the affinity of the meiocyte cytoplasm for basic dyes during prophase was noted by various earlier cytologists (e.g. Py, Ref. 3), but Painter [5] appears to have been the first to recognise that this is due to a reduction in RNA. Painter's observations were made on *Rhoeo discolor*; more recently the fall in cytoplasmic RNA has been confirmed in other monocotyledons [6] and in dicotyledons [7, 8], and there is every reason to believe that it is a general concomitant of meiosis in the anther. Its magnitude has been estimated by extraction methods (Table 1) and by cytophotometry (Fig. 1, p. 33 of this volume).

Ribosome counts from the cytoplasm of *Lilium* meiocytes show that the reduction in cytoplasmic RNA is due to a decline in the ribosome population (Table 1). As the cells enter the prophase

Table 1. CHANGES IN TOTAL RNA OF FIXED CELLS, CYTOPLASMIC RIBOSOME NUMBERS, NUCLEOLOIDS AND CHROMOSOME-ASSOCIATED RNA DURING MEIOSIS IN *Lilium henryi*. FROM DICKINSON AND HESLOP-HARRISON [16]

Meiotic stage	RNA per cell (μμg)	Adenine/ Guanine	Ribosome count/ unit vol.	Nucleoli	Nucleo- loids
Leptotene	60	ND	15·4	1–4	0
Zygotene	60	0·99	9·8	1–2	0
Pachytene	31	1·24	2·9	1–2 major, 'cap' type, +0–3 super- numeraries	0
Diplotene	29	1·67	1·0	1–2 major spheri- cal + a few small supernumeraries	0
Diakinesis	29	1·67	0·6	1 (rarely 2) major + a few supernumeraries; losing basiphilia	0
Metaphase I	57	ND	9·2	none visible	0
Anaphase I	ND	0·95	9·8	several among separating chromo- some groups	15–30
Dyad	ND	ND	ND	up to 20 per nucleus; wide size range	15–30
Metaphase II	114·6	ND	ND	none visible	few visible
Telophase II	ND	ND	ND	1–2 major, up to 10 smaller supernumeraries	10–20 per cell
Young spore	ND	1·03	ND	1–2 major, de- clining numbers of supernumeraries	6–10 per spore; numbers declining

Quantitative data from Ref. 6; standard errors omitted. ND, data not available.

there is an abundance of free ribosomes and an appreciable, although not large, amount of ribosomal endoplasmic reticulum. A sub-stantial fall in ribosome number follows during the zygotene–pachytene interval, coinciding in time with the loss of basiphilia detectable cytochemically. A further indication that the reduction is due to ribosome elimination is given by the fact that the A/G value for the total cell RNA rises as RNA content declines (Table 1).

The ribosome elimination does not appear to be complete, at least to the extent that this can be judged from electron micrographs. A few configurations perhaps interpretable as polysomes persist, although the cytoplasm is quite strikingly 'empty' during the pachytene–metaphase I interval.

So radical a modification in the character of the cytoplasm as that occurring during the meiotic prophase must obviously be preceded and accompanied by substantial changes in cell metabolism. Enzyme studies on extracted meiocytes have not hitherto given any notable indication of this [9], but cytochemical observations on meiocytes *in situ* do indeed show that there are substantial rises in the activity of various lytic enzymes in mid-prophase. The trend in acid phosphatase activity in the prophase meiocytes of *Cosmos bipinnatus* is illustrated in Fig. 1, p. 33 of this volume [8]; corresponding trends have been detected in cytochemically detectable ribonuclease. The magnitude of the changes are such that it may now be confidently predicted that they will be observable in extraction experiments, given suitable technique. In the last connection, it may be noted that not only is the timing of sampling important to catch what is undoubtedly a transient phenomenon, but also the need to avoid contamination, since the cytochemical results show that the trends in enzyme activity in the enveloping tapetal layers are quite different from those in the meiocytes (see also Ref. 9).

The ribosome population is restored during or immediately after the meiotic mitoses, and this process is seemingly intimately linked with the behaviour of the nucleolus during meiosis, which we may now consider.

THE NUCLEOLAR CYCLE

It has long been known that the meiotic nucleolar cycle differs markedly from the mitotic [10], and it now seems likely that this is but another facet of a general difference in RNA metabolism. Meiocytes enter prophase with one or several nucleoli, depending upon genotype; where there are more than one, the number is reduced at the time of synapsis. Thereafter there is usually an increase in volume [11], and often a change in shape due to the flattening of the nucleolus towards one pole [12–14]. The spherical shape is restored at diplotene, but both protein and RNA stainability are usually lost at the end of diakinesis, and the parental cell nucleolus is often not discernible at all after the congression of the bivalents at metaphase I. There are, however, cases where the mother cell nucleolus does persist through the meiotic divisions [13].

In spite of the increase in nucleolar size noted by Lin [11], there is good reason to believe that there is little or no synthesis of nucleolar RNA after late leptotene. This has been shown for the meiocytes of *Zea mays* by the autoradiographic study of Das [15], who found no incorporation of ^{3}H-uridine in the nucleoli after leptotene. Chromosome-associated synthesis was, however, found to rise during prophase to a maximum in diplotene, declining again through diakinesis (Table 2).

Table 2. SILVER GRAIN COUNTS FROM AUTORADIOGRAPHS OF SECTIONS OF *Zea mays* ANTHERS LABELLED WITH ^{3}H-CYTIDINE AND ^{3}H-URIDINE AT DIFFERENT MEIOTIC STAGES, SHOWING CHROMOSOMAL AND NUCLEOLAR INCORPORATION AND CHROMOSOMAL/NUCLEOLAR RATIO. FROM DAS [15]

Meiotic stage	Chromosomal incorporation	Nucleolar incorporation	Chromosomal/ Nucleolar
Premeiotic interphase	27·4	10·8	2·5
Preleptotene–leptotene	12·8	2·7	4·7
Pachytene	26·8	0·9	29·8
Diplotene	9·9	0·2	49·5
Diakinesis	6·2	0·1	62·0

Nucleolus-like bodies appear again at the time of the meiotic mitoses, but in both formation and fate they differ strikingly from those of somatic cells. Although they contain both RNA and protein and are structurally similar to nucleoli, they do not appear to be formed at the nucleolus-organising regions of chromosomes, and lie in the cytoplasm during part of the division cycle [16]. These 'nucleoloids' have been illustrated and described in various early accounts, and they have been reported in a sufficiently wide sample of species to suggest that they may be a characteristic feature of meiosis in the anther. Among the more notable references, those of Latter on *Lathyrus odoratus* [12], Frankel on *Fritillaria* species [13], Håkansson and Levan on *Pisum sativum* [17] and Lindemann on *Bellevalia romana* and *Agapanthus umbellatus* [18] may be mentioned.

The nucleolus cycle in *Lilium* is summarised in Table 1. Although the nucleoloids are evidently not formed at nucleolus-organising regions of specific chromosomes, small nucleolus-like particles are visible in association with the anaphase I chromosomes in lily, and these bodies may correspond to those described in *Zea mays* by Walters [19]. Some are likely to be the precursors of the nucleoloids of telophase I, while some are undoubtedly incorporated in the telophase nuclei. Again, in anaphase II several nucleolus-like bodies can be distinguished in the spindle and among the chromosomes, and once more some remain dispersed in the cell while others are retained in the telophase nuclei. The analytical data of Table 1 show that a substantial synthesis of a high-GC type of RNA occurs towards the end of the prophase in lily, and it would be in keeping with the cytological observations if this were to take place intranuclearly during diplotene and diakinesis, contributing eventually to the nucleoloids. The observations of Das (Table 2) indicate a chromosome-associated synthesis in diplotene in *Zea mays*, and it is noteworthy that plant diplotene chromosomes have some lampbrush-like features [20]. Because of the possible effects of the callose meiocyte wall on the movement of tracers mentioned in a later section, there is some uncertainty about the interpretation of autoradiographic studies of RNA synthesis after diakinesis. However, some reports have given evidence of a late prophase synthesis, notably that of Taylor and Moses on *Tradescantia paludosa* [21].

The analytical results given for *Lilium henryi* in Table 1 also suggest that a further synthesis of a ribosomal type of RNA occurs in the dyad period (see also Ref. 22), and this could be related to the subsequent appearance of nucleoloids during meiosis II. If the dyad is effectively a closed system after the sealing of the callose wall, this synthesis would be at the expense of precursors already present in the meiocyte, and it may not always be detectable in tracer experiments.

THE NUCLEOLAR CYCLE AND RIBOSOME REPOPULATION

In the light of the correlations set out in Table 1, it seems possible that the nucleoloids may represent the main mechanism through which the ribosome population of the cytoplasm is restored during late meiosis after the prophase elimination [16]. According to this interpretation, the nucleoloids are packeted ribosomes or ribosome sub-units, which are dispersed, or assembled and dispersed, after exclusion from the nucleus at each successive division.

The observations must be extended to a wider range of species before the pattern seen in lily can be accepted as being a general one in flowering plants, but some similarities with events in other groups may be noted. In the maturation of the amphibian oocyte, 1000 or so extrachromosomal nucleoli are formed during the extended diplotene in association with DNA synthesised at the nucleolus-organising region of nucleolar chromosomes at pachytene. This production of additional DNA is interpreted as representing a multiplication of the cistrons concerned with rRNA synthesis.

The significance of this gene amplification is related to the need to achieve massive RNA synthesis during egg maturation without disturbing the normal meiotic behaviour of the chromosomes [23]. A similar requirement would seem to be present in angiosperm microsporogenesis, given that a rapid restoration of the ribosome populations of the young spores is necessary for them to support protein synthesis in the period immediately succeeding meiosis [24]. Part of the prophase DNA synthesis reported by Hotta and Stern [25] could be concerned with a multiplication of the cistrons concerned with rRNA synthesis in a manner analogous with that seen in the amphibian oocyte.

THE CYTOPLASMIC MEMBRANES

In *Lilium*, radical changes overtake the cytoplasmic membranes of the meiocyte concurrently with the prophase decline in ribosome number [26]. The leptotene meiocytes possess moderate amounts of ribosomal endoplasmic reticulum of a normal, somatic-cell type. As the ribosome elimination phase begins towards the end of zygotene, plate-like profiles are progressively replaced in electron micrographs by circular profiles of paired membranes, until ultimately all the cytoplasmic membranes present this appearance. The absence of any other aspects shows that the membranes form complete spheres, and not cylinders or convoluted tubes. The spheres are, moreover, often disposed in a concentric manner, giving a very characteristic multimembraned body. The closure of the membranes brings about a compartmentation of the cytoplasm, and the multimembraned bodies are frequently seen to include cytoplasmic structures—plastids, mitochondria and spherosomes [26, 27]. The increasing density of the ground substance through successive shells suggests that the protein concentration is higher in the sample of cytoplasm trapped in the centre of the multi-membraned bodies than outside.

Ribosomes are lost from the membranes during the period of compartmentation. Electron micrographs do not show any marked unit-membrane organisation; but in material fixed through standard glutaraldehyde procedures with uranyl acetate and lead hydroxide post-staining, dark, granulate material is seen to be associated with the membranes. The cytoplasmic membranes return to a more normal form concurrently with the disappearance of ribosomes. In *Lilium*, dyad stages customarily show both plate-like profiles of endoplasmic reticulum and residual multimembraned bodies in the cytoplasm. The latter evidently lose their sheaths progressively, one peeling away at a time. At the time of release of the spores from the meiotic tetrads the endoplasmic reticulum is indistinguishable from that of a somatic cell.

Membrane metamorphoses comparable with those seen in the lily meiocyte have not been reported from other species, and it is possible that the pattern of behaviour is not a general one. However, some change in the character of the endoplasmic reticulum is certainly to be expected in all species as a concomitant of the prophase ribosome elimination.

THE ORGANELLES

The extensive light-microscopic studies of the behaviour of the cytoplasmic organelles of the meiocyte made by Guilliermond's school, and particularly by Py [3], have already been mentioned. In six angiosperm species belonging to different families, both monocotyledonous and dicotyledonous, Py reported that the organelles constituting the 'chondriome' increased in number in the premeiotic period, and that during meiosis they entered a so-called 'granular phase' which persisted from zygotene until the time of tetrad cleavage. During the granular phase the elements of the chondriome were small and undifferentiated, and there was no starch storage in the plastids. The conclusion of the granular phase was marked by the resumption of starch storage and organelle multiplication, with the appearance of elongated 'chondrioconts' —chains of mitochondria. More recently, the light-microscopic observations of Steffen and Landmann [28] on *Impatiens glandulifera* have confirmed quantitatively the absence of plastid multiplication during the meiotic prophase.

The structural changes in the organelles during meiosis have been clarified by electron microscopic studies, and it is now possible to interpret Py's 'granular phase' as being one of drastic de-differentiation in both mitochondria and plastids. An early account for *Tradescantia paludosa* was given by Bal and De [29], who found that the mitochondria were small, spherical and without cristae in the mid-prophase. Marumaya [30] has confirmed this observation for *T. paludosa* and has given a fuller account of the mitochondrial cycle. The elongated mitochondria of premeiotic stages give place to spherical ones by pachytene, and aspects suggestive of division are no longer seen. The mitochondria then progressively lose internal structure, until by meiosis I they are essentially without cristae, being recognisable only as spherical bodies of diameters around 0·4 μm. Re-differentiation follows with the development of the internal membrane system, until by pollen mitosis the mitochondria have a normal form with well-defined cristae. Division figures are common after the period of re-differentiation.

The behaviour of the plastids through meiosis has been followed in detail in *Lilium longiflorum* [31]. In the premeiotic period, profiles suggestive of division stages are frequent; starch is present, and also a moderately developed lamellar system. In leptotene, the starch is eliminated, division aspects decline in number, and the lamellar system regresses. Division ceases by zygotene, and with the transition to pachytene the plastids reach maximum structural simplification. They are then irregularly spherical or ellipsoidal, with few or no discernible ribosomes and a severely reduced lamellar system. This condition prevails through diplotene and diakinesis. An unusual and possibly unique structural feature appears at interphase, consisting of an association between a membranous component, interpreted as a flattened tubule or ribbon, and a cluster of particles of diverse sizes. This membrane–particle association is probably present in all plastids from the dyad to the late tetrad. Its significance cannot be specified, but the fact that it is consistently present suggests that it may be concerned with the re-differentiation of the plastids, which proceeds from late tetrad onwards. This re-differentiation involves the restoration of the ribosome population, the development of a lamellar system, and the reappearance of starch. As it proceeds, the plastids become amoeboid, and replication begins once more.

Although detailed fine-structural information about the behaviour of plastids and mitochondria in the meiotic prophase is available for only two species, both monocotyledons, aspects of the de-differentiated organelles have been illustrated casually in various accounts of plant microsporogenesis dealing with other species. Moreover, the electron microscopic observations can be reconciled so readily with those of Py [3] and other light-microscopic cytologists that there is an adequate basis for the generalisation that the organelles regularly undergo a cycle of de-differentiation and re-differentiation in the course of meiosis in the angiosperm microsporocyte, although they show genetic continuity throughout.

CORRELATION OF THE CYTOPLASMIC AND ORGANELLAR CHANGES

There seems good reason to believe that the de-differentiation of the plastids and mitochondria is closely correlated in time with the change in the ribosome population of the cytoplasm. The ribosome elimination takes place during the zygotene–pachytene interval, and this is also the period of the principal plastid metamorphosis in *Lilium*, as well as the time when, according to Muramaya [30], the mitochondria of *Tradescantia* metamorphose into a spherical form and then lose their cristae. The re-differentiation of the organelles is similarly correlated with the ribosome restoration phase, beginning at the time of meiosis I.

The correlation with the nuclear events of meiosis is obviously also quite close. It may be noted that if the ribosome cycle is closely interlocked with the nucleolar cycle, as suggested in an earlier section, then this correlation is presumably obligate, because as here interpreted the formation of supernumerary nucleoli at the time of the meiotic mitoses depends upon pachytene–diplotene syntheses.

Before we examine the possible significance of these findings, the role of the cytoplasm in the interaction between different meiocytes within the anther loculus merits some comment.

PLASMODESMATA AND CYTOPLASMIC CONNECTIONS

At the onset of meiosis, the meiocytes within an anther loculus are linked by plasmodesmata of a normal, somatic-cell, type. The cells pass into synchrony in this state before the meiotic S-period of preleptotene, but thereafter profound changes in the intercellular connections are initiated. During leptotene and zygotene, massive cytoplasmic channels are developed between neighbouring cells, and in many species the interconnections extend throughout the whole sporogenous tissue of the anther so that a kind of syncytium is produced [32, 33]. The formation of the channels proceeds concurrently with the synthesis of the callose special mother cell wall, which begins in leptotene.

Although the initiation of synchronous behaviour in the meiocyte nuclei does not depend upon the existence of the syncytial organisation, there is reason to believe that the sharing of a common cytoplasm is an important factor in *maintaining* close synchrony through the meiotic stages; because in species where the syncytium does not extend throughout the loculus, neighbouring groups of interconnected meiocytes lose synchrony as the prophase advances [34]. As with the nuclear events, the cytoplasmic and organellar changes of prophase are closely synchronised throughout the sporogenous syncytium. The cytoplasmic interconnections of mid-prophase are, however, usually severed at the time of the meiotic mitoses, and the formation of further callose walls between the spores isolates them within the tetrad. After the severance of the links, synchrony in both nuclear and cytoplasmic behaviour is progressively lost.

In summary, the changes in the walls and cellular interconnections in the sporogenous tissues are such as to produce during mid-prophase an unusually high level of intercellular communication, and later, after the completion of meiosis, one which is unusually low. The evidence from a wide range of species suggests that this is the general situation in flowering plants, and the same behaviour is known in pteridophytes [35]. The most reasonable rationalisation of what is observed is to suppose that it is advantageous for there to be free movement of nutrients and possibly also information between the cells of the sporogenous tissue in prophase, while after the meiotic segregation there is a premium upon the isolation of the meiotic products. Arguments favouring the idea that the isolation of the spores in the tetrad period is required for their assertion of genetic independence have been given elsewhere [33, 36]. Here it may be noted that there is now direct evidence that while materials readily enter the sporogenous tissue in prophase stages, the penetration of certain molecules is retarded after the closure of the meiocyte wall [37, 38, 47; this volume, p. 119]. The fact that nucleosides may not readily enter the tetrad [37] is a factor to be taken into account in appraising the results of tracer experiments.

CONCLUSIONS

It was argued above largely on theoretical grounds that a far-reaching cytoplasmic reorganisation must take place concurrently with meiosis in the angiosperm anther, a reorganisation related to the sporophyte–gametophyte transition. The foregoing review shows that beyond doubt major changes do occur, in both cytoplasm and organelles, and that they are closely locked to the chromosomal cycle of meiosis. The conclusion seems inescapable that these changes are different manifestations of the predictable general reorganisation—the cytoplasmic 'spring clean', long postulated by Mather (e.g. Ref. 39) as likely to be associated with sexual reproduction. It is not possible yet to say why a wholesale removal of the ribosomes of the diploid mother cell should be a concomitant of diplophase–haplophase transition, but certainly such an event does not contradict the hypothesis that a re-standardisation of the cytoplasm is necessary to provide the appropriate environment for the later expression of gametophytic potentials. What remains faintly surprising is the scale of the operation; were the target merely to expunge diplophase information carried in the form of persistent mRNA, then the elimination of the ribosomes would seem to contain an element of over-kill.

The fact that the organelles are recognisable at all times through meiosis shows that the lineage of those present in the spores extends back without break to the diploid mother cell. There clearly are no phases of elimination and subsequent restoration from the nuclear envelope as envisaged for the fern egg [40, 41]. However, the regression of structure in the mitochondria and plastids suggests that in the organelles also some process of re-standardisation is carried through. This is obviously related to the cytoplasmic change, but its significance remains obscure at present, unless it be that here also some persistent relic of diplophase control has to be expunged.

The correlation in time between the changes in the cytoplasm and organelles and the chromosomal cycle of meiosis would be comprehensible were the extranuclear events consequent upon phase-specific gene transcriptions. According to this interpretation, the 'autodigestive' process of zygotene–pachytene would be dependent upon gene transcription in leptotene–zygotene manifested in the rise in hydrolase activity in the cytoplasm, while, as suggested earlier, the later ribosome restoration would in turn be dependent upon rRNA synthesis in diplotene–diakinesis. The implication here is that the changes in the constitution of the cytoplasm are obligatorily linked—at least in the normal course of events—to the meiotic chromosomal cycle, which presumably is such as to permit gene transcription only at certain intervals. The test here will be whether the time correlations observed in the comparatively small number of species so far studied are truly universal.

If the function of the cytoplasmic and organellar changes associated with meiosis be indeed to produce a kind of *tabula rasa* for the haploid spore nucleus, the question arises: When does the event occur which determines what inscriptions that nucleus will make? It so happens that we are already close to having an answer to this, albeit from indirect evidence: the young spore at the time of release from the tetrad is probably not committed to any particular developmental pathway, but is free to adopt one of three—it may develop as a male gametophyte, the normal behaviour; or it may under some circumstances form a female gametophyte; or yet again it may pass through the appropriate embryonic stages to produce a sporophyte. 'Pollen-grain embryo sacs' have been induced by temperature and other treatments in various species [42, 43], while Guha and Maheshwari [44] showed that haploid sporophytes may arise from spores of *Datura innoxia* in anthers cultured *in vitro* on suitable media. Nitsch in this volume (p. 234) gives a brief account of the same phenomenon in *Nicotiana tabacum* [45], and emphasises that the potential for sporophyte development is present only ?t a particular period—in fact, that covering the young spore interval.

The conclusion that the young spore is essentially uncommitted at the time of release from the tetrad must mean that in normal development influences impinge upon it which direct it inevitably into the male-gametophytic pathway. One may speculate that these arise in the tapetum, or are already present in the thecal fluid; but as yet we are essentially without evidence on the point. What is needed now are experiments on isolated spores cultures *in vitro* on media of various compositions and with the indicated kinds of addenda—thecal fluid, tapetal extracts and nucellar extracts.

We may note that if the cytoplasmic re-standardisation is a necessary prerequisite for the transition from sporophyte to gametophyte, then events must occur concurrently with meiosis in the ovule corresponding to those in the anther, in anticipation of the inception of the embryo sac. There have been no adequate studies of the cytoplasm during meiosis in the megasporocyte, so we lack information on the point; with the firm indications of what to expect, however, evidence should not be too difficult to obtain.

While the transition from sporophyte to gametophyte is normally correlated in time with meiosis, we have seen that the evidence of apospory shows that the two events are not necessarily linked. The aposporous development of the male gametophyte seems not to have been noted in angiosperms, and it may be doubted whether it would be recognisable as such in view of the reduced nature of this generation. Aposporous development of the female gametophyte is of course well known, as it is for the bisexual gametophyte of ferns. Given that the cytoplasmic reorganisation is essential for the inception of the gametophyte, it is pertinent to inquire how this is managed in cells other than meiocytes. The expectation is that the aposporous mother cells of embryo sacs and other gametophytes should pass through ribosome elimination and restoration phases, and presumably also through the cycle of organellar de- and re-differentiation. Furthermore, one might predict that these changes should be instituted only in cells which are isolated from organised diplophase tissue. There are already indications of this: discussing control of the pteridophyte life cycle, Bell [46] points out that aposporous gametophytes arise from fern leaves only when the tissues are near death and the intercellular correlations are interrupted. Clearly fine-structural and cytochemical investigation of aposporous mother cells could be highly profitable in further efforts aimed at elucidating the basis of the alternation of generations.

REFERENCES

1 GUILLIERMOND, A., 'Sur l'évolution du chondriome pendant la formation des grains de pollen de *Lilium candidum*', *C.r. hebd. Séanc. Acad. Sci., Paris*, **170**, 1003–6 (1920)
2 GUILLIERMOND, A., 'Recherches sur l'évolution du chondriome pendant le development du sac embryonaire et des cellules-mères des grain de pollen dans les Liliacées et sur la significance des formations ergastoplasmiques', *Ann. Sci. nat., Bot.*, **6**, 1–52 (1924)
3 PY, G., 'Recherches cyologiques sur l'assise nourricière des microspores et les microspores des plantes vasculaires', *Rev. Gen. Bot.*, **44**, 316–68; 369–413; 450–62; 484–510 (1932)
4 WAGNER, N., 'Évolution du chondriome pendant la formation des grains de pollen des Angiosperms', *Biologia gen.*, **3**, 15–66 (1927)
5 PAINTER, T. S., 'Cell growth and nucleic acids in the pollen of *Rhoeo discolor*', *Bot. Gaz.*, **105**, 58–68 (1943)
6 MACKENZIE, A., HESLOP-HARRISON, J. and DICKINSON, H. G., 'Elimination of ribosomes during meiotic prophase', *Nature, Lond.*, **215**, 997–9 (1967)
7 SAUTER, J. J. and MARQUART, H., 'Die Rolle des Nukleohistons bei der RNA–und Proteinsynthese während die Mikrosporogenese von *Paeonia tenuifolia* L.', *Z. Pfl. Physiol.*, **58**, 126–37 (1967)
8 KNOX, R. B., DICKINSON, H. G. and HESLOP-HARRISON, J., 'Cytochemical observations on changes in RNA content and acid phosphatase activity during the meiotic prophase in the anther of *Cosmos bipinnatus* Cav.', *Acta bot. neerl.*, **19**, 1–6 (1970)
9 LINSKENS, H. F., 'Die Änderung des Protein und Enzym-Musters während der Pollenmeiose und Pollenentwicklung', *Planta*, **69**, 79–91 (1966)
10 GATES, R. R., 'Nucleoli and related nuclear structures', *Bot. Rev.*, **8**, 337–409 (1942)
11 LIN, M., 'Chromosome control of nuclear composition', *Chromosoma*, **7**, 340–70 (1955)
12 LATTER, J., 'The pollen development of *Lathyrus odoratus*', *Ann. Bot.*, **40**, 277–314 (1926)
13 FRANKEL, O., 'The nucleolar cycle in some species of *Fritillaria*', *Cytologia*, **8**, 37–47 (1937)
14 MOENS, P. B., 'The structure and function of the synaptinemal complex in *Lilium longiflorum* sporocytes', *Chromosoma*, **23**, 418–51 (1968)
15 DAS, N. K., 'Inactivation of the nucleolar apparatus during meiotic prophase in corn anthers', *Expl Cell Res.*, **40**, 360–4 (1965)
16 DICKINSON, H. G. and HESLOP-HARRISON, J., 'The ribosome cycle, nucleoli and cytoplasmic nucleoloids in the meiocytes of *Lilium*', *Protoplasma*, **69**, 187–200 (1970)
17 HÅKANSSON, A. and LEVAN, A., 'Nucleolar conditions in *Pisum*', *Hereditas*, **28**, 436–40 (1942)
18 LINDEMANN, R., 'Vergleichende cytologische Unterschungen an den Liliaceen *Bellevalia romana, Agapanthus umbellatus* und *Lilium regale*', *Planta*, **48**, 2–28 (1956)

19 WALTERS, M. S., 'Ribonucleoprotein structures in meiotic prophase of *Zea mays*', *Heredity, Lond.*, **23**, 39–47 (1968)

20 GRUN, P., 'Plant lampbrush chromosomes', *Expl Cell Res.*, **14**, 619–21 (1958)

21 TAYLOR, J. H. and MOSES, M. J., 'Desoxypentose nucleic acid synthesis during microsporogenesis in *Tradescantia*', *Expl Cell Res.*, **9**, 474–88 (1955)

22 HOTTA, Y. and STERN, H., 'Inhibition of protein synthesis during meiosis and its bearing on intracellular regulation', *J. Cell Biol.*, **16**, 259–79 (1963)

23 GALL, J. G., 'The genes for ribosomal RNA in oogenesis', *Genetics Supplement*, **61**, 121–32 (1969)

24 HESLOP-HARRISON, J., 'Ribosome sites and S gene action', *Nature, Lond.*, **218**, 90–1 (1968)

25 HOTTA, Y., ITO, M. and STERN, H., 'Synthesis of DNA during meiosis', *Proc. natn. Acad. Sci. U.S.A.*, **56**, 1184–91 (1966)

26 DICKINSON, H. G., *Studies on Microsporogenesis*, Thesis, University of Birmingham (1969)

27 HESLOP-HARRISON, J. and DICKINSON, H. G., 'A cycle of spherosome aggregation and disaggregation correlated with the meiotic divisions in *Lilium*', *Phytomorphology*, **17**, 195–9 (1967)

28 STEFFEN, K. and LANDMAN, W., 'Chondriosomen-, Sphärosomen- und Proplastidenzahlen in Beziehung zu hen Differenzierungsvorgängen bei der Antherenentwicklung', *Planta*, **51**, 30–48 (1958)

29 BAL, A. K. and DE, D. N., 'Developmental changes in the submicroscopic morphology of cytoplasmic components during microsporogenesis in *Tradescantia*', *Devl Biol.*, **3**, 241–54 (1961)

30 MARUMAYA, L., 'Electron microscopic observations of plastids and mitochondria during pollen development in *Tradescantia paludosa*', *Cytologia*, **33**, 482–97 (1968)

31 DICKINSON, H. G. and HESLOP-HARRISON, J., 'The behaviour of plastids during meiosis in the anther of *Lilium longiflorum* Thunb.' (In press)

32 HESLOP-HARRISON, J., 'Cell walls, cell membranes and protoplasmic connections during meiosis and pollen development'. In: *Pollen Physiology and Fertilisation* (Ed. H. F. LINSKENS), 39–47, North-Holland Publishing Co., Amsterdam (1964)

33 HESLOP-HARRISON, J., 'Cytoplasmic connections between angiosperm meiocytes', *Ann. Bot.*, **30**, 221–30 (1966)

34 HESLOP-HARRISON, J., 'Microsporogenesis in some triploid orchid hybrids', *Ann. Bot.*, **17**, 540–9 (1953)

35 LUGARDON, B., 'Sur l'existence de liaisons protoplasmiques entre les cellules-mères des microspores de Ptéridophytes au cours de la prophase hétérotypique', *C.r. hebd. Séanc. Acad. Sci., Paris*, **267**, 593–6 (1968)

36 HESLOP-HARRISON, J., 'Cytoplasmic continuities during spore formation in flowering plants', *Endeavour*, **25**, 65–72 (1966)

37 MACKENZIE, A. and HESLOP-HARRISON, J., 'Autoradiography of soluble (2–^{14}C)-thymidine derivatives during meiosis and microsporogenesis in *Lilium* anthers', *J. Cell Sci.*, **2**, 387–400 (1967)

38 KNOX, R. B., DICKINSON, H. G. and HESLOP-HARRISON, J., 'Cytochemical observations on changes in RNA content and acid phosphatase activity during the meiotic prophase in the anther of *Cosmos bipinnatus* Cav.', *Acta bot. neerl.*, **19**, 1–6 (1970)

39 MATHER, R., 'Genes and cytoplasm in development', *Encycl. Plant Physiol.*, **15**, 41–67 (1965)

40 BELL, P. R., FREY-WYSSLING, A. and MÜHLETHALER, K., 'Evidence for the discontinuity of plastids in the sexual reproduction of a plant', *J. Ultrastruc. Res.*, **15**, 108–121 (1966)

41 BELL, P. R. and MÜHLETHALER, K., 'The degeneration and reappearance of mitochondria in the egg cells of a plant', *J. Cell. Biol.*, **20**, 235–48 (1964)

42 STOW, I., 'On the female tendencies of the embryo-sac-like giant pollen grains of *Hyacinthus orientalis*', *Cytologia*, **1**, 417–39 (1930)

43 NAITHANI, S. P., 'Chromosome studies in *Hyacinthus orientalis*. III. Reversal of sexual state in the anthers of *Hyacinthus orientalis* var. Yellow Hammer', *Ann. Bot., N.S.*, **1**, 369–77 (1937)

44 GUHA, S. and MAHESHWARI, S. C., 'Development of embryoids from pollen grains of *Datura* in vitro', *Phytomorphology*, **17**, 454–61 (1967)

45 NITSCH, J. P., this volume, 234

46 BELL, P. R., 'The experimental investigation of the pteridophyte life cycle', *J. Linn. Soc., Bot.*, **56**, 188–202 (1959)

47 KNOX, R. B. and HESLOP-HARRISON, J., 'Direct demonstration of the angiosperm meiotic tetrad using a fluorogenic ester', *Z. Pfl. Physiol.*, **62**, 451–9 (1970)

Cytoplasmic RNA and Enzyme Activity During the Meiotic Prophase in *Cosmos bipinnatus**

R. B. Knox, *Department of Botany, Australian National University, Canberra, Australia*
H. G. Dickinson, *Department of Botany, University College, Gower Street, London, England*
J. Heslop-Harrison, *Institute of Plant Development, University of Wisconsin, Madison, Wisconsin, U.S.A.*

Changes in the content of acetic-alcohol fixable RNA and in enzyme activity have been followed in meiocyte cytoplasm during the meiotic prophase in *Cosmos bipinnatus*, Compositae, by quantitative and semi-quantitative cytochemical methods. The composite capitulum provides a uniquely favourable object for cytochemical studies of meiosis in the anther, since the radius of a single sectioned capitulum provides a time axis, a fact which greatly reduces errors due to variation in section thickness and processing procedures.

  The RNA content of the meiocyte cytoplasm was estimated by cytophotometry of 6-μm sections after pyronin Y [1] or azure B staining. The dimensions of the meiocytes at successive periods through meiosis were ascertained, volumes were computed and total relative RNA content per cell was calculated from the cytophotometric values for concentration per unit volume. The changes in RNA content in the interval from preleptotene (corresponding to the premeiotic S-period) to diplotene are seen in Fig. 1. After an initial rise which may not be significant, the RNA content per meiocyte falls abruptly between zygotene and pachytene to a value less than half of that measured at leptotene. This decline in fixable cytoplasmic RNA corresponds to that recorded in the meiocytes of *Lilium* and *Trillium* (Ref. 2; see p. 18 of this volume) on the basis of microchemical estimates using the Edström technique. The timing of the decline—in the zygotene–pachytene interval—is the same; and the magnitude of the change also corresponds closely, notwithstanding the difference in measurement technique and the
_____
* Abstract.

32

fact that with the monocotyledonous species estimates were of total fixable cell RNA, while with *Cosmos* they were of the cytoplasm alone. The mid-prophase diminution of meiocyte RNA has recently been noted by Sauter and Marquardt [3] in another dicotyledon, *Paeonia tenuifolia*.

Ribosome counts in *Lilium* show that the loss in cytoplasmic RNA is largely—perhaps entirely—accounted for by ribosome

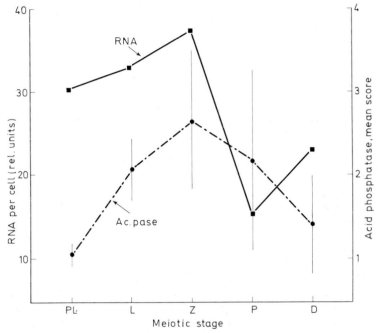

Figure 1. Changes in RNA content and acid phosphatase activity in glutaraldehyde-fixed meiocytes of Cosmos bipinnatus. *The RNA curve is derived from pyronin microdensitometry, and the acid phosphatase curve is based upon visual scoring of preparations in which enzyme activity was revealed with 2-naphthyl-thiol phosphate as substrate*

elimination [2]; it may be surmised that this is also true for *Cosmos*, although fine-structural evidence is not yet available. In an attempt to detect changes in enzyme activity that might be associated with ribosome dissolution, the prophase meiocytes of *Cosmos* have been surveyed for lysosome enzymes, particularly acid phosphatase and ribonuclease. Observations were made both on unfixed, freeze-sectioned capitula, and on capitula fixed in 2·8% glutaraldehyde buffered at pH 7·4. Acid phosphatase activity was detected by the methods of Hanker *et al.* [4] and Barka and Anderson [5], and ribonuclease by a modification of the lead salt method of Enwright, Frye and Atwal [6; see also Ref. 7].

The change in acid phosphatase activity detectable in fixed meiocytes with 2-naphthyl-thiol phosphate as substrate is shown in Fig. 1. Notwithstanding the inherently high variability in the estimates, the trend is clear enough: activity is low in the prelepto-tene meiocytes, increases to a peak in zygotene, and then declines somewhat to pachytene with a steeper fall subsequently to diplotene. Observations made on freeze-sectioned fresh anthers using α-naphthyl acid phosphate as substrate gave a similar result, with a peak of activity in zygotene; however, the activity detectable in pachytene was higher than in the fixed material, perhaps indicating that enzyme was lost from the later-stage meiocytes in processing. The observations on ribonuclease activity were less clear-cut because of higher non-specific staining and the fact that the pro-cedure gave some reaction in the cell nucleus at all stages. Never-theless, the trend matched that seen with acid phosphatase: peak cytoplasmic activity appeared in zygotene, with a subsequent decline in pachytene and diplotene. The distribution of the reaction products in the cytoplasm with both acid phosphatase and ribo-nuclease suggested that some of the enzyme activity might be associated with particles; the effect could, however, result from technique artefacts.

These results show that there are phase-specific changes in cyto-chemically detectable enzyme activity in the meiocyte cytoplasm during prophase. The functional significance of 'acid phosphatase' and 'ribonuclease' activity detected in this manner cannot be specified, but the fact is that peak activities do occur just before, and during, the period when cytoplasmic RNA content is declining spectacularly. This may be taken to indicate that the changes observed are connected with ribosome breakdown process.

REFERENCES

1 MOSS, G. I., 'A cytochemical study of DNA, RNA and protein in the developing maize anther', *Ann. Bot.*, **31**, 545–72 (1967)

2 MACKENZIE, A., HESLOP-HARRISON, J. and DICKINSON, H. G., 'Elimination of ribosomes during meiotic prophase', *Nature, Lond.*, **215**, 997–9 (1967)

3 SAUTER, J. J. and MARQUARDT, H., 'Die Rolle des Nukleohistons bei der RNA-und Proteinsynthese während die Mikrosporogenese von *Paeonia tenuifolia* L.', *Z. Pfl. Physiol.*, **58**, 126–37 (1967)

4 HANKER, J. B., SEAMAN, A. R., WEISS, L. P., UENO, H., BERGMAN, R. A. and SELIGMAN, A. M., 'Osmiophilic reagents: new cytochemical principle for light and electron microscopy', *Science, N.Y.*, **146**, 1039–43 (1964)

5 BARKA, T. and ANDERSON, P. J., 'Histochemical methods for acid phosphatase using hexazonium pararosanilin as coupler', *J. Histochem. Cytochem.*, **10**, 741–53 (1962)

6 ENWRIGHT, J. B., FRYE, F. L. and ATWAL, O. S., 'Ribonuclease activity of peripheral leucocytes and serum in rabies-susceptible and rabies-refractory mice', *J. Histochem. Cytochem.*, **13**, 515–17 (1965)

7 KNOX, R. B. and HESLOP-HARRISON, J., 'Pollen-wall proteins: localization and enzymic activity', *J. Cell Sci.*, **6**, 1–37 (1970)

Histones, RNA and Protein Synthesis in Pollen Cells of *Paeonia**

Jörg J. Sauter, *Institute of Forest Botany, University of Freiburg, West Germany*

Basic proteins associated with DNA may function as regulators of gene action. The mechanism of regulation, however, is not yet clarified. It is known that (*a*) RNA synthesis proceeds mainly in the 'diffuse chromatin', the dense chromatin being inactive with respect to DNA-directed RNA synthesis; (*b*) lysine-rich histones cross-linking the chromatin fibrils are most probably responsible for the formation of the dense, inactive chromatin; and (*c*) active and inactive chromatin do not differ in the amount of lysine-rich histone extractable with biochemical methods. Therefore, it has been supposed that the regulation of gene action is not achieved by quantitative but by qualitative alteration of the DNA-associated histone—for instance, by acetylation (for literature, see Ref. 1). Acetylation modifying DNA–histone interactions is thought to influence the condensation degree of chromatin and thus the template activity of the chromatin for RNA synthesis.

The aim of the present study was to test whether these mainly biochemical findings could also be proved with the cytochemical technique. With the alkaline Fast green (FG) staining method only the unaltered (non-acetylated) histones can be stained which are considered to be active in suppressing DNA–RNA transcription. Consequently, it is expected that there would be a difference in histone-staining behaviour of nuclei from cells with highly suppressed and from cells with highly activated RNA and protein synthesis. To test this assumption, we used young pollen grains from *Paeonia tenuifolia* L. in the post-pollen-mitosis stage. During this period the vegetative and the generative pollen cells differ extremely in RNA and protein synthesis, as we demonstrated previously in

* Abstract.

36

cytochemical and autoradiographic studies [2, 3], the vegetative cell being extremely active and the generative cell nearly inactive.

In this system, which thus seemed to be suitable for studying the above-mentioned findings, histones have been investigated cytochemically by the alkaline FG staining method. Controls were performed to distinguish between DNA- and RNA-associated histones as well as between arginine-rich and lysine-rich histones (for details, see Ref. 1). Additionally, DNA, RNA and arginine-containing proteins have been stained by the Feulgen, the gallocyanin chromalum and the Sakaguchi methods, respectively.

Summarising, we obtained the following most striking result (for details see Ref. 1).

In the nucleus of the generative cell there is a very high content of DNA-associated histone which was found to be of the lysine-rich type and to be present in an unaltered and FG-stainable form; at the same time, this nucleus is highly condensed and the cell is nearly inactive with respect to RNA and protein synthesis.

On the other hand, in the nucleus of the vegetative cell there is nearly no DNA-associated lysine-rich histone in a stainable form, which indicates that it is either absent or, more probably, present but in an altered (acetylated?) form; in this nucleus the chromatin is highly decondensed, and the cell consistently shows a conspicuous RNA and protein synthesis.

These cytochemical findings agree well with the biochemical results mentioned above, which supports the view that the DNA-associated histones of the lysine-rich type are involved in the condensation and decondensation of chromatin and by this in controlling release of genetic information. The changes in the degree of condensation may be achieved by differential acetylation of the lysine-rich histone.

Furthermore, in our cytochemical study another histone associated with RNA has been detected in the cytoplasm of the vegetative cell which is very rich in arginine and most probably forms part of ribosomes.

REFERENCES

1 SAUTER, J. J., 'Cytochemische Untersuchung der Histone in Zellen mit unterschiedlicher RNS- und Protein-Synthese', Z. Pfl. Physiol., **60**, 434–49 (1969 a)
2 SAUTER, J. J., 'Autoradiographische Untersuchungen zur RNS- und Proteinsynthese in Pollenmutterzellen, jungen Pollen und Tapetumzellen während der Mikrosporogenese von Paeonia tenuifolia L.', Z. Pfl. Physiol., **61**, 1–19 (1969 b)
3 SAUTER, J. J. and MARQUARDT, H., 'Die Rolle des Nucleohistons bei der RNS- und Proteinsynthese während der Mikrosporogenese von Paeonia tenuifolia L.', Z. Pfl. Physiol., **58**, 126–37 (1967)

Pollen Development and the Pollen Grain Wall

The Role of the Tapetum During Microsporogenesis of Angiosperms

P. Echlin, *Botany School, University of Cambridge, England*

INTRODUCTION

In writing this paper I have concentrated on the reports in the literature which have appeared in the 12 months preceding December 1969; earlier work has recently been reviewed and commented on in papers by Echlin and Godwin [1] and Heslop-Harrison [2]. A general survey of the development of the tapetum in lower vascular plants is given by Foster and Gifford [3].

The tapetum is the innermost wall layer of the microsporangium of the anther, and is the tissue in closest contact with the developing pollen grains. Typically, the tapetum is composed of a single layer of cells characterised by densely staining protoplasts and prominent nuclei. The tapetal cells in the developing anthers of angiosperms are often enlarged and may be multinucleate or polyploid. Early work had suggested that these nuclear divisions were endomitotic, but Maheshwari [4] had shown that in spite of certain abnormalities the tapetal nuclei generally undergo normal mitosis and that amitosis does not occur. The tissue plays an essential nutritive role in the formation of the pollen grains, for all the food materials entering the sporogenous cells must either pass through it or be metabolised by it.

Shortly after its differentiation in a stamen primordium, an embryonic anther as seen in cross-section consists of a mass of ground meristematic tissue. Although the entire sub-epidermal layer of the young anther may be potentially sporogenous, it is more usual that sporangium initiation is restricted to separate areas which come to correspond to the corners of the anther when seen in cross-section. At each of these regions a discrete number of hypodermal cells undergo periclinal divisions, and this archesporial tissue gives rise to primary parietal cells to the outside and primary

41

sporogenous cells to the inside. The primary sporogenous cells in turn give rise to the sporogenous tissue which matures to form the pollen grains, and the primary parietal cells undergo repeated periclinal and anticlinal divisions to form the several layers of the spore wall and the tapetum.

There is still doubt (see Maheshwari, Ref. 4) whether the entire tapetum arises from the primary parietal cells or only a large part of it. Because of the similarity in appearance of the tapetal and sporogenous cells in the light microscope, earlier workers had considered that part of the tapetum may have arisen by sterilisation of the outer cells of the sporogenous tissue. However, the general consensus of opinion now is that the tapetal cells are of parietal origin.

The significant point of these earlier studies is that they show that both the tapetal cells and the pollen grains have developed from the same archesporial tissue. The mature pollen grains are haploid cells which have arisen from diploid pollen mother cells; and the tapetal cells are diploid, frequently polyploid, cells which have also arisen from diploid cells.

Two major types of tapetum are recognised, based on their behaviour during microsporogenesis: the glandular or secretory type of tapetum where the cells remain in their original position but progressively become more disorganised and finally undergo complete autolysis, and the amoeboid tapetum, which is characterised by an early loss of the tapetal cell walls and the intrusion of the tapetal protoplasts among the developing pollen grains, followed by the fusion of these protoplasts to form a tapetal periplasmodium.

ULTRASTRUCTURE AND ONTOGENY OF THE TAPETUM

There are numerous reports in the literature which refer to the development of the tapetum at the level of the light microscope. It is not intended to discuss this aspect further but to concentrate on the more recent studies which have used the electron microscope. For the sake of clarity and ease of understanding it is proposed to separate the development of the tapetum from the development of Ubisch bodies, although it is quite clear that both processes are intimately connected and that tapetum development is, in turn, closely associated with the development of the microspores which they envelop and nourish. Little reference will be made to the ontogeny of pollen other than to relate the maturation of the microspore to the developmental stage of the tapetum. Other morphological and physiological variations in the development of the tapetum will also be dealt with separately.

Development of the amoeboid or periplasmodial tapetum

In a brief report by Heslop-Harrison [5] a description is given of an acetolysis-resistant membrane which invests the tapetum and developing microspores in certain Compositae. Although no electron micrographs are given which pertain to the development of the tapetum, Heslop-Harrison finds that the tissue remains in a parietal position until the breakdown of the meiotic tetrads to release the individual microspores. Following this, the tapetal cells dissociate from one another, become amoeboid, and extend into the anther loculus. The longitudinal walls slowly break down and the whole tissue changes into a plasmodium which extends into the spore mass.

A more detailed study is given in the recently published work by Mepham and Lane [6], who have investigated the fine structure and histochemistry of the development of the periplasmodial tapetum in *Tradescantia bracteata*. In contrast to what appears in the secretory tapetum, the periplasmodial tapetum of *Tradescantia* possesses an organised and apparently functional ultrastructure, and as development proceeds the cells undergo a reorganisation rather than a degeneration.

In the premeiotic stage of the sporogenous tissue, the tapetal cells contain abundant and prominent organelles. One interesting feature is the appearance of raphides, which are found free in the cytoplasm and not associated with any particular organelle. The tapetal protoplasm becomes vacuolated and numerous dictyosome-derived vesicles appear at the cell periphery. Some of these vesicles appear enclosed within elements of the endoplasmic reticulum and are similar in some respects to the multi-vesicular bodies described by Echlin *et al.* [7]. There is some evidence that these vesicles in *Tradescantia* are discharged from the cell, as Mepham and Lane show the presence of similar vesicles between the cell membrane and the cell wall. It is at this stage that lysis of the thin tapetal walls occurs, and the authors consider that this is probably due to the release of hydrolytic enzymes from the dictyosome-derived vesicles. Tapetal wall dissolution is fairly rapid, and is followed by a 'wave of lysis' which extends into the anther loculus and may even remove some of the sporogenous cell walls. At about this time the callose special cell walls are initiated around the pollen mother cells which will undergo meiosis to form the pollen tetrads, but in some instances the lysis is so rapid that a few meiocytes remain temporarily bereft of a cell wall.

By the end of this stage of development the tapetal cells, which are characterised by deeply staining polysaccharides, infiltrate between the pollen mother cells. On the basis of histochemical and enzymatic studies, Mepham and Lane consider that the cell walls of the tapetum and the sporogenous tissue are largely pectic in composition, whereas the cells of the middle layer, endothecium and the epidermis are cellulosic.

During pollen mother cell meiosis long extensions of the tapetal cell membrane extend into the anther loculus and fuse where they make contact between the mass of sporogenous tissue. Eventually the tapetal cytoplasm engulfs each pollen mother cell, apparently isolating each of them by investing membranes. Towards the end of meiosis the new periplasmodial tapetal cytoplasm undergoes a 'reorganisation' which is accompanied by characteristic changes within the cytoplasm. Callases, which are considered to be derived from dictyosome activity within the tapetal cytoplasm, enzymatically divest the tetrads of their callose coat, and long amoeboid-like periplasmodial processes penetrate between the individual pollen grains. Eventually the pollen grains, like the pollen mother cells which engendered them, are enclosed within vacuoles of the tapetal cytoplasm.

The tapetal cell metabolism appears then to change, from one associated primarily with carbohydrate formation and degradation, to the synthesis of lipid material. Lipid globules appear within the plastids and are progressively extruded from them into the tapetal cytoplasm. Parallel cytochemical studies confirm the presence of unsaturated lipids and phospholipids within the tapetal periplasmodium. During the whole of this phase in which pollen grain wall development proceeds, the tapetal cell organelles are readily recognisable and there is morphological evidence of protein synthesis.

Towards the end of pollen grain wall development the tapetal plasmodium becomes progressively more hydrated and the cells vacuolated, and at pollen grain mitosis the tapetal plastids show evidence of polysaccharide synthesis. As the time of anthesis approaches, each pollen grain is surrounded by a thin layer of tapetal cytoplasm, and the tonoplast appears to be in close contact with the outermost parts of the pollen grain wall. Just prior to anthesis, changes in the anther cuticle and connective give rise to extensive dehydration of the tapetal cytoplasm and result in the deposition of the periplasmodium as tryphine on the surface of the pollen grain. This tryphine can be shown to contain both polysaccharide granules and lipid globules, both of which are considered to be distinguishable from the lipid material which is extruded from the pollen grains themselves. The significance of this close relationship which develops between the tapetum and the pollen grain wall at this stage of development has already been discussed by Mepham and Lane [8, 9], Godwin [10] and Echlin [11].

In their study of the development of the tapetum in *Tradescantia* Mepham and Lane have demonstrated that the tissue maintains a high degree of organisation until just prior to anthesis. Because of the close association of the tapetum with the pollen grains, the authors consider it 'could exercise a selective influence over the transport of material into the pollen'. They do not consider that nutrients for pollen growth are derived *from* the tapetum but concede that they may pass *through* the tissue. They also consider that a continuity could develop between the tapetal membrane systems and those in the pollen grain, and that such a connection would facilitate the rapid and easy transfer of nutrients. The full significance of the *extrusion* of lipid material from within the pollen grains is not at present clearly understood.

The development of the secretory tapetum

Considerably more work has been carried out on this type of tissue, and for a summary of the significance of these studies prior to 1968 reference should be made to the papers by Echlin and Godwin [1, 12, 13]. For the sake of completeness of this present record, a brief description will be given of the sequence of events which occur during the development of the secretory tapetum in *Helleborus foetidus*, together with some more recent work on other species.

The earliest recognisable stage is of sporogenous tissue surrounded by actively dividing tapetal cells. These cells have prominent nuclei and nucleoli which tend to be in the half of the cell nearest to the developing sporogenous tissue. In contrast to the findings of Mepham and Lane, the cell organelles are not readily recognisable, although it is possible to see mitochondria and a few plastids. The tapetal cytoplasm also contains a few dictyosomes with peripherally associated vesicles, together with profiles of endoplasmic reticulum. Starch grains are transitorily present in the tapetal cells at this stage.

The tapetal cell walls are relatively thin and appear to be composed of middle lamellae with a small amount of cellulosic primary wall. Polyvesicular bodies are associated with the tapetal–sporogenous tissue cell walls and with the tapetal–endothecium walls.

At the pollen mother cell stage the tapetal–pollen mother cell wall has become much thicker and the tapetal cytoplasm appears more electron-dense, owing to the increased number of ribosomes and pro-Ubisch bodies (see below). The thickening of the tapetal–anther cavity cell wall is irregular, more cellulose being deposited on the tapetal side of the wall. At the time of the meiotic division of the pollen mother cell, the tapetal nuclei are greatly increased in size and the endoplasmic reticulum is now more apparent. By the tetrad stage prominent plasmodesmata can be seen interconnecting the tapetal cells. The cytoplasm contains fewer organelles, with the exception of the dictyosomes, which are highly active, and the cell membrane appears highly convoluted.

At this same stage of development in *Ipomoea purpurea* [14], the tapetal cells reach the peak of maturity, and by the late tetrad stage the cells have already begun to show sign of autolysis. The cell walls are much thinner and eventually disappear, and the protoplasts retract by a process which appears to be associated with the discharge of large vesicles at the cell surface.

In *Beta vulgaris* [15] the endoplasmic reticulum is very contorted and is a prominent feature within the cell. Until this stage the development has been as in *Helleborus* except that the cell organelles were a more prominent feature of the cell cytoplasm during early growth.

At the critical phase of wall initiation the tapetum in *Helleborus* begins to show signs of degeneration. The tapetal cell membrane is highly convoluted and many of the enclosures made by the convulotions are equivalent in size and appearance to the electron-transparent vesicles immediately below the surface, which themselves appear to be derived from the dictyosomes. The endoplasmic reticulum is now recognisable and both smooth and rough profiles may be seen running approximately parallel to the cell surface. There is an increase in the degree of vacuolation within the cell; and although there is a continual decrease in the number of recognisable organelles, the nucleus remains a prominent but increasingly more lobed structure. The common wall between tapetum and anther cavity appears considerably less electron-dense, and there is a space between it and the cell membrane. In *Beta* the tapetal cell walls now degenerate and the cell organelles are liberated into the anther locule. These organelles then undergo further deterioration and the cytoplasmic components of the tapetal cells are interdispersed among the microspores until the onset of microspore vacuolation.

By the stage of primary exine deposition in *Helleborus* the tapetum can be seen to have undergone further deterioration. All the cell walls have virtually disappeared and only a thin layer remains. The remainder of the cytoplasm appears fairly normal with the prominent nuclei and nucleoli occupying a more central position in the cell, surrounded by endoplasmic reticulum, ribosomes and dictyosomes. As wall development proceeds through to the secondary exine deposition and intine formation, the tapetum becomes progressively more disorganised. The mitochondria and plastids are the first organelles to disappear, and this is accompanied by an increase in the vesiculation within the cytoplasm. Other organelles disappear, yet the cell membrane together with the nucleus and endoplasmic reticulum remains intact. During the final stages of anther development the tapetal cytoplasm is completely disorganised. The limiting membrane is ruptured and very few recognisable organelles are found in the residue. A similar cytoplasmic degeneration is seen in *Ipomoea*, and in *Beta* the tapetal organelles are also unrecognisable in the fluid of the anther loculus. Hoefert [15] considers, on morphological evidence alone, that the breakdown products of the tapetal organelles may be incorporated into the microspore. In *Helleborus* this material covers the pollen grains and frequently fills the interbacular cavities.

The recent study by Marquardt, Barth and Von Rahden [16] has concentrated on the behaviour of the nucleus during tapetal development. During the early stages the cytoplasm is rich in ribonucleic acid and protein. This is followed by a phase of mitotic divisions without cell wall formation during which the ploidy of the cells increases to 16 with an accompanying sixteenfold increase in deoxyribonucleic acid. This stage is accompanied by the endoplasmic reticulum becoming far more distended and vesiculate.

During the final stages of development there is an increase in the ribonucleic acid content of the cells, and they show many invaginations in the cell membrane which are interpreted as secretory structures.

It can thus be seen that the two types of tapetum develop in quite different ways, the principal difference being the apparent functional longevity of the periplasmodial type of tapetum compared to the short-lived secretory tapetum. As might be expected, the timing of the various changes in the secretory tapetum shows some small variations; but the general trend is that whereas the development of the tapetum goes through an initial phase of synthesis followed by a process of degeneration, the developing pollen grains go through a continued synthetic stage. It is interesting to note that in the secretory type of tapetum the stage of maximum pollen grain wall development coincides with the stage of tapetal degeneration.

DEVELOPMENT OF THE UBISCH BODIES

One of the distinctive features of the secretory tapetum is that during the tetrad stage its cells often acquire particles of sporopollenin on the inner locular faces. Workers in the Cambridge group were the first to definitively describe the ultrastructural ontogeny of these bodies, and referred to them as Ubisch bodies a term in general use to describe these structures as seen in the light microscope. Heslop-Harrison [2] considers this term inappropriate, claiming that they were observed and commented on before the original description by von Ubisch. He prefers to use the term 'orbicles' or 'plaques'. The three terms will be used interchangeably, as the correct term now appears to be a problem of semantics and personal choice rather than scientific accuracy.

Ubisch bodies are spheroidal structures found in the anthers of many genera of angiosperms, both monocotyledons and dicotyledons, and some gymnosperms. They occur in large numbers on the walls of the tapetal cells, especially those lining the anther loculus. They are generally only a few micrometres in diameter, but frequently fuse into larger compound aggregates in the later stages of their development. In some instances such aggregates constitute platelets extending far across the tapetal cell surface and crossing the cell boundaries. The walls of the coating of Ubisch bodies apparently consist of sporopollenin, the main constituent of the mature microspore exine, and a material of considerable durability and characteristic resistance to acetolysis. The mature wall is usually thick in proportion to the total size of the Ubisch body and, although generally uniform in thickness, it is sometimes verrucose or spinose, so exhibiting a greater or lesser similarity to the pollen grain exine. It would seem that Ubisch bodies may represent the capacity for organising similar but less complex exinous systems by smaller cytoplasmic units. The earlier studies on Ubisch bodies show that they are produced by the glandular or secretory type of tapetum and not by the amoeboid. With the final autolysis of the tapetal cells in the later stages of sporogenesis, the Ubisch bodies tend to lie irregularly upon the remnants of the tapetum next to the fibrous endothecium, and among the maturing pollen grains.

The following is a brief description of Ubisch body formation as seen in *Helleborus foetidus*. For further details of the process together with summaries of earlier work, reference should be made to the paper by Echlin and Godwin [1]. Since the publication of this study some other investigations have been made; although they do not substantially alter the account which follows, reference will be made to these investigations where appropriate.

As indicated earlier the appearance and development of Ubisch bodies is closely associated with the ontogeny of the tapetum. In *Helleborus*, at the sporogenous cell stage, a number of medium-electron-dense bodies are found distributed throughout the tapetal cytoplasm. These bodies have a limiting membrane and are referred to as pro-Ubisch bodies, as subsequent studies show that they develop into the Ubisch bodies proper. As anther development proceeds, the pro-Ubisch bodies increase in number, particularly in that sector of the cell nearest the anther cavity. At the time of late tetrad formation the pro-Ubisch bodies appear to be surrounded by a zone of ribosomes which radiate from them like the spokes of a wheel. The limiting membrane of the pro-Ubisch bodies is usually discontinuous at those places which correspond to the insertion of the rays of radiating ribosomes. During the formation of recognisable exinous elements in the pollen grain wall, profiles of endoplasmic reticulum are seen in close association with the pro-Ubisch bodies, and there appears to be open continuity between the lumen of the pro-Ubisch bodies and the cisternae of the endoplasmic reticulum; the material in both is of similar electron-density and structure. This appearance suggests that both the ribosomes and endoplasmic reticulum are involved in the metabolism of material in the pro-Ubisch body.

Heslop-Harrison and Dickinson [17] found similar medium-electron-dense structures in *Lilium*, but were unable to demonstrate either a limiting membrane around the pro-Ubisch bodies or evidence that their formation was associated with the endoplasmic reticulum and ribosomes. As in the study on *Helleborus*, Heslop-Harrison and Dickinson commented on the close similarity between pro-Ubisch bodies and spherosomes. In the description by Carniel [18] of Ubisch body formation in *Oxalis*, clearly delimited pro-Ubisch body-like structures can be seen in the tapetal cells at this stage, but they are not associated, in the early stages at least, with any sub-cellular components.

The work by Risueno *et al.* [19] on *Allium* to a certain extent corroborates the findings in *Helleborus*. These workers find that the endoplasmic reticulum widens in certain areas and fills with an 'extraordinary electron dense material', and this material is considered to represent the nucleus of the Ubisch bodies. No initial association is seen between pro-Ubisch bodies and ribosomes. As development proceeds, the cisternum of the endoplasmic reticulum continues to dilate and increasing amounts of electron-dense material continue to fill the space. The growth in size of the Ubisch bodies, as these are now called, is by lamellar apposition of membranes. The apparent increase in electron-density of these bodies compared to the pro-Ubisch bodies in *Helleborus*, *Lilium* and *Oxalis* is probably due to the use of permanganate as a fixative rather than glutaraldehyde and osmium.

In spite of some conflicting reports in the earlier literature, there now appears to be general agreement that pro-Ubisch bodies are extruded from the tapetal cytoplasm into the anther loculus. In *Helleborus* it has been possible to show that the pro-Ubisch bodies become incorporated into, and eventually pass through, the tapetal cell membrane. In no instance has it been possible to find a break in the membrane, and it is suggested that during extrusion the membrane surrounding the pro-Ubisch body fuses with the tapetal membrane. Once the pro-Ubisch bodies have been extruded from the tapetal cytoplasm into the space between the membrane and the cell wall, they rapidly become invested with a layer of electron-dense material which appears in some instances to be deposited on both sides of an electron-transparent layer of unit membrane dimensions. This electron-dense material is only deposited after the pro-Ubisch body has been released from the cytoplasm and has never been observed to occur while the body is still within the cell. The same process of extrusion and coating of pro-Ubisch bodies is seen in *Lilium* [17], although occasionally the extrusion of the pro-Ubisch body results in a persistent stalk linking the pro-Ubisch body to the tapetal cell. Heslop-Harrison and Dickinson consider that this would account for the tapetal origin of the unit membrane surrounding each Ubisch body. However, these workers do not find any association of lamellae or electron-transparent layers with the deposition of electron-dense material around the Ubisch bodies. In *Oxalis* the morphological equivalents of pro-Ubisch bodies move to the cell membrane and accumulate in groups, and the deposition of sporopollenin occurs after their extrusion into the anther loculus. It is at this stage in the development of Ubisch bodies in *Oxalis* that strands of endoplasmic reticulum may be seen associated with the pro-Ubisch bodies prior to their extrusion from the cell. No growth-lamellae are seen and Carniel considers that the tubular internal structures of the Ubisch bodies originate by dissolution of the original material of the pro-Ubisch bodies. In *Helleborus* the gradual disappearance of this material is a result of increased deposition of sporopollenin.

In *Allium* the dilated cisternae of the endoplasmic reticulum containing the electron-dense Ubisch bodies furrow the whole of the tapetal cytoplasm, interconnecting with one another and eventually opening into the anther loculus. Risueno *et al.* show

THE CONTRIBUTION OF THE TAPETUM TO THE DEVELOPING POLLEN GRAIN

Early studies had shown that the callose special wall which forms around the pollen mother cell and remains in position until the release of the tetrads is an effective barrier to the interchange of materials between tapetum and developing microspore. Recent studies by Southworth (p. 115, this volume) have shown that both phenylalanine and leucine would not penetrate the callose wall. As callose is a β-1,3-glucan, it was not too surprising that Southworth was able to show that glucose was incorporated into the callose wall. Acetate was incorporated into the wall at all stages of growth, whereas mevalonate was not taken up by the microspores. However, it cannot be discounted that both acetate and glucose may both have been taken up into the wall by incorporation of their breakdown products.

The tapetum appears to make little or no contribution to either the cellulosic primexine or the primary exine, both of which are formed while the microspores are invested with the impenetrable layer of callose. Also, neither the secretory nor the periplasmodial type of tapetum has entered the phase of decline, associated with the possible general release of metabolites into the anther cavity. The possibility cannot be excluded, however, that small molecules —trioses, for example—may not pass from tapetum to developing microspore.

In all the plants so far examined, during the phase of secondary exine deposition, the tapetum passes through a phase of greater or lesser degeneration accompanied by a phase of Ubisch body extrusion in those species which possess these curious structures. It is difficult to concede that the tapetum is not making some contribution to the developing pollen grain wall; and although there is some circumstantial evidence that this is so, definitive proof is still lacking. It is presumed that any such material passing from the tapetum to the microspores would be some precursor of, or unpolymerised, sporopollenin, and one may envisage three ways in which this material may be incorporated into structures within the anther cavity.

(1) Solubilised precursors could be deposited in an orderly fashion, such as on pre-existing sites, thin lamellae or tapes on and within the exine of the pollen grain wall.

(2) The precursor could be deposited in the form of Ubisch bodies.

(3) The precursor could be deposited as small granules into or

on to a pre-existing surface, as appears to be the case in the final stages of wall development in *Helleborus* [13].

Brooks and Shaw [24] have substantial evidence that the pollen grain exine is derived by an oxidative polymerisation of carotenoids and carotenoid esters present in the anther. Heslop-Harrison [25, 26] has followed the development of the pollen grain wall and the maturation of the tapetum of *Lilium longiflorum* in relation to the synthesis of carotenoids as measured by their absorbance at 450 nm. He does not consider that these carotenoid pigments contribute to sporopollenin synthesis, as sporopollenin deposition is completed before these pigments are released. He does concede, however, that although coloured carotenoids do not appear to be involved in sporopollenin synthesis in *Lilium*, this does not preclude colourless precursors of related molecules being involved.

In a more recent paper Brooks and Shaw [27] consider that the Ubisch bodies described by Echlin and Godwin are polymerised carotenoid material. Unlike Echlin and Godwin, Brooks and Shaw consider that these bodies are laid down on the microspore when it is released from the tetrad and 'adhere to the surface in a moderately mobile manner', suggesting some form of incomplete polymerisation. In their studies on *Lilium henryi* the pro-Ubisch bodies are initially laid down on the microspore in close surface proximity to each other; but as the pollen grain matures and expands, so the bodies move apart, resulting in the filigree surface appearance of the pollen grain. Brooks and Shaw claim that this concept would account for pollen wall development in other species, but it is not at present clear how this would operate in *Helleborus* and *Lilium longiflorum*, where the bulk of exine deposition appears to be by accretion around thin membranous structures, followed by lamellar apposition.

In any event, the developing pollen grains must obtain nourishment from some source, as in some species there is up to a thirtyfold increase in volume without an apparent increase in hydration. In the species which have so far been investigated there does not appear to be any substantial amount of reserve material in the pollen grain cytoplasm at or even before this massive deposition of exine. Echlin (unpublished data) has good evidence from the studies on *Helleborus* that there is a massive re-appearance of starch grains in the cytoplasm *after* exine deposition at about the time of pollen tube germination. This would support the contention that the pollen grain has not lost its synthetic ability, as it may now not obtain material from the tapetum, which at this late stage is completely disintegrated. If the materials do not come in one form or another from the tapetum, the only other source of metabolites is the microspores themselves or the degenerating callose, but this is only conjecture, as at present we lack any evidence regarding the ultimate fate of the breakdown products of this substance. Wiermann and his co-workers [28–30] have followed the changes in the content of flavonols, anthocyanins and carotenoids during microsporogenesis and related their findings to investigations carried out in the electron microscope. They found in the garden tulip 'Apeldoorn' that at the time of pollen intine deposition the tapetum changed from a large physiologically active structure to completely disintegrated cells. In both the garden tulip and *Narcissus pseudonarcissus* the major synthesis of flavonolglycosides and carotenoids occurred at about the time of microspore tetrad separation while the tapetum was functionally active, whereas the major appearance of anthocyanins coincided with the degeneration of the tapetum.

Mepham and Lane [6, 8, 9] consider that in *Tradescantia* the exine is secreted, controlled and polymerised by the microspore itself and that the tapetum makes no direct contribution to it, as appears to be the case in all other plants so far investigated. They do not rule out the possibility that the pollen grain absorbs precursors or nutrients from the tapetum, but are insistent that these nutrients are not derived from the degenerative breakdown of the tapetum but are merely transported through it. They consider that the lamellae seen traversing the lipid globules are in fact membranes of the tapetal endoplasmic reticulum preserved in the lipid exudate. It is difficult to differentiate satisfactorily between the lipid material thought to be extruded by the pollen grain cytoplasm and the periplasmodial lipid deposited at the pollen grain surface. However, the report by Mepham and Lane represents the first piece of work on the amoeboid tapetum, which they have shown to be very different from development of the secretory tapetum. Much of their evidence is based on the presence of lipid in the interbaculoid cavities, and it is difficult to see how this exinous material passes out from pollen grain cytoplasm. The recent study of Horvat [31] on the formation and development of the exine in *Tradescantia paludosa* shows that there are abundant lipid deposits in the interbacular cavity, although there are no comparable deposits in the pollen grain cytoplasm.

An interesting paper by Flynn in this volume (p. 121) shows the presence of microtubules in the pollen wall in *Nuphar* and tubules between the exine and the tapetal surface in *Aegiceras*. The microtubules in *Nuphar* are perpendicular to the cell surface and are about 10·0 nm in diameter. The 'extramural' microtubules in *Aegiceras* are the same size and extent from the exine surface to the tapetal surface. These microtubules provide the route for the transfer of materials from sporophyte to gametophyte without encountering a membrane barrier. The intraexinous microtubules seen in *Nuphar* appear similar in some respects to the membranous profiles seen in the lipid exudate of mature *Tradescantia* pollen, which Mepham and Lane consider to penetrate the endexine and become continuous with the endoplasmic reticulum of the pollen grain cytoplasm. Such extraexinous microtubules would support the as yet undemonstrated contention of Mepham and Lane that a route for the transit of materials from within the pollen grain to the outside does exist in *Tradescantia*. The extraexinous microtubules in *Aegiceras* would provide a route for the transfer of material from the tapetum to the pollen grain wall, as originally postulated by Rowley, Mühlethaler and Frey-Wyssling [32].

POLLENKITT FORMATION

The formation and deposition of *Pollenkitt* around the pollen grains of *Lilium*, together with a resumé of earlier reports on the appearance of this substance, has already been described by Heslop-Harrison [17, 25, 33], and it will only be necessary to summarise his findings and conclusions.

The *Pollenkitt* is an oily layer found on the outside of mature pollen grains from many insect-pollinated species, and appears to arise from secretions by the tapetum. Heslop-Harrison was able to show that this material is mainly lipid and contains carotenoids, the principal carotenoid in *Lilium longiflorum* being α-carotene-5,6-epoxide. The work of Brooks and Shaw (described earlier in this paper) suggests that carotenoids are concerned in sporopollenin synthesis but, as stated previously, Heslop-Harrison was able to show that the maximum appearance of coloured carotenoids in the anther occurred well after the period of maximum exine deposition.

The carotenoids first appear as highly osmophilic globules in the tapetal cytoplasm at about the time of tetrad break-up, but do not appear to be released into the thecal cavity until the dissolution of the tapetum. It is interesting to note that in the same species Heslop-Harrison was able to demonstrate that Ubisch body formation and release occurs during the period of maximum exine deposition of the pollen grain and is completed before the *Pollenkitt* is released. Although Mepham and Lane do not refer to it as such, it would appear that *Pollenkitt* deposition occurs in *Trade-*

scantia. These workers observe that lipidophilic globules are extruded from plastids into the periplasmodial cytoplasm and that this lipid material is eventually deposited on the pollen grain surface. They consider that this material is separate and distinguishable from the lipid exuded from within the pollen grain.

The biological function of the *Pollenkitt* is far from certain. It has been suggested that it may act as an insect-attractant, or as a protection against the damaging effects of ultra-violet radiation. Similarly, the sticky nature of the *Pollenkitt* would serve as an adherent to the insect's body, and because of its hydrophobic nature might even be associated with the dispersal of the pollen grains. Heslop-Harrison [34] suggests that the *Pollenkitt* may be functioning as the pollen-borne substances involved in sporophytic incompatibility systems between pollen and stigma.

TRYPHINE FORMATION

In many respects tryphine is similar to *Pollenkitt*. The major difference is that whereas *Pollenkitt* appears to be principally hydrophobic lipids and species-specific carotenoids, the tryphine appears to be a complex mixture of hydrophilic substances and is derived from the breakdown of the tapetal cells during the final stages of anther maturation. It is possible to recognise cytoplasmic elements in the tryphine, notably strands of endoplasmic reticulum and, in *Helleborus*, pro-Ubisch bodies. The material is only transitory and large deposits are not apparent on the surface of mature pollen grains. This may be a consequence of the dehydration of the anther at anthesis, which would result in a dramatic decrease in the bulk of material, which by its nature would be quite hydrophilic.

It would be interesting to identify tryphine chemically, and then see whether both tryphine formation and *Pollenkitt* formation occur in the same species. Although a detailed survey is lacking, it appears that tryphine formation occurs in both insect- and wind-pollinated plants. The function of tryphine is obscure, although some of the functions ascribed to *Pollenkitt* may equally well apply. It is just as likely, however, that it has no function, and it may represent the final dissolution of the tapetal cells.

CONCLUSIONS

Although a considerable amount of work has been carried out describing the ultrastructure and ontogeny of the tapetum of about a dozen species of flowering plants, there is a remarkable paucity of information about the biochemistry and physiology of this tissue. It is becoming more apparent that the tapetum does in fact contribute material to the developing microspore, but there are no data as to *what* is contributed and how this contribution is effected.

It is now the turn of the biochemist and the cytochemist to provide answers to some of the remaining questions. Following the lead on anther culture by Nitsch, it should now be possible to incorporate labelled precursors into anthers and follow the course of this material during development. Specifically, it should be possible to introduce labelled glucose into the callose which surrounds the developing microspore and to follow the fate of this material after tetrad break-up. This should provide some information regarding the possible contribution to exine development by the material derived from callose dissolution. Similarly, the use of labelled acetate or mevalonate may provide data concerning both the formation of sporopollenin precursors in the tapetum and the pollen grain, and the subsequent deposition and polymerisation of this material either at the pollen grain surface or in the form of Ubisch bodies.

REFERENCES

1 ECHLIN, P. and GODWIN, H., 'The ultrastructure and ontogeny of pollen in *Helleborus foetidus* L. I. The development of the tapetum and Ubisch bodies', *J. Cell. Sci.*, **3**, 161–74 (1968)

2 HESLOP-HARRISON, J., 'Pollen wall development', *Science, N.Y.*, **161**, 230–7 (1968)

3 FOSTER, A. A. and GIFFORD, E. M., *Comparative Morphology of Vascular Plants*, Freeman, San Francisco (1959)

4 MAHESHWARI, P., *An Introduction to the Embryology of Angiosperms*, McGraw-Hill, New York (1950)

5 HESLOP-HARRISON, J., 'An acetolysis-resistant membrane investing tapetum and sporogenous tissue in the anthers of certain Compositae', *Can. J. Bot.*, **47**, 541–2 (1969)

6 MEPHAM, R. H. and LANE, G. R., 'Formation and development of the tapetal periplasmodium in *Tradescantia bracteata*', *Protoplasma*, **68**, 175–92 (1969)

7 ECHLIN, P., GODWIN, H., CHAPMAN, B. and ANGOLD, R., 'The structure of polyvesicular bodies associated with cell walls in developing anthers of *Ipomoea purpurea* (L.) Roth.', *VIth Int. Cong. Elec. Micros.*, *Kyoto*, Vol. II, 317–18 (1966)

8 MEPHAM, R. H. and LANE, G. R., 'Exine and the role of the tapetum in pollen development', *Nature, Lond.*, **219**, 961–2 (1968)

9 MEPHAM, R. H. and LANE, G. R., 'Role of the tapetum in the development of *Tradescantia* pollen', *Nature, Lond.*, **221**, 282–4 (1969)

10 GODWIN, H., 'Pollen exine formation', *Nature, Lond.*, **220**, 389 (1968)

11 ECHLIN, P., 'Development of the pollen grain wall', *Ber. dt. bot. Ges.*, **81**, 461–70 (1969)

12 ECHLIN, P. and GODWIN, H., 'The ultrastructure and ontogeny of pollen in *Helleborus foetidus* L. II. Pollen grain development through the callose special wall stage', *J. Cell. Sci.*, **3**, 175–86 (1968)

13 ECHLIN, P. and GODWIN, H., 'The ultrastructure and ontogeny of pollen in *Helleborus foetidus* L. III. The formation of the pollen grain wall', *J. Cell. Sci.*, **5**, 459–77 (1969)

14 GODWIN, H., ECHLIN, P. and CHAPMAN, B., 'The development of the pollen grain wall in *Ipomoea purpurea* (L.) Roth.', *Rev. Palaeobot. Palynol.*, **3**, 181–95 (1967)

15 HOEFERT, L. L., 'Ultrastructure of *Beta* pollen. I. Cytoplasmic constituents', *Am. J. Bot.*, **56**, 363–8 (1969)

16 MARQUARDT, H., BARTH, O. M. and VON RAHDEN, U., 'Zytophotometrische und elektronenmikroskopische Beobachtungen über die Tapetumzellen in den Antheren von *Paeonia tenuifolia*', *Protoplasma*, **65**, 407–21 (1968)

17 HESLOP-HARRISON, J. and DICKINSON, H. G., 'Time relationships of sporopollenin synthesis associated with tapetum and microspores in *Lilium*', *Planta*, **84**, 199–214 (1969)

18 CARNIEL, K., 'Licht und elektronenmikroskopische Untersuchung der Ubischkorperentwicklung in der Gattung *Oxalis*', *Öst. bot. Z.*, **114**, 490–501 (1967)

19 RISUENO, M. C., GIMENEZ-MARTIN, G., LOPEZ-SAEZ, J. F. and GARCIA, M. I. R., 'Origin and development of sporopollenin bodies', *Protoplasma*, **67**, 361–74 (1969)

20 GHERARDINI, G. L. and HEALEY, P. L., 'Dissolution of outer wall of a pollen grain during pollination', *Nature, Lond.*, **224**, 718–19 (1969)

21 KNOX, R. B. and HESLOP-HARRISON, J., 'Cytochemical localization of enzymes in the wall of the pollen grain', *Nature, Lond.*, **223**, 92–4 (1969)

22 CARNIEL, K., 'Uber die Kornchen-Schict in den Pollensäcken von *Gnetum gnemon*', *Öst. bot. Z.*, **113**, 368–74 (1966)

23 BANERJEE, U. C., 'Ultrastructure of the tapetal membranes in grasses', *Grana palynol.*, **7**, 365–77 (1967)

24 BROOKS, J. and SHAW, G., 'Chemical structure of the exine of pollen grains and a new function for carotenoids in nature', *Nature, Lond.*, **219**, 532–3 (1968)

25 HESLOP-HARRISON, J., 'Tapetal origin of pollen coat substances in *Lilium*', *New Phytol.*, **67**, 779–86 (1968)

26 HESLOP-HARRISON, J., 'Wall development within the microspore tetrad of *Lilium longiflorum*', *Can. J. Bot.*, **46**, 1185–92 (1968)

27 BROOKS, J. and SHAW, G., 'The post-tetrad ontogeny of the pollen wall and the chemical structure of the sporopollenin of *Lilium henryi*', *Grana palynol.*, **8**, 227–34 (1968)

28 WIERMANN, R., 'Untersuchungen zum Phenylpropanstoffwechsel des Pollens. I. Übersicht über die bei Gymnospermen und Angiospermen isolierten flavonoiden Verbindungen', *Ber. dt. bot. Ges.*, **81**, 3–16 (1968)

29 WIERMANN, R., 'Untersuchungen zum Phenylpropanstoffwechsel des Pollens. II. Über die Veranderungen des Flavonol und Anthocyaningehaltes während der Mikrosporogenese,' *Planta*, **88**, 311–20 (1969)

30 WIERMANN, R. and WEINERT, H., 'Untersuchungen zum Phenylpropanstoffwechsel des Pollens. III. Über die Auspigmentierung während der Endphasen der Mikrosporogenese in den Antheren der Darwintulpen-Hybride "Apeldoorn" ', *Z. Pfl. Physiol.*, **61**, 173–83 (1969)

31 HORVAT, F., 'Contribution a la connaissance d'l'ultrastructure des parois du pollen de *Tradescantia paludosa* L.', *Grana palynol.*, **6**, 416–34 (1966)

32 ROWLEY, J. R., MÜHLETHALER, K. and FREY-WYSSLING, A., 'A route for the transfer of material through the pollen grain wall', *J. biophys. biochem. Cytol.*, **6**, 537–8 (1959)

33 HESLOP-HARRISON, J., 'Anther carotenoids and the synthesis of sporopollenin', *Nature, Lond.*, **220**, 605 (1968)

34 HESLOP-HARRISON, J., 'Ribosome sites and S-gene action', *Nature, Lond.*, **218**, 90–1 (1968)

Dictyosome Development During Microsporogenesis in *Canna generalis**

J. J. Skvarla, *Samuel Roberts Noble Laboratory of Electron Microscopy, Department of Botany and Microbiology, University of Oklahoma, Norman, Oklahoma, U.S.A.*
A. G. Kelly, *Department of Biology, Lycoming College, Williamsport, Pennsylvania, U.S.A.*

The Golgi apparatus undergoes profound changes in form in the course of microsporogenesis in *Canna generalis*. In the micro-sporocytes before the meiotic divisions, the Golgi constitutes several interassociated dictyosomes. Each dictyosome consists of a stack of cisternae 0·5 μm or less in diameter (Figs. 1–3). Tubular proliferations approximately 300 Å wide extend from the periphery of the cisternae and are united in networks of tubules. These tubules can be traced considerable distances from the central regions of the cisternae and there is evidence of interconnection between those of neighbouring dictyosomes (Figs. 4–5); some aspects suggest that large groups of dictyosomes may be interconnected in this manner (Figs. 6 and 7).

In both ontogeny and function the Golgi apparatus and the endoplasmic reticulum (ER) are closely associated [1, 2], and there may be physical links between the two systems [3]. Both laminar ER and tubular ER are present in the *Canna* microsporocytes, the former mainly rough-surfaced and the latter either rough or smooth. The smooth tubular form of the ER frequently appears to be associated with the dictyosomes. The mid-regions of the cisternae in the dictyosomes are lamellate, in contrast with the tubular peripheral regions, and the lamellate part appears to become progressively more extensive from one face to the other (Fig. 8). On the broader face, where the tubular elements are more exten-sive, there are indications of continuity with the ER. However, when the membranes are closely packed, the distinction between dictyosome and ER is not readily made [1].

* Abstract.

Ribosome numbers are low in the microsporocyte at this stage. Those near the dictyosomes are mostly situated in the vicinity of the smaller face; this association is the more conspicuous because of the paucity of ribosomes, and there is an obvious possibility of some functional interaction.

The microsporocytes show an extraordinary frequency of morphological 'twins' in the dictyosome population (Figs. 9, 11 and 12). The members of a pair always seem to be at equivalent stages of development, with similar dimensions and conformations. In cross-sectional view the members of a pair often appear as mirror images of each other (Fig. 9). Such views might be interpreted as resulting from the sectioning of the arms of U-shaped or urn-shaped dictyosomes, but this seems improbable, because sections of extremely reflexed dictyosomes have never been seen.

In meiotic prophase stages there is no clear morphological evidence of secretory activity by the dictyosomes, but with the beginning of the deposition of the last pollen wall layer, the intine, prolific vesiculation begins (Figs. 13–15). This is in accord with other studies which suggest that Golgi vesicles contribute to the intine [4]. The vesicles from the dictyosomes are often larger and more abundant around the less extensive face, which suggests that the cisternae on this face are more active in secretory functions than are those of the other face.

In the mature pollen, after completion of wall formation, the secretory activity subsides, and the dictyosomes appear as discontinuous entities, with few associated vesicles (Fig. 16).

Numbers of cisternae per dictyosome were estimated in counts of all sectioned views at three developmental stages. Results were as follows:

	Mean number of cisternae	Range
Microsporocytes	8·0	3–14
Late microspore	5·7	3–9
Mature pollen	5·3	3–7

There is thus a progressive fall in the numbers of cisternae as microsporogenesis progresses.

It may be assumed that the minimal number of three cisternae persists so long as the dictyosome maintains its integrity. The pattern of variation suggests that there is an accumulation of additional, non-vesiculating cisternae in early development, after which extensive vesiculation again reduces the number of cisternae. The

ELECTRON MICROGRAPHS OF MICROSPOROCYTES AND SPORES OF *Canna generalis*. FIXATION WITH ACROLEIN-GLUTARALDEHYDE-OsO_4; POST-STAINING OF SECTIONS WITH URANYL ACETATE AND LEAD CITRATE. THE ORIGINAL MAGNIFICATIONS ARE GIVEN; ALL FIGURES REPRODUCED AT $\frac{3}{5}$

Figures 1–3. Sections of individual dictyosomes in face (Fig. 1), side (Fig. 2) and oblique (Fig. 3) views. Note the highly tubular nature of cisternae periphery. Magnification: Fig. 1, ca. ×41000; Fig. 2, ca. ×33600; Fig. 3, ca. ×30000

Figures 4–5. Sections through two interassociated dictyosomes in face and side views, respectively. Ribosomal aggregates are evident between interassociated tubules. Magnification: Fig. 4, ca. ×31000; Fig. 5, ca. ×40000

Figures 6–7. Linear arrays of interassociated dictyosomes. Ribosomes are present in cross-sections of membranes in cytoplasm and on less extensive faces of dictyosomes. Magnification: Fig. 6, ca. ×35000; Fig. 7, ca. ×42000

Figure 8. Cross-section of dictyosome illustrating accumulation of tubules at more extensive face. Magnification: ×40000

Figure 9. Cross-section of dictyosome 'twins' displaying mirror-image effect. Magnification: ca. ×29000

Figures 10–12. Sequence of views illustrating formation of dictyosome 'twins'. Magnifications: Fig. 10, ca. ×41000; Fig. 11, ca. ×41000; Fig. 12, ca. ×41000

Figure 13. Face view of dictyosome showing secretory vesicle production from peripheral tubules. Magnification: ca. ×19000

Figures 14–15. Secretory vesicles appear most abundant at less extensive face of dictyosomes. Ribosomes are also abundant at less extensive faces. Magnification: Fig. 14, ca. ×21000; Fig. 15, ca. ×35000

Figure 16. Mature pollen showing dictyosome at low level of activity. Magnification: ca. ×38000

(Figures 8–16 on p. 66)

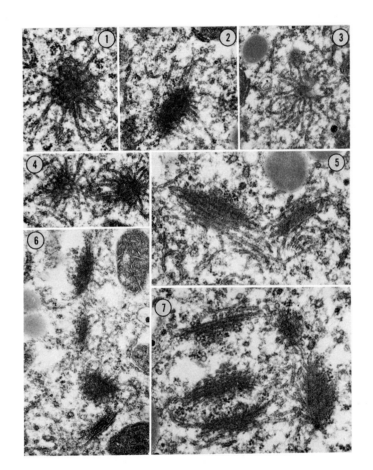

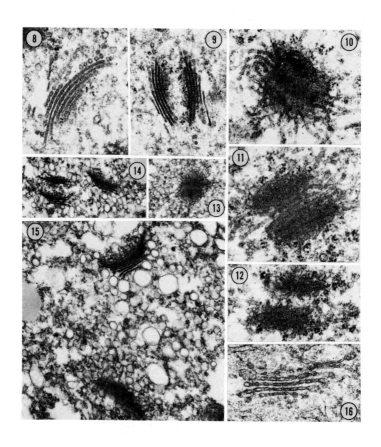

added cisternae may originate as convergent tubules of smooth ER, and the consistent appearance of tubular elements at the more extensive face indicates that this is the forming face. If correct, this would imply that existing cisternae serve as a locus for the assembly and fusion of ER-derived elements to produce new cisternae. This interpretation gains some support from the fact that the associated ribosomes are clustered near the opposite, less extensive, face, where vesiculation is prolific, an indication of greater functional activity.

Formation of the dictyosome in this fashion would account for the interconnections, and also for the elaborate form of the Golgi apparatus when seen after isolation and negative staining [5–7]. However, this interpretation does not account for the 'twinning' seen in the microsporocyte. It is suggested that the dictyosomes continue to grow by fusion of ER elements until they attain a certain size, whereupon they fragment into 'daughter' dictyosomes by vertical cleavage. A more convincing argument for this point could be made if images were available showing dictyosomes approximately twice the size of the individual twins, and of the actual fragmentation. These have not yet been observed; but it could be that the divisions are more or less synchronous in the microsporocytes, in which case the critical period could be missed easily enough.

This work was supported by National Science Foundation grant No. GB 6768.

REFERENCES

1 ESSNER, E. and NOVIKOFF, A. B., 'Cytological studies on two functional hepatomas: Interrelationships of endoplasmic reticulum, Golgi apparatus and lysosomes', *J. Cell Biol.*, **15**, 289–312 (1962)
2 NOVIKOFF, A. B., ESSNER, E., GOLDFISCHER, S. and HEUS, M., 'Nucleoside–phosphatase activities of cytomembranes'. In: *The Interpretation of Ultrastructure* (Ed. R. J. C. HARRIS), 149–92, Academic Press, New York (1962)
3 PALADE, G. E., 'The endoplasmic reticulum', *J. biophys. biochem. Cytol.*, **2**, Suppl., 85–98 (1956)
4 HESLOP-HARRISON, J., 'Pollen wall development', *Science, N.Y.*, **161**, 230–7 (1968)
5 CUNNINGHAM, W. P., MORRE, D. J. and H. H. MOLLENHAUER, 'Structure of isolated plant Golgi apparatus revealed by negative staining', *J. Cell Biol.*, **28**, 169–79 (1966)
6 CUNNINGHAM, W. P. and MORRE, D. J., 'Interassociation of dictyosomes to form plant cell Golgi apparatus', *J. Cell Biol.*, **27**, 68A (1965)
7 CUNNINGHAM, W. P., 'Tubular connections between dictyosomes and forming secretory vesicles in plant Golgi apparatus', *J. Cell Biol.*, **29**, 373–6 (1966)

Pollen Grain and Sperm Cell Ultrastructure in *Beta**

L. L. Hoefert, *United States Department of Agriculture Research Station, Salinas, California, U.S.A.*

Pollen development after mitosis involves the following stages: (*a*) formation of the generative cell wall, (*b*) attachment of the generative cell wall to the pollen grain intine, (*c*) separation of the generative cell from the intine, (*d*) free generative cell, (*e*) spindle-shaped generative cell, (*f*) generative cell division, and (*g*) development of the two sperm cells. In *Beta* all these events occur before anther dehiscence and shedding of pollen.

The first mitotic division resembles normal mitosis except that the plane of division is radial. The generative cell chromosomes occupy a peripheral position, while the vegetative chromosomes are centrally located in the dividing microspore. A curved cell plate develops and attaches to the pollen grain intine. The unequal mitosis results in a smaller generative cell enclosed by a wall, and a larger vegetative cell. A similar sequence of generative wall formation has been described in *Dactylorchis* [3] and in *Endymion* [1]. In monitor sections the generative cell wall of *Beta* stains with ruthenium red, periodic acid–Schiff's reagent, resorcin blue and aniline blue, which indicates the presence of pectic substances, carbohydrate and callose. The generative cell increases in size with a concomitant decrease in thickness of its wall, and separates from the pollen grain intine. Microtubules are present inside the generative cell wall from the time of its inception, and are aligned parallel to the wall surface. The free generative cell assumes an ellipsoidal shape prior to the second mitotic division within the pollen grain. In the ellipsoidal generative cell, microtubules are aligned with the long axis of the cell. The generative cell wall at this stage appears either to disappear or to be greatly diminished in extent. Monitor sections show no staining of the generative wall for carbohydrate or

* Abstract.

68

pectic substances. Gorska-Brylass [2] observed a loss of callose staining in the generative cell wall at the ellipsoidal stage in *Chlorophytum*, *Hyacinthus* and *Tradescantia*.

The second mitotic division occurs wholly within the two limiting membranes of the generative cell. Nuclear division does not appear to differ from normal mitosis but the details of cytokinesis have not been recorded. Therefore the mechanism by which the membranes develop around the two sperm cells, and the means of dispersal of cytoplasm between the two cells, are matters that require further study.

The two sperm cells in *Beta* are spindle-shaped and surrounded by two membranes. Each sperm cell contains its own cytoplasm and each displays cytoplasmic microtubules with a pole-to-pole orientation. A similar orientation of microtubules occurs in protozoan parasites [4], and it is thought to have some function in movement of the protozoan through the host cytoplasm.

REFERENCES

1 ANGOLD, R. E., 'The formation of the generative cell in the pollen grain of *Endymion non-scriptus* L.', *J. Cell Sci.*, **3**, 573–8 (1968)
2 GORSKA-BRYLASS, A., 'Temporary callose wall in the generative cell of pollen grains', *Naturwissenschaften*, **54**, 230–1 (1967)
3 HESLOP-HARRISON, J., 'Synchronous pollen mitosis and the formation of the generative cell in massulate orchids', *J. Cell Sci.*, **3**, 457–66 (1968)
4 SHEFFIELD, H. G. and MELTON, M. J., 'The fine structure and reproduction of *Toxoplasma gondii*', *J. Parasit.*, **54**, 209–26 (1968)

Histochemistry and Ultrastructure of Pollen Development in *Podocarpus macrophyllus* D. Don*

I. K. Vasil and H. C. Aldrich, *Department of Botany, University of Florida, Gainesville, Florida, U.S.A.*

In the early stages of development of the microsporophyll the cells are rectangular with a profusely vacuolated cytoplasm and numerous amyloplasts, and plasmodesmatal connections are present. The archesporial cells are differentiated in the hypodermis on either side of the provascular strand; they show increased basiphilia and have fewer amyloplasts. This is the condition during the dormant phase, which lasts about ten months. Thereafter the archesporial cells undergo repeated periclinal divisions to produce 5–6 layers of cells, including 1–2 layers of tapetum. The sporogenous cells are characterised by a profusion of vesicles in their cytoplasm at this stage, and show little PAS reaction. As they enter the meiotic prophase, the cytoplasm withdraws from the wall and the plasmodesmata disappear. No cytomictic or plasma channels are formed, a fact consistent with the asynchrony observed in the developing microspore mother cells. The nucleus enlarges greatly in early prophase, and the pores in the nuclear envelope are conspicuous. Synaptinemal complexes are formed in zygotene, in the manner now familiar from many other species.

During prophase the tapetal cells are characterised by abundant ribosome and polysome populations, amyloplasts and Golgi activity. The outer tangential walls of the tapetal cells break down from the middle lamella inwards, leaving the middle lamella and the outer part of the wall, that of the adjacent sporangium wall cell, intact. The inner tangential and radial walls of the tapetal cells are eventually broken down completely.

* Abstract.

As the plasmalemma of the microspore mother cells withdraws from the parent wall, mildly PAS-positive fibrillar material remains, presumably a breakdown product of the original wall. A thin wall is now secreted *de novo* at the plasmalemma (Fig. 1). This shows weak fluorescence in the aniline blue–UV test, indicating that callose

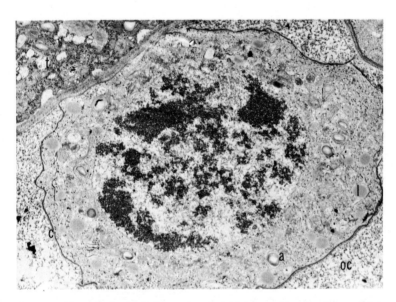

Figure 1. *Microspore mother cell in early meiosis showing a thin callose-like wall (c), remains of the original cellulose wall (oc) and part of a tapetal cell (t). a, amyloplast; l, lipid. Magnification: × ca.3300*

may be present. However, the massive callose special wall seen in most other species of seed plants is not found in *Podocarpus*.

Long, thin, three-layered structures of an intensely electron-dense material appear in the cytoplasm of the microspore mother cells in early meiosis (Fig. 1). These are generally associated with Golgi cisternae and/or endoplasmic reticulum, and gradually accumulate near the plasmalemma (Fig. 2). It is possible that the material is a sporopollenin precursor.

While the parenchymatous cells of the cone axis are packed with starch, there is very little in the microspore mother cells and tapetal cells during meiosis, and this is true also of the tetrads and young microspores. During the later development of the male gametophyte most of the starch disappears from the cone axis, and massive accumulation of protein, starch and lipid globules begins in the cytoplasm of the pollen grains.

The tapetal nuclei and cytoplasm show signs of intense activity from meiosis II until after tetrad formation. Droplets of sporopollenin aggregate on the outer surface of the tapetal plasmalemma.

The air-sacs are formed after the deposition of the sexine but while the four microspores are still associated in the tetrad. To

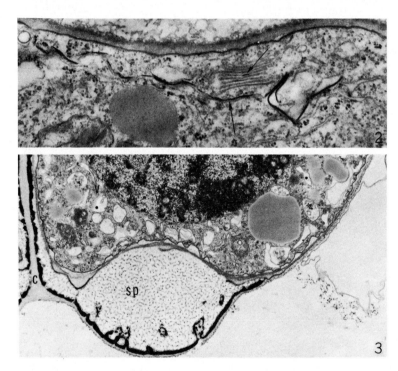

Figure 2. Portion of the peripheral cytoplasm and newly formed callose-like wall from a microspore mother cell in early meiosis. Bands of electron-dense material are seen on the endoplasmic reticulum and Golgi cisternae (arrows). Magnification: ×22000
Figure 3. Parts of two microspores from a tetrad showing their attachment by a thin layer of callose-like material (c), an air-sac filled with a network of sporopollenin (sp) and formation of vesicles from the nuclear envelope (v). Magnification: ×8400

begin with, a fine reticulate network of sporopollenin-like material fills the air-sacs (Fig. 3), but this is later lost. Occasional bands of microtubules are seen along the plasmalemma and nuclear envelope of the young spore, along with coated vesicles in the peripheral cytoplasm. There is a profusion of vesicles originating from Golgi lamellae; the nuclear surface is marked by convoluted invaginations (Fig. 3), and some vesicles may have their origin here. There is as yet no evidence to suggest what materials are being transferred by this process.

Several layers of granular material are secreted outside the spore plasmalemma, and an uneven layer of sporopollenin is deposited between this material and the callose-like wall. The first depositions occur in the regions of the probacula. A continuous layer, representing the sexine, is deposited all around the spore except at the lower or proximal end between the air-sacs, the zone where the tube emerges during germination. The sexine is very electron-dense to begin with, but some density is lost as it matures. Sculpturing of the sexine is completed before release of the spores from the tetrad.

The tapetal protoplasts degenerate rapidly after the dissolution of the spore tetrads. The degeneration takes place while the plasmalemma is still intact. Cytoplasm and most cell organelles disappear, but the sporopollenin plaques are persistent.

A thick continuous layer of sporopollenin is deposited between the sexine and the plasmalemma of the microspore, and on the outside it shows at least five distinct electron-transparent and laminated layers over which new sporopollenin deposition takes place. This is the laminated portion of the nexine or endexine, the nexine 1. A continuous layer formed within is the nexine 2. In *Podocarpus* a third nexine layer, nexine 3, is laid down during the final stages of pollen maturation. The wall of the air-sacs is formed of sexine only (Fig. 3).

As the microspore nucleus undergoes mitosis, the intine is deposited between the plasmalemma and nexine. This layer is very thin in the upper or distal portion of the grain where the prothallial cells are formed. All the nuclei of the male gametophyte are clearly demarcated from one another by their own cytoplasm and plasmalemma, in addition to inconspicuous cell walls traversed by plasmodesmata. The cytoplasm of the pollen grain is ultimately filled with a large number of lipid globules, starch grains and a granular electron-dense material, which is proteinaceous in nature. As the pollen grains mature, the small granules coalesce to form larger masses.

The surface of the mature exine in the neighbourhood of the air-sacs is marked by minute undulations caused by a large number of small but smooth projections, seen in scanning electron micrographs. A number of crater-like depressions appear on the surface of the air-sacs.

The line of demarcation between the distal face of the grain and the air-sacs is markedly different in sculpturing from the rest of the pollen surface, bearing a number of thick finger-like processes. The slit-like area from which the tube emerges is clearly seen from the proximal pole.

Detailed accounts of this work are to appear in *Journal of Ultrastructure Research* and *Protoplasma*.

The Pollen Wall: Structure and Development

J. Heslop-Harrison, *Institute of Plant Development, University of Wisconsin, Madison, Wisconsin, U.S.A.*

THE MATURE WALL: STRATIFICATION AND SURFACE FEATURES

Before the advent of electron microscopy, the investigation of the stratified and often elaborately patterned wall of the angiosperm pollen grain depended upon optical microscopy of whole and sectioned grains and upon the staining reactions of the strata. One of the most effective approaches was the so-called LO (Lux-Obscurus) method of Erdtman [1], which depends upon the interpretation of the diffraction images produced during downward focusing through the wall. In conjunction with the study of sectioned walls and the use of ultra-violet microscopy, this method led to a satisfactory understanding of the main morphological features of wall stratification, which has now been confirmed in its essentials and extended into much greater detail by the electron microscopy of ultra-thin sections. The structural studies have been complemented by microchemical methods, and differences in the chemical composition of the strata have been exposed by staining and other procedures. The importance of this factor in the classification of wall layers has been stressed especially by Faegri [2]. Yet a third approach to interpreting the stratigraphy is offered by ontogenetic studies, to be discussed in some detail in this chapter.

With several sources of criteria available, more than one valid classification of the layers of the pollen wall can be made. Erdtman [3] has suggested the morphological terminology set out in Fig. 1. We will adopt Erdtman's classification here as being widely known from his extensive publications.

In general, the stratification of angiosperm pollen grain walls can be understood in terms of the layers defined in Fig. 1. Variations occur in the relative thickness of the various components; for

75

example, there are some grains where the nexine 2 is enormously thick, and there appear to be others in which it is absent. Subdivisions may occur within the various layers to produce aspects that can be of considerable taxonomic significance. Thus, Larson, Skvarla and Lewis [4] and Skvarla and Turner [5], extending the earlier work of Stix [6] on composite pollen, showed that the foot layer or nexine 1 may be detached in some species from the bases of the bacula, and that the latter may be connected by a second foot layer or *stutzmembran*, separated from the nexine 1 by a *cavus*. Also exemplified in the Compositae is the development of bacula in more than one layer. An 'internal tectum' may be produced below the level of the tectum proper, and this may form a continuous stratum

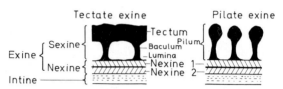

Figure 1. *Terminology of pollen wall layers, based upon that of Erdtman* [1]. *The characteristic material of the exine is sporopollenin; the intine is pectocellulosic. In the mature, untreated grain, fibrillar material derived from the primexine is present in the lumina of the tectate exine and between the columns of the pilate exine*

or be present over only part of the surface of the grain. In still more elaborate examples several layers of tecta may be formed, sometimes in such complexity that the numbers cannot be distinguished.

Pores and slits form conspicuous features of the pollen wall. These are defined as regions where the full exine stratification is absent or modified, the area of the aperture being covered only by the intine and layers corresponding to the nexine, usually without bacula, or with bacula over only part of the area. Papillae or more distinct elongated processes may arise from the nexine over the aperture region, and not infrequently the central part is found to be thickened somewhat, forming an operculum, lens-shaped in section. In certain pollen grains, as in those of the grasses, the pore region is surrounded by an annulus derived by the thickening of the nexine 2. In other pollen types the annulus is bordered in turn by a region of modified sexine structure, so that an elaborate 'aperture mechanism' exists.

Where a tectum is not formed, the pilae, or the muri resulting from the fusion of their heads, may be distributed to form a distinctive pattern. The tectum may ensheath the baculate area completely, or be only partial. Even where the tectum is complete, the distribution of the bacula seen through it may be characteristic for a particular taxon. The surface of the tectum itself may reveal a wealth of structural detail useful for classificatory purposes, in the distribution and dimensions of perforations, ridges, striae, spines, spinules, warts, papillae and the like.

Shortly before final maturation of the pollen grains, surface coatings of various kinds are acquired, derived from secretory or breakdown products of the tapetum. These coatings have been variously named as *perine, tryphine* and *Pollenkitt.* Practically all pollens carry some tapetal debris at the time of dispersal, and this may have little or no functional significance. There is no doubt, however, that in many species materials applied to the exine during the final phase of activity of the tapetum fulfil important roles. They may, for example, serve to cement grains together, to aid dispersal through the agency of insects by giving the grains characteristic colours and odours, and perhaps also to protect the protoplast from radiation damage during dispersal. It is now also known that enzymes may be transferred to the exine surface and even into its cavities during pollen maturation (p. 171), and there is a possibility that incompatibility substances are also surface-borne in some species.

CHEMISTRY OF THE MICROSPORE WALLS

Chemically, the primary division in the wall of the pollen grain is between the intine and the exine.

The *intine* has long been known to be pectocellulosic [7], and it seems that its constitution differs in no essential features from that of the primary wall of somatic cells. Intines of different species vary greatly in thickness; in some the cellulosic layer is too thin to be observed with the optical microscope, while in others, such as certain *Carex* species, it occludes practically the whole lumen of the pollen grain. The cellulose of the intine shows a microfibrillar structure electron microscopically. The microfibrils are oriented parallel to the surface of the grain, but their disposition in this plane is usually random, even in non-spherical grains, except sometimes in the vicinity of elongated apertures. Because of this distribution of the microfibrils, the intine shows negative spherical birefringence when viewed in section with the polarising microscope; that is to say, the maximum extinction plane is tangential. The intine often appears lamellated in electron micrographs, and at least in some pollens this is due to the interbedding of the cellulose layers with layers of protein (p. 172).

The most striking characteristic of the material of the exine is its remarkable resistance to degradation by both physico-chemical and biological agencies. The general class of resistant exine materials was given the name 'sporopollenin' by Zetzsche [8]. In the preparation of pollen for palynological work it is customary to remove contaminating materials from the exine by acetolysis [9]. Sporopollenin is the only class of plant wall polymer to survive this treatment without solvation or radical structural changes, so acetolysis has achieved something of the status of a microchemical test for this type of compound. Sporopollenin is degraded by chromic acid and by 2-amino-ethanol [10].

Recent work on the chemistry of sporopollenin is reviewed in this volume by Brooks and Shaw (p. 99). Here we may note that notwithstanding the broad chemical similarity of sporopollenins from different sources, the name defines a range of compounds. During the early development of the exine of lily, progressive chemical change is shown by varying reactivity to stains. Resistance to acetolysis is acquired early in the formation of the sexine, but the material of the wall is not chemically identical with that of the mature exine at this time, and, hence, it has been referred to as protosporopollenin [11]. There is also a differentiation within the mature exine itself, since, as recognised by Faegri [2], the last exine layer to be formed, the nexine 2 or 'endexine', responds differently from the layers formed earlier to stains such as acid fuchsin. Taken together with the fact that exines from different species vary in their responses to stains, these observations suggest that 'sporopollenin' is variable either in the relative proportions of different moieties, or in the degree of cross-linking, or in the extent of the masking of reactive groups.

Variations in the physical state of sporopollenin may also account for some of the differences in staining responses. In several early electron microscopic studies of exine structure, distinctions were drawn between homogeneous and lamellated sporopollenin. Afzelius, Erdtman and Sjöstrand [12] noted lamellae of various dimensions in the outer part of the spore wall of *Lycopodium clavatum*, and concluded from the distribution and thickness that they could represent aggregates of 'elementary lamellae' of ca. 60 Å. In later work Afzelius [13, 14] found the inner part of the exines of various gymnosperms and angiosperms to be lamellated, again with lamellae of thickness 50–60 Å, while the outer part was amorphous-granular. The lamellated stratum corresponded to the nexine 2 of Erdtman's terminology (Fig. 1), and the amorphous-granular to the sexine plus nexine 1. Afzelius suggested that both structural forms were composed of 50–60 Å granules, and calculated that such granules could contain 50 'units' of the empirical formula proposed by Zetzsche and Vicari [15] for the sporopollenin of *Lycopodium*.

Many subsequent fine-structural studies have revealed a lamellated structure in part or all of the nexine 2 of mature grains, although it is now clear that the age of the grain and the type of preparative technique, including the pretreatment given and the types of electron stains used, greatly affect the image obtained.

These fine-structural studies show that the distinction made by Faegri [2] between the outer, basic fuchsin-staining 'ectexine' (= sexine + nexine 1) and an inner 'endexine' (= nexine 2) with little affinity for the stain is in part a distinction between the strata of the exine with the homogeneous type of sporopollenin, and strata of a more persistently lamellated type. The distinction is also apparent in optical properties of the strata. Sitte [16] demonstrated negative spherical form birefringence in the exines of various flowering plants, and Stix [17] was able to show for *Echinops banaticus* that this was contributed by the inner layer, corresponding to the nexine 2. These observations confirm the supposition that this layer has a concentrically lamellated structure. In *Echinops* bacula and tectum revealed positive spherical form birefringence, which indicates that in these parts of the exine the submicroscopic structural elements are radially arranged.

As we shall see, these variations in the state of the sporopollenin in various parts of the exine probably reflect differences in the developmental history of the strata. The first-formed parts, bacula, and tectum where present, reveal a lamellated structure at the time of their inception, with the lamellae directed radially; the nexine is formed later by the apposition of tangentially oriented lamellae. After further consolidation of the sexine, the lamellated appearance can no longer be detected electron microscopically, although it seems that in species like *Echinops banaticus* the basic anisotropy continues to be revealed in birefringence properties. The nexine 2, in contrast, never appears to consolidate as completely as the sexine, so that the lamellated structure is often visible electron microscopically in the mature grain.

Some reports [18, 19] have indicated that at the time of its earliest deposition sporopollenin may exist in a stranded form, the strands being ca. 50 Å in diameter. According to Rowley, these strands are initially aggregated in bundles in the exine of *Poa annua*, but as consolidation occurs the stranded appearance is lost and the sporopollenin becomes homogeneous.

WALL DEVELOPMENT

Several nineteenth-century cytologists were attracted by the developmental problems presented by pollen walls, but the most significant early work was that of Beer [20, 21]. Beer's paper of 1905 was concerned with various plants of the Onagraceae, and that of 1911 with *Ipomoea purpurea*. This latter species possesses a large pollen grain with an elaborately sculptured wall, and Beer was able to trace its development in some detail. He confirmed the important point, known to Strasburger and other earlier cytologists, that the main features of exine patterning are laid down while the microspores are still enclosed within the callose wall of the tetrad, which he referred to as the 'special mother cell wall'. He recognised that

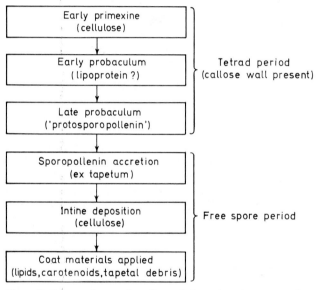

Figure 2. Scheme of events in pollen wall formation, time axis reading downwards [65]

the individual spore walls are formed independently and in isolation from parental cells. Realising that this made it unlikely that parental tissues exerted a direct control over exine patterning, Beer suggested that the haploid spore nucleus governed the morphogenetic processes in the wall, and recorded the presence of 'kinoplastic fibrils' running from the nuclear membrane towards the cell wall during the patterning-determining period. These fibrils were found to disappear after the release of the spores from the tetrad. During the subsequent period of exine growth the early-imprinted pattern is preserved, and Beer considered that this did not require further intervention of the protoplast.

In the event, it proved that the investigation of wall ontogeny could not be taken much further than Beer had pressed it until the development of electron microscopy and ultra-thin section technique almost half a century later. Several fine-structural studies of wall development have now been published, of varying degrees of detail, and the taxonomic range of the species examined has been considerable.

Most of the published accounts are in substantial agreement so far as the major features of wall ontogeny are concerned, so there is a reasonable basis for generalisation, but there are various divergences in interpretation, traceable to both differences in technique and actual diversity in the material. The following review is based primarily upon observations of wall formation in *Lilium longiflorum*; the main events in this species are summarised in Fig. 2.

The tetrad period

As Beer confirmed, the tetrad period is the critical one for the institution of wall patterning. In lily the cleavages which give the tetrad are successive. The wall produced after meiosis bisects the meiocyte to give the dyad configuration, and the two walls formed following meiosis II complete the subdivision to give the tetrad. As we have seen, these walls are of callose and are not penetrated by any plasmodesmata. In the lily tetrad the spores are usually disposed in the same plane, the walls formed after meiosis II being parallel in the two halves of the dyad. In some species with successive cleavage these walls are regularly at right angles to each other, and this configuration appears occasionally in lily. The plane of the walls is, of course, related to the orientation of the spindles of the second meiotic division. In species like *Helleborus foetidus*, the simultaneous cleavage of the meiocyte to give the tetrad is preceded by the formation of secondary phragmoplasts between each of the four daughter nuclei; cell plates form across the equator of each, and callose is deposited upon these [22, 24, 25]. The product is a tetrad in which the spore nuclei lie at the points of a tetrahedron.

The cleavage pattern seems to have no significance for the succeeding stages of wall formation, except to the extent that the orientation of the spores within the tetrad affects the disposition of some elements of patterning (p. 95). The first evidence of the individual spore walls appears very soon after the completion of the cleavage process. In lily, as in other described species, fibrillar material is formed between the plasmalemma and the investing callose wall, at first irregularly and later more uniformly. The plasmalemma often appears highly convoluted at this time; this appearance may, however, be a fixation artefact, since some preparations reveal a smooth plasmalemma in close apposition with the wall material [23, 24].

This first-formed wall has been named the *primexine* [23, 25] because of its role as a precursor of the exine proper. The finely fibrillar material forms the matrix of the primexine, and its solubility and other properties indicate that it is cellulose. Very shortly after the beginning of primexine formation two aspects of the future exine pattern appear: the prospective aperture regions are defined and the locations of the future bacula are established.

In species in which the apertures of the mature grain are covered only by extensions of the nexine, the aperture regions are set off in the early period of wall development as areas where the plasmalemma of the spore remains in direct contact with the callose of the tetrad wall; they are therefore zones where the cellulose matrix material is not deposited. It is probable that in grains where a baculate operculum is formed, primexine matrix material is laid down over the central region of the aperture; however, developmental studies have yet to be conducted on pollens of this type.

In at least some species the endoplasmic reticulum of the spore is intimately concerned with the establishment of pore and colpus sites. The spherical pollen grains of the Caryophyllaceae are pocked with numerous circular pores, uniformly distributed over the surface. In the early primexine stage each pore-site is marked out by a plate of endoplasmic reticulum, closely apposed to the plasmalemma, which in turn lies directly against the callose of the tetrad wall [23, 25]. In *Helleborus foetidus*, studied by Echlin and Godwin [26b], the aperture takes the form of a furrow. Here again it is defined during the early development of the primexine by the apposition of a plate of endoplasmic reticulum to the plasmalemma, which itself remains in contact with the callose tetrad wall while the matrix material of the primexine is being deposited elsewhere. The pollen grain of *Helleborus foetidus* is tricolpate; Dickinson [68] has recently shown that the same association between a plate of endoplasmic reticulum and the prospective aperture site may be seen also in the monocolpate grain of *Lilium longiflorum*. It seems indisputable, then, that in these species the cytoplasmic membranes do discharge an important morphogenetic role.

The second manifestation of pattern-determination is in the

localisation of the future bacula. Very soon after the initiation of primexine development, the fibrillar matrix material is seen to be traversed by radially directed rods, the *probacula*. The electron microscopic appearance of the probacula at this early stage is critically dependent upon preparative technique. With permanganate fixation they appear granular-amorphous, with a particle size around 50–60 Å, and this appearance is common also with osmic fixation, especially where the fixative has penetrated the callose tetrad wall poorly. With the best current technique, glutaraldehyde pre-fixation followed by osmication, the probacular material appears lamellated. The lamellations show a repeat distance of around 100 Å, and the response to post-stains such as uranyl acetate and lead citrate suggests that there may be a lipoprotein component, albeit one which is more unstable than that found in the unit membranes of the cytoplasm [27]. The lamellae arise directly from the plasmalemma, and in lily the trunks of favourably sectioned bacula reveal aggregates of 6–12. Their disposition is at present unclear; each baculum could be made up of several flat or pleated plates, of concentric cylinders, or of a single spirally disposed or pleated ribbon.

As with the prospective aperture region it is often—but seemingly not invariably—possible to associate the sites of the probacula with structures in the neighbouring layers of cytoplasm. In early accounts of *Silene pendula* and *Saponaria officinalis* [23, 25] and *Parkinsonia aculeata* [28], elements of the endoplasmic reticulum were found to be associated with the feet of the probacula, lying appressed to the plasmalemma. Skvarla and Larson [29] illustrated a particularly conspicuous association in *Zea mays*, where tubules of endoplasmic reticulum run radially towards the bases of the probacula. These demonstrations of cytoplasmic membranes associated with the probacula have been mostly with permanganate-fixed material, and in several studies using other fixations the relationship has not been found (e.g. Ref. 30).

In lily, vesicular components often appear in the vicinity of the burgeoning probacula, but here again a regular association with cytoplasmic membranes is not apparent. In the very youngest stages there does, however, seem to be some relationship between the sites of the probacula and concentrations of bodies of the dimensions and staining qualities of ribosomes. These frequently appear in polysome-like configurations. Should these bodies indeed be ribosomes, their appearance at this time may be significant, since the normal ribosome complement of the cytoplasm is not restored completely until late in the tetrad period (see this volume, p. 19).

When the probacula are first formed, they are isolated from each other both above and below. The pattern they form is thus one of isolated rods, filling the space between the aperture sites where there are several, or extending over the greater part of the spore surface when only one aperture is present. If the callose wall is removed mechanically at this time, the naked spores show evidence of pattern by optical microscopy, but this pattern is removed when the wall is exposed to cellulose digestion [12]. This fact indicates that the integrity of the pattern depends upon cellulose at this stage of development and that no other wall polymer has structural continuity over the whole surface of the spore.

Two changes quickly ensue. Connections develop between the heads of the probacula, both above and below the level of the primexine matrix, and new materials are injected into them. In lily the heads of the probacula expand laterally to form flattened caps. Since the probacula are disposed in rows, the caps eventually fuse to give a raised reticulum. Similar changes in tectate exines connect the heads in two dimensions, forming a continuous roof over the primexine matrix.

Chemical change is shown by a progressive alteration in response to electron stains and in ultrastructural appearance. With both permanganate and osmic impregnation, and with such post-stains as lead citrate, the probacula reveal increasing electron-opacity, and eventually all evidence of lamellated structure disappears. These changes in the probacula are accompanied by increasing resistance to degradation treatments. With the linking of the heads of the probacula, the patterned component of the primexine gains structural integrity, and it can be extracted intact from the tetrads by acetolysis. The product is a net, which in its structure clearly foreshadows the organisation of the mature exine [11].

The foot layer of nexine 1 is established by plate-like outgrowths from the bases of the probacula and by apposition of lamellae arising at the plasmalemma [27]. In its early stages of growth this layer does not survive acetolysis; then as it thickens and becomes more or less continuous it acquires resistance [11]. While this development proceeds, the heads of the probacula and the connections between them become rounded, so that the appearance of the muri of the mature grains gradually emerges.

All of this development occurs within the callose tetrad wall, and it is not accompanied by any appreciable growth in volume.

The free-spore period

The spores are released from the tetrad by the rapid dissolution of the callose wall. This dissolution is mediated by a callase, and the process has been studied in *Cucurbita* by Eschrich [31]. The enzyme appears in the locular fluid over only a comparatively short period of time and is therefore phase-specific. The first products of the breakdown of the callose wall are β-1,3-linked oligosaccharides such as laminaribiose and laminaritriose, and these were detected by Eschrich in the anther fluid for a period while the structure of the tetrad walls was being eroded. Thereafter these sugars were broken down to glucose. In some species the middle lamellae laid down during the tetrad cleavage division resist digestion for a while, and the geometrical figure they form, referred to by Beer [21] as a 'triradiate lamella', persists among the young spores.

In all species where measurements have been made, the spores on release from the tetrad undergo a rapid increase in size. In lily the growth in volume is of the order of 2·8 times within the 24 h following the break-up of the tetrads.

This increase in volume means a considerable thinning out of material of the premexine. The matrix material is shredded and dispersed, both in species like lily, where it lies exposed between the muri, and in those like *Silene pendula*, where it is enclosed in the cavities beneath the tectum. The residuum remains visible in both scanning and transmission electron micrographs [11, 65]. It can be shown by the use of solvents that this residue is largely cellulose, and its continued presence may be significant during the final phase of pollen maturation, when surface materials are added [66].

The sporopollenin of the patterned part of the primexine is also thinned out during the expansion period, and this has been described in several accounts from that of Beer [20], dealing with species of the Onagraceae, onwards. Changes in the overall shape of pollen grains occur as the wall is stretched; grains may become more spherical, or change from a triangular to an ellipsoidal form, or even from a spherical to an ellipsoidal form, as in various grasses. These shape modifications must represent the consequences of a substitution of the constraints imposed by the callose tetrad wall by others set by the mechanical properties of the exine itself. Larson and Lewis [32] have suggested that the expansion of the spores of *Parkinsoniana aculeata* after release from the tetrad may not be purely dependent upon stretching of the wall, since invaginations are present in the primexine which are taken up during growth following the dissolution of the tetrad wall.

While a thinning out of the sporopollenin part of the wall has been recorded in several species, there are others where such a change is barely recognisable, apparently because there is a very rapid accretion of new materials during the release period. For the young spores of lily the interval following the disappearance of the tetrad wall is one of substantial exine growth, and there seems little doubt that this is dependent upon precursor materials from the tapetum (p. 92). Banerjee, Rowley and Alessio [24] have examined the changes in dimensions of the various components of the exine of *Sparganium androcladium* during the period following release from the tetrad, and have shown that in this species also there must be a rapid deposition of sporopollenin over this interval. The diameter doubles between the late tetrad stage and the time when intine is apparent in the young spore. In this same period the bacula show no decrease in diameter, although their average height is reduced by 30%; the tectum thins out by about 20% and the nexine 1 by about 10%. The expectation would be that the fourfold increase in surface area would reduce the thickness of the exine layers by some 75%, so there is evidently much concomitant growth. As with *Parkinsonia*, part of the initial expansion of the nexine 1 is accommodated by pleats present in the tetrad, which are pulled out on release. The elements of the sexine are drawn apart as the area increases, and Banerjee *et al.* suggest that this results in changes in some structural features of the exine mentioned below.

The early period of spore expansion beginning with the break-up of the tetrads is followed by an interval of relatively slower growth, the rate gradually declining until a maximum volume is reached some time before dehiscence. The sexine of the lily spore continues to show an accretion of sporopollenin through the early part of this interval of growth, indeed until the dissolution of the tapetum, and the data of Banerjee *et al.* for *Sparganium* show that there must be an increase here also. During this last period of exine growth, the electron-staining properties of the sporopollenin alter progressively, the electron-density after osmication declining. The change is presumably due to the masking of reactive groups through increase of cross-linking.

In some species the accretion of sporopollenin does not continue to keep pace with growth in surface area, so that the last period of expansion in the grains is accompanied by an attentuation of the exine [28, 32].

Where they have provided enough detail to judge, fine-structural accounts of wall growth during the free-spore period confirm without exception Beer's proposition that the major features of sexine pattern are determined in the early tetrad period, and that subsequent changes involve in the main a development of structural features already present. This conclusion is a significant one, since it implies that the basic morphogenetic processes are worked out in the isolated spore very shortly after tetrad cleavage, and that what follows does not involve the participation of cytoplasm in a pattern-determining role. In an earlier period of work on the growth of spore walls this matter provoked controversy [20]. Fitting [33] supposed that growth in the spore walls of *Selaginella* did not depend upon protoplasmic contacts, and his views were supported by Campbell [34] and Denke [35]. Lyon [36], on the other hand, argued that there was no need to invoke 'action at a distance' in relation to protoplasmic control of wall growth in *Selaginella*. A distinction critical to this issue was made by Beer [21], who realised that his own observations on spore wall development in *Oenothera* and *Ipomoea* meant that while the initial morphogenetic processes did require protoplasmic control, later growth need not. Given this situation, it would seem entirely possible for growth to continue in the walls of spores after the degeneration of the protoplast, as recorded by Tischler [37] for male sterile *Mirabilis jalapa*. The requirements would be an existing template, produced before the demise of the cell contents, and a medium—the thecal fluid— charged with the necessary precursors, and possibly also enzymes, for sporopollenin assembly.

The conclusion that sexine growth during the free-spore period does not involve any new morphogenetic activity at the level where the structural interpretation of genetical information is initiated does not imply that the pattern of the mature sexine is no more than an enlarged version of that expressed in the primexine. There can be no doubt that differential growth takes place following release from the tetrad, so that changes in the proportions of various elements of the sexine result. Such changes may perhaps be explained in allometric terms, but it is difficult to avoid the conclusion that in some spores different growth 'potentials' are built in to the primexine which ensure that the accretion of sporopollenin during the early free-spore period is not uniform. In many strongly sculptured grains, such as those of the Compositae, eminences on the surface of the tectum extend disproportionately to produce spines and spinules [37]. This type of growth is commonly associated with an invasive tapetum. There is a marginal possibility that the tapetum here *is* acting as a primary morphogenetic agent, but the behaviour is more readily reconciled with the general character of exine growth by the assumption that particular parts of the surface of the tectum are preferred areas for sporopollenin deposition because of the existence of different catalytic or other conditions in these areas at the time of release from the tetrad [38].

Nexine 2

Deposition of the nexine 2 (endexine) begins either just before the break-up of the tetrads or shortly afterwards. Like the sexine and nexine 1, the nexine 2 is composed of a material of the general class of sporopollenin, but, as we have seen (p. 79), it shows different staining properties and may present a different appearance electron microscopically at maturity. There is no reasonable doubt that the material of the nexine 2 originates within the spore itself, and there is abundant evidence that it accumulates through the apposition of lamellae. These lamellae in section reveal an electron-transparent plate variously estimated as 40–50 Å in thickness, on the surface of which sporopollenin is deposited. The lamellae bear some resemblance to those seen in the very young probacula, but they are oriented at right angles, being tangential rather than radial.

The apposition of lamellae in the nexine 2 has been described in *Anthurium* spp. by Rowley and Southworth [40], in *Ipomoea purpurea* by Godwin, Echlin and Chapman [41], in *Lilium longiflorum* by Dickinson and Heslop-Harrison [27] and in *Helleborus foetidus* by Echlin and Godwin [26c]. The lamellae arise at or very near to

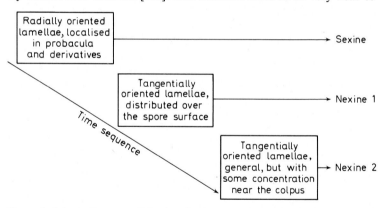

Figure 3. Scheme for extracellular lamella formation in the developing exine. The lamellae provide the sites for sporopollenin deposition. The strata on the right are initiated by the production of lamellae formed at or near the plasmalemma in the sequence shown on the left [67]

the plasmalemma, and are at first without sporopollenin coatings. Sporopollenin then accumulates on the two faces, and increases in thickness as the lamellae are pressed outwards towards the nexine 1 by the new ones originating within.

In many species the plates of sporopollenin coalesce throughout much of the nexine 2 as this process continues, and indications of the parent lamellae disappear so that the layer as a whole appears homogeneous. However, there are exines in which lamellated structure persists in the nexine 2, and, as we have seen, the lamellation may be revealed electron microscopically by certain pretreatments even when the usual appearance is homogeneous, while the birefringence properties of the nexine 2 may also expose its basically lamellated structure.

It seems likely that the parent lamellae concerned in sporopollenin deposition in the nexine 2 are homologous with those of the young probacula and the developing nexine 1. If this be so, then it may be concluded that sporopollenin deposition—at least in its initial stages—follows the same basic plan wherever and whenever it occurs in the spore wall. The sequence for lily is as in Fig. 3 [27]. It is significant that a similar mode of origin has been described for the sporopollenin of liverwort spores by Horner [42]. Rowley and co-workers [40, 43] have pointed to the dimensional agreement between the parent lamellae and cytoplasmic unit membranes. The lamellae are not, however, likely to be of the nature of unit membranes. Those formed in the early tetrad period appear to rise directly from the plasmalemma, forming the trunk of the probaculum (p. 83); such right-angle junctions between unit membranes are unknown, and on physico-chemical grounds unlikely to occur except transiently, whatever interpretation of membrane structure be accepted. Furthermore, while unit membranes fix readily for electron microscopy with a range of fixatives, the lamellae of the probacula are more labile. A lipoprotein composition is not, however, excluded, if response to electron stains be accepted as a criterion.

Intine growth

The deposition of the innermost wall of the young spore, the intine, begins soon after the release of the spores from the tetrad; in lily, for example, there is evidence of the presence of intine material over parts of the plasmalemma almost from the beginning of the development of the nexine 2. There would appear to be a spatial conflict inherent in a process requiring the simultaneous growth of two walls one within the other, and how this is resolved is not immediately evident; but it seems that the lamellae for the nexine 2 are actually passed through the young, unconsolidated intine fabric [44].

In all species studied, the intine becomes compacted first in the vicinity of the apertures. Thereafter the zone of growth extends around the full circumference of the grain. As Sitte [7] showed, the material deposited contains microfibrillar cellulose, with a matrix material of pectic substances and hemicelluloses. The microfibrils lie with their long axes directed more or less tangentially, but in the plane of the wall they are oriented randomly, the only departure being in the neighbourhood of elongated apertures.

In most preparations of the lily spore during early intine growth the plasmalemma is seen to be separated from the innermost layers of cellulose microfibrils, the intervening volume being filled with a loose amorphous or very finely fibrillar material. The effect is probably a fixation artefact, due to the delicately poised osmotic situation in the spore in this early period of independence, but it does indicate that the wall material at the surface of the plasmalemma is poorly compacted at this time.

Intine growth involves the activity of dictyosomes in the adjacent layers of cytoplasm, in the manner now well established for the growth of cellulose walls of somatic cells. Vesicles of the 'coated' type are frequently to be seen apposed to the plasmalemma, and the derivation of these from neighbouring dictyosomes seems to be in little doubt [44]. The role of microtubules in intine thickening is, however, enigmatic. Microtubules are never very frequent in the spore cytoplasm, and where they are associated with plasmalemma during the period of cellulose accretion in the intine they occur only in groups of two or three; there is never any question of a population as dense as that frequently seen during the formation of primary and secondary walls in somatic tissues. This is probably to be expected if the primary role of the microtubule in wall growth is in the establishment of microfibril orientation; the random distribution of microfibrils over much of the intine surface shows that no orienting mechanism is in fact at work.

In some intines thickening proceeds in waves, so that the wall becomes stratified or lamellated. The strata are often marked off by interrupted layers of granular or filamentous inclusions, as in *Silene pendula* [23]. These inclusions have been interpreted as cytoplasmic residues, or even persistent protoplasmic filaments. The lamellation and the included particles or filaments are often most clearly visible in the intine in the vicinity of the apertures, especially in species like those of the Caryocaraceae described by Barth [45], where the intine is enormously thickened below the colpi. It is now clear that the included material is protein and that the intine is the site of several hydrolytic enzymes [46, 47]. The significance of the intine proteins is discussed elsewhere in this volume (p. 171).

THE TAPETUM DURING WALL FORMATION

It has long been supposed that the tapetum must play some part in the fashioning of the spore walls, either through its presumed function as a centre of synthesis in the anther or, more directly, through participation in the morphogenetic processes concerned in wall growth. In the anthers of many plants with the parietal type of tapetum, granules of material with chemical properties similar to those of sporopollenin accumulate on the inner faces of the tapetal cells and persist there until pollen dispersal. These granules assume the form of flat plates, cups or hollow spheroids, and range in size from 6–8 μm in maximum dimension to below the resolution limit of the optical microscope. Attention was first drawn to them by Rosanoff in 1865 [48] as occurring around the polyads of mimosoid pollen, and in 1919 and 1922 Mascré [49, 50] described resistant particles on the tapetal membranes of species of Solanaceae and Boraginaceae. Schnarf [51] observed their presence in several other families, including the Liliaceae. Krjatchenko, describing the development of the granules in *Lilium croceum* [52], suggested that they arose from lipid droplets secreted by the tapetal cells, droplets possibly originating from mitochondria. Additional descriptive accounts were given by Ubisch [53], Kosmath [54] and Py [55]. In various recent re-investigations of the tapetal sporopollenin particles they have been referred to as 'Ubisch bodies'. The name is inappropriate, since Ubisch was not their discoverer. They will be referred to here as orbicules.

Early fine-structural observations of tapetal orbicule development were recorded for *Poa annua* by Rowley [56], and for *Silene pendula* and *Cannabis sativa* by Heslop-Harrison [57, 23]. Rowley described the development of the orbicules by the accretion of sporopollenin around 'pro-Ubisch' bodies, considered to be lipid in nature. Similar precursor bodies were reported as being formed within the protoplasts of the tapetal cells of *Cannabis* [57]; following the earlier suggestion of Krjatchenko, the possibility that the bodies might be derived from mitochondria was considered, but more recent work has indicated that this interpretation is erroneous. Carniel [58] has shown that the pro-orbicular bodies of *Oxalis* are lipid in nature, and that they accumulate within the tapetal cells near the plasmalemma and are then passed to the cell surface, where they acquire a sporopollenin coating.

This origin has been confirmed in its essentials in work on *Populus* and *Salix* [59] and *Lilium longiflorum* [67], and especially in the comprehensive study of *Helleborus foetidus* by Echlin and Godwin [26a], who have given evidence bearing on many details of the process. However, there remain several debatable points, such as, for example, the role of the endoplasmic reticulum in the formation of the pro-orbicular bodies, and the question whether they are wholly extruded from the tapetal cells or remain appended by extensions of the plasmalemma.

The significance of the tapetal sporopollenin depositions remains obscure, but that there is no direct relationship with exine development is shown by the timing and sites of orbicule formation. In *Silene pendula*, *Lilium longiflorum* and *Oxalis* spp. sporopollenin appears at the tapetal surface early in the tetrad period, revealing from the first an electron-density after osmication much higher than that of the developing probacula, and more comparable with that revealed by the late probacula and the exine during its period of

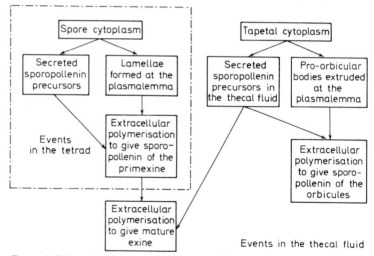

Figure 4. Scheme for the deposition of sporopollenin in the tapetal orbicules and exine, based on the evidence from genera with a tapetum of the parietal, secretory type [67]

consolidation and growth following release of the spores from the tetrads. In these species, then, the sporopollenin synthesis associated with the tapetum begins before that in the exine. However, the deposition of the sporopollenin on the orbicules starts only after they have been extruded from the tapetal cells into the space below the degenerating walls. This presumably means that, as with the exine, sporopollenin polymerisation occurs only on surfaces external to the cell; for as Echlin and Godwin [26a, b, c] have stressed, it is never found on membranes or other structures within the cytoplasm.

Taken together these various observations suggest that the formation of the orbicules and the synthesis of sporopollenin in the exine may be related as in the scheme of Fig. 4. In species with an invasive type of tapetum a similar scheme would apply, except that less prominence would need to be given to the tapetum-associated sporopollenin synthesis. Orbicules are produced in smaller numbers or not at all in these species [51, 39], although a sheath of a material with some of the properties of sporopollenin is produced on the outer faces of the tapetal cells [61].

CONTROL OF EXINE PATTERNING

The generation of complex exine patterns often with high taxonomic specificity raises morphogenetic problems of unusual interest, concerning both the location of genetic control and the manner in which it is exerted. The sculptured wall itself is produced round haploid spores, yet within the environment of the anther, an organ of the diploid parent. Is the control exerted by the spore nucleus or is it imposed by the sporophyte? In the latter case what is the executive agent? Is it the sporophytic tissue in the vicinity during the actual process of wall growth? Or are cytoplasmic determinants handed on through meiosis from the parent meiocyte?

Since the events leading to exine pattern determination occur within the tetrad, and since the spores during the critical period are physically isolated from neighbouring tissues, the readiest conclusion is that the whole process is under the control of the haploid genome, in interaction only with microenvironment within the tetrad, which presumably can be resolved into a system of constrictive wall forces and perhaps diffusion gradients of small molecules. This proposition would imply that the haploid spore nucleus is activated very rapidly after meiosis, since primexine formation begins very soon after the cleavage following meiosis II. A normal ribosome population is not restored by this time, but it is significant that what look like polyribosomes do become associated with the burgeoning probacula (p. 83). It is indeed as if one of the earliest preoccupations of the spore cytoplasm were to establish a protein-synthetic system concerned with this aspect of wall growth. What cannot yet be specified is the source of mRNA, whether this originates in the spore nucleus, or persists from the parent meiocyte.

Godwin [62] has discussed genetic evidence concerning the control of pollen wall characteristics, and has pointed out that segregation of different exine types in one and the same tetrad has never been recorded. Evidence of segregation of some kind would certainly be expected if exine features were exclusively under the control of the spore nucleus. Godwin has also mentioned the relevant evidence from the development of genetically deficient spores. Imperfect meiosis in triploids, aneuploids and structural heterozygotes produces spore nuclei which are genomically unbalanced. This is commonly revealed in the failure of pollen mitosis in a proportion of the cells of each tetrad of polyad or in later gametophytic malfunction. Were normal exine development to depend upon the haplophase genomes, it is to be expected that deficiencies in unbalanced spores would be reflected in blocked or modified exine development. Yet in many cases the genetically unbalanced and essentially unviable spores of polyads do show normal exine ontogeny, the only modifications being such as might be associated with variations in the volume of cytoplasm bequeathed from the parent cell.

The extreme situation is seen where cytoplasmic enclaves are formed at the time of cleavage which are entirely without nuclei. Since the cleavage walls are normally formed in relation to phragmoplasts, and since phragmoplast formation requires the presence of some nuclear structure, be it only one chromosome, it is unusual for enucleate cells to be found even in polyads produced after highly disturbed meioses. However, this has been reported as occurring regularly in *Acanthus spinosus* [63], where nucleus-free cytoplasmic buds are set off immediately following meiosis II. These buds round up within the callose wall of the polyad and go on to form an essentially normal exine.

Needless to say, evidence of this kind lends powerful support to the view that the spore nuclei are not concerned in the execution of exine patterning. If polysomes are active in this process, they could hardly have accepted messengers after the tetrad cleavage, if in consequence of the cleavage they become stranded in enucleated cytoplasm. The possibilities remaining are that control is exerted by the neighbouring sporophytic tissues, or that determinants are transmitted from the parental meiocyte.

The sporophytic tissue in closest proximity to the spores throughout exine ontogeny is the tapetum, but it seems in the highest degree improbable that it plays any essential pattern-*determining* role. As we have seen, the critical events occur in the very young tetrad, at a time when the spores are enclosed within the callose wall, isolated physically and also to some extent chemically from the tapetum and its secretion products. Penetration by tapetal cytoplasm, or by diffusing macromolecular determinants—if such be supposed to exist—is probably excluded. Moreover, the fine-structural evidence reviewed in earlier paragraphs is good enough to assure us that the living spore protoplast is intimately involved in the initial establishment of wall pattern. Whatever may be the tapetal role in the later stages of pollen growth, it cannot be a factor intervening in any direct manner during the critical tetrad period.

The proposition we are left with is that the establishment of exine patterning depends upon the existence in the spore cytoplasms of determinants synthesised by the parent meiocyte. Such determinants could take the form of protein sub-units, capable of initiating patterning processes perhaps by mutual association whenever a favourable environment emerges in the spores; or of persistent messengers carrying the potential for synthesising such proteins, again when a suitable environment becomes available. These alternatives have the same genetical, but different biochemical, implications. The genetical implications are that the 'read out' of exine information precedes meiosis, and that the extranuclear carrier of the transcribed, or perhaps transcribed *and* translated, information is not subject to destruction during the cytoplasmic clean-up process described in another paper in this volume (p. 16). Obviously the interpretations differ in respect to biochemical detail in that one requires protein synthesis as a prerequisite for the initiation of the patterning processes while the other does not. The possible importance of the presence of polyribosome-like aggregates in the vicinity of the wall during the patterning phase has already been mentioned. It should be possible to determine experimentally whether localised protein synthesis does occur in the very young tetrad, and whether exine patterning is delayed or modified if this is selectively blocked.

In conclusion, it may be noted that the spore environment is an important factor in controlling the orientation and disposition of various exine features. This point has recently been re-emphasised by Godwin [61], and several earlier authors, especially Drahowzal [62] and Wodehouse [63], discussed aspects of it in detail. Observations of early exine ontogeny in many species have revealed constant relationships between the locations of spores, colpi and other aspects of exine pattern with the axis of the young spore of the tetrad. As various dicotyledonous examples show rather clearly, the contact faces between spores within the tetrad are evidently important points of alignment for the pores of triporate grains. The more complex situations examined by Wodehouse [63] reveal similar regularities, and a most informative interpretation of the situation in *Pinus* has recently been published by Martens, Waterkeyn and Huyskens [64].

The conclusion may be drawn that the pattern-generating processes within the spore, while necessarily reflecting the capacities of the genome so far as detail is concerned, are dependent upon cellular interactions for their initiation and early orientation.

REFERENCES

1 ERDTMAN, G., ' "LO-analysis" and "Welcker's rule" ', *Svensk bot. Tidskr.*, **50**, 1–7 (1956)
2 FAEGRI, K., 'Recent trends in palynology', *Bot. Rev.*, **22**, 639–64 (1956)
3 ERDTMAN, G., 'Sporoderm morphology and morphogenesis. A collocation of data and suppositions', *Grana palynol.*, **16**, 318–23 (1966)
4 LARSON, D. A., SKVARLA, J. J. and LEWIS, C. W., 'An electron microscopic study of exine stratification and fine structure', *Pollen Spores*, **4**, 233–46 (1962)
5 SKVARLA, J. J. and TURNER, B. L., 'Systematic implications from electron microscopic studies of composite pollen', *Ann. Mo. bot. Gdn*, **53**, 220–56 (1966)
6 STIX, E., 'Pollenmorphohologische Untersuchungen an Compositen', *Grana palynol.*, **22**, 41–114 (1960)
7 SITTE, P., 'Untersuchungen zur submikroskopischen Morphologie der Pollen- und Sporenmembranen', *Mikroskopie*, **8**, 290–9 (1953)
8 ZETZSCHE, F., 'Sporopollenine'. In: *Handbuch der Pflanzen-analyse* (Ed. G. KLEIN), Vol. 3, Wein (1932)
9 ERDTMAN, G., 'The acetolysis method', *Svensk bot. Tidskr.*, **54**, 561–4 (1960)
10 ROWLEY, J. R. and FLYNN, J. J., 'Single-stage carbon replicas of microspores', *Stain Technol.*, **41**, 287–90 (1966)
11 HESLOP-HARRISON, J., 'Wall development within the microspore tetrad of *Lilium longiflorum*', *Can. J. Bot.*, **46**, 1185–92 (1968)
12 AFZELIUS, B. M., ERDTMAN, G. and SJÖSTRAND, F. S., 'On the fine structure of the outer part of the spore wall of *Lycopodium clavatum* as revealed by the electron microscope', *Svensk bot. Tidskr.*, **48**, 155–61 (1954)
13 AFZELIUS, B. M., 'On the fine structure of the pollen wall in *Clivia minuta*', *Bot. Notiser*, **108**, 138–45 (1955)
14 AFZELIUS, B. M., 'Electron microscopic investigations into exine stratification', *Grana palynol.*, **1**, 22–37 (1956)
15 ZETZSCHE, F. and VICARI, H., 'Untersuchungen über die Membran der Sporen und Pollen, II. *Lycopodium clavatum* L.', *Helv. chim. Acta.*, **14**, 58–62 (1931)
16 SITTE, P., 'Polarisationsmikroskopische Untersuchungen an Sporodermen', *Z. Naturforsch.*, **14**, 575–82 (1959)
17 STIX, E., 'Polarisationsmikroskopische Untersuchungen am Sporoderm von *Echinops banaticus*', *Grana palynol.*, **5**, 289–97 (1964)
18 ROWLEY, J. R., 'Nonhomogeneous sporopollenin in microspores of *Poa annua*', *Grana palynol.*, **3**, 3–20 (1963)
19 ROWLEY, J. R., 'Stranded arrangement of sporopollenin in the exine of microspores of *Poa annua*', *Science, N.Y.*, **137**, 526–8 (1962)
20 BEER, R., 'On the development of the pollen grain and anther of some Onagraceae', *Beih. bot. Zbl.*, **19**, 286–313 (1905)
21 BEER, R., 'Studies in spore development', *Ann. Bot.*, **25**, 199–214 (1911)
22 WATERKEYN, F., 'Les parois microsporocytaires de nature callosique', *Cellule*, **62**, 225–55 (1962)
23 HESLOP-HARRISON, J., 'Ultrastructural aspects of differentiation in sporogenous tissue', *Symp. Soc. exp. Biol.*, **17**, 315–40 (1963)
24 BANERJEE, U. C., ROWLEY, J. R. and ALESSIO, M. L., 'Exine plasticity during pollen grain maturation', *J. Palynol.*, 70–89 (1965)
25 HESLOP-HARRISON, J., 'An ultrastructural study of pollen wall ontogeny in *Silene pendula*', *Grana palynol.*, **4**, 7–24 (1963)

26a ECHLIN, P. and GODWIN, H., 'The ultrastructure and ontogeny of pollen in *Helleborus foetidus* L. I. The development of the tapetum and Ubisch bodies', *J. Cell Sci.*, **3**, 161–74 (1968)

26b ECHLIN, P. and GODWIN, H., 'The ultrastructure and ontogeny of pollen in *Helleborus foetidus* L. II. Pollen grain development through the callose special wall stage', *J. Cell Sci.*, **3**, 175–86 (1968)

26c ECHLIN, P. and GODWIN, H., 'The ultrastructure and ontogeny of pollen in *Helleborus foetidus* L. III. The formation of the pollen grain wall', *J. Cell Sci.*, **5**, 459–77 (1969)

27 DICKINSON, H. G. and HESLOP-HARRISON, J., 'Common mode of deposition for the sporopollenin of sexine and nexine', *Nature, Lond.*, **220**, 927–8 (1968)

28 LARSON, D. A. and LEWIS, C. W., 'Pollen wall development in *Parkinsonia aculeata*', *Grana palynol.*, **3**, 21–7 (1963)

29 SKVARLA, J. J. and LARSON, D. A., 'Fine structural studies of *Zea mays* pollen. I: Cell membranes and exine ontogeny', *Am. J. Bot.*, **53**, 1112–25 (1966)

30 HORVAT, F., 'Contribution a la connaisance de l'ultrastructure des parois du pollen de *Tradescantia paludosa* L.', *Grana palynol.*, **6**, 416–34 (1966)

31 ESCHRICH, W., 'Untersuchungen über den Ab-und Aufbau der Callose', *Z. Bot.*, **49**, 153–218 (1961)

32 LARSON, D. A. and LEWIS, C. W., 'Fine structure of *Parkinsonia aculeata* pollen. I. The pollen wall', *Am. J. Bot.*, **48**, 934–43 (1961)

33 FITTING, H., 'Bau und Entwicklungsgeschichte der Makrosporen von *Isoetes* und *Selaginella*', *Bot. Ztg*, **58**, 107–64 (1900)

34 CAMPBELL, D. H., 'Studies on the gametophyte of *Selaginella*', *Ann. Bot.*, **16**, 419–28 (1902)

35 DENKE, P., 'Sporenentwicklung bei *Selaginella*', *Beih. bot. Zbl.*, **12**, 182–9 (1902)

36 LYON, F., 'The spore coats of *Selaginella*', *Bot. Gaz.*, **40**, 285–95 (1905)

37 TISCHLER, G., 'Zellstudien an sterilen Bastardpflanzen', *Arch. Zellforsch.*, **1**, 33–151 (1908)

38 HESLOP-HARRISON, J., 'Scanning electron microscopic observations on the wall of the pollen grain of *Cosmos bipinnatus* (Compositae)', *Proc. Engis Stereoscan Colloquium, 1969*, 89–96 (1969)

39 HESLOP-HARRISON, J., 'The origin of surfaces feature of the pollen wall of *Tagetes patula* as observed by scanning electron microscopy', *Cytobios*, **2**, 177–86 (1969)

40 ROWLEY, J. R., and SOUTHWORTH, D., 'Deposition of sporopollenin on lamellae of unit membrane dimensions', *Nature, Lond.*, **213**, 703–4 (1967)

41 GODWIN, H., ECHLIN, P. and CHAPMAN, B., 'The development of the pollen grain wall in *Ipomoea purpurea* (L.) Roth', *Rev. Palaeobotan. Palynol.*, **3**, 181–95 (1967)

42 HORNER, H. T., LERSTEN, N. R. and BOWEN, C. C., 'Spore development in the liverwort *Riccardia pinguis*', *Am. J. Bot.*, **53**, 1048–64 (1966)

43 ROWLEY, J. R. and DUNBAR, A., 'Sources of membranes for exine formation', *Svensk bot. Tidskr.*, **61**, 49–64 (1967)

44 HESLOP-HARRISON, J., 'Some fine structural features of intine growth in the young microspore of *Lilium henryi*', *Port. Acta Biol.*, **10**, 235–40 (1968)

45 BARTH, O. M., 'Estudos morfologicos dos polens em Caryocaraceae', *Rodriguesia*, **25**, 351–428 (1966)

46 KNOX, R. B. and HESLOP-HARRISON, J., 'Cytochemical localization of enzymes in the wall of the pollen grain', *Nature, Lond.*, **223**, 92–4 (1969)

47 KNOX, R. B. and HESLOP-HARRISON, J., 'Pollen wall proteins: localization and enzymic activity', *J. Cell Sci.*, **5**, 1–15 (1970)

48 ROSANOFF, S., 'Zur Kenntniss des Baues und der Entwicklungsgeschichte des Pollens der Mimoseae', *J. wiss. Bot.*, **4**, 441–50 (1865)

49 MASCRÉ, M., 'Sur le rôle de l'assise nourricière du pollen', *C.r. hebd. Séanc. Acad. Sci., Paris*, **168**, 1214 (1919)

50 MASCRÉ, M., 'Sur l'étamine des Borraginées', *C.r. hebd. Séanc. Acad. Sci., Paris*, **175**, 987–8 (1922)

51 SCHNARF, K., 'Kleine Beiträge zur Entwicklungsgeschichte der Angiospermen. IV. Über das Verhalten des Antherentapetums einiger Pflanzen', *Öst. bot. Z.*, **72**, 242–5 (1923)

52 KRJATCHENKO, D., 'De l'activité des chondriosomes pendant le développement des grains de pollen et des cellules nourricières du pollen dans *Lilium croceum* Chaix', *Revue gén. Bot.*, **37**, 193–211 (1925)

53 UBISCH, G. V., 'Zur Entwicklungsgeschichte der Antheren', *Planta*, **3**, 490–5 (1927)

54 KOSMATH, L., 'Studien über das Antherentapetum', *Öst. bot. Z.*, **76**, 235–41 (1927)

55 PY, G., 'Recherches cytologiques sur l'assise nourricière des microspores et les microspores des plantes vasculaires', *Revue gén. Bot.*, **44**, 316–78, 379–413, 484–512 (1932)

56 ROWLEY, J. R., 'Ubisch body development in *Poa annua*', *Grana palynol.*, **4**, 25, 36 (1963)

57 HESLOP-HARRISON, J., 'Origin of exine', *Nature, Lond.*, **195**, 1069–71 (1962)

58 CARNIEL, K., 'Licht- und elektronenmikroskopische Untersuching der Ubisch-körperenticklung in der Gattung *Oxalis*', *Öst. bot. Z.*, **114**, 490–501 (1967)

59 ROWLEY, J. R. and ERDTMAN, G., 'Sporoderm in *Populus* and *Salix*', *Grana palynol.*, **7**, 517–67 (1967)

60 HESLOP-HARRISON, J., 'An acetolysis-resistant membrane investing tapetum and sporogenous tissue in the anthers of certain Compositae', *Can. J. Bot.*, **47**, 541–2 (1969)

61 GODWIN, H., 'The origin of the exine', *New Phytol.*, **67**, 667–76 (1968)

62 DRAHOWZAL, G., 'Beiträge zur morphologie und entwicklungsgeschichte der pollenkorner', *Öst. bot. Z.*, **85**, 241–69 (1936)

63 WODEHOUSE, R. P., *Pollen Grains*, McGraw-Hill, New York (1935)

64 MARTENS, P., WATERKEYN, L. and HUYSKENS, M., 'Organization and symmetry of microspores and origin of intine in *Pinus sylvestris*', *Phytomorphology*, **17**, 114–18 (1967)

65 HESLOP-HARRISON, J., 'Pollen wall development', *Science, N.Y.*, **161**, 230–7 (1968)

66 HESLOP-HARRISON, J., 'Tapetal origin of pollen coat substances in *Lilium*', *New Phytol.*, **67**, 779–86 (1968)

67 HESLOP-HARRISON, J. and DICKINSON, H. G., 'Time relationships of sporopollenin synthesis associated with tapetum and microspores in *Lilium*', *Planta*, **84**, 199–214 (1969)

68 DICKINSON, H. G., 'Ultrastructural aspects of primexine formation in the microspore tetrad of *Lilium longiflorum*', *Cytobiologie*, **1**, 437–49 (1970)

Recent Developments in the Chemistry, Biochemistry, Geochemistry and Post-tetrad Ontogeny of Sporopollenins Derived from Pollen and Spore Exines

J. Brooks and G. Shaw, *School of Chemistry, University of Bradford, England*

INTRODUCTION

When pollen grains are treated successively with organic solvents, hot aqueous alkalis and acids, the contents of the grain are removed and a wall remains which is virtually nitrogen-free. Further treatment of this with warm 85% phosphoric acid for several days removes cellulose and other polysaccharides, leaving a residual exine which, frequently but not always, is morphologically intact. The exine is completely stable to further treatment with reagents of the above type but will succumb to strong oxidising agents. The reported dissolution of micro amounts of exines [1] in ethanolamine at 100°C results, we believe, from a base-catalysed aerial oxidation to simpler, more soluble compounds and on any macro scale we find that little or no measurable dissolution occurs. The exine is chemically composed of a substance named sporopollenin [2], a unique and novel biopolymer [3] which is characterised by resistance to acetolysis, high stability to anaerobic biological, and non-oxidative chemical, attack, but is chemically unsaturated and as a consequence of this readily undergoes oxidation. The diversity of form and structure of pollen and spore exines and their extraordinary resistance to decay and strong chemicals form the basis of modern palynology.

The work of Zetzsche* et al. [4] in the 1930s and our own more

* Fritz Zetzsche was born in Berlin on 25 March 1892. He studied chemistry at the University of Berlin and in 1919 he was appointed assistant to Professor Tambor at the University of Berne, Switzerland. In 1920 he became Reader at the Institute of Organic Chemistry in Berne and in 1929 he was appointed Extraordinary Professor for Special Organic Chemistry at the University. He remained at Berne until 1936, when he was called back to the University of Berlin, where he continued his researches until his death on 1 May 1945 during the Fall of Berlin.

recent work [5–7] has clearly established that sporopollenins comprise a distinct and unique group of substances. Zetzsche [4] showed that sporopollenins from different sources had similar empirical formulae, and that they were highly unsaturated, contained C-methyl and hydroxyl groups (Table 1) and after oxidation gave a mixture of simple dicarboxylic acids (C_2–C_6) as the major products. None of these results, however, gave any indication of the monomeric substances from which the polymer was derived.

Table 1. ANALYSES OF SOME SPOROPOLLENINS. DATA OF ZETZSCHE et al. [4]

Material	Formula	C-Me/mol	OH/mol
Sporopollenins			
Lycopodium clavatum	$C_{90}H_{144}O_{27}$	2·04	15
Equisetum arvense	$C_{90}H_{144}O_{31}$		13
Ceratozamia mexicana	$C_{90}H_{148}O_{31}$	4·08	
Corylus avellana	$C_{90}H_{138}O_{22}$		11
Picea excelsa	$C_{90}H_{144}O_{26}$		
Picea orientalis	$C_{90}H_{144}O_{25}$		14
Phoenix dactylifera	$C_{90}H_{150}O_{23}$	3·45	
Pinus sylvestris	$C_{90}H_{144}O_{24}$	1·70	13
Taxus baccata	$C_{90}H_{138}O_{26}$	2·74	
Fossil Sporopollenins			
Tasmanin	$C_{90}H_{136}O_{17}$	3·00	
Geiseltalpollenin	$C_{90}H_{129}O_{19}S_7N$	3·40	
Lange sporonin	$C_{90}H_{82}O_{17}N$	4·50	

The many and varied advances in techniques for the separation and structure determination of organic compounds which have been introduced into chemistry during the past two decades encouraged us to re-examine the problem of sporopollenin structure. It soon became clear, however, that the difficulties faced by Zetzsche and his co-workers were not to be overcome easily. These difficulties were especially concerned with the lack of suitable chemical or biochemical methods for degrading the sporopollenin molecule in a manner that would give sufficiently large structurally valuable molecular units. Since acids, alkalis and other powerful chemical reagents were without action, we had to rely almost exclusively on oxidising agents [6]. Because of their availability in quantity and high sporopollenin content (see Table 2 for comparative data) our initial studies were devoted largely to *Lycopodium clavatum* spore and *Pinus sylvestris* pollen exines, and used ozone as the most valuable degradative reagent [6]. Ozonisation of exines from which cellulose had not been removed left a morphologically intact intine composed of almost pure cellulose. These cellulose intines,

Table 2. COMPOSITION OF THE WALL OF SOME MODERN POLLEN GRAINS

Material	% Cellulose	% Sporopollenin
Lycopodium clavatum	2·7	23·4
Pinus sylvestris	6·0	23·8
Pinus montana	7·1	23·7
Pinus contorta	8·0	18·2
Pinus radiata	10·0	20·0
Pinus uliginosa	8·5	18·1
Pinus mughus	9·7	17·3
Sambucus nigra	1·7	12·2
Typha angustifolia	5·3	11·3
Alnus glutinosa	5·1	10·7
Alnus incana	1·8	8·8
Chenopodium album	4·4	11·2
Betula verrucosa	3·2	8·2
Cichorium intybus	5·1	9·8
Populus alba	4·2	5·1
Chamaenerion angustifolium	2·8	5·1
Lilium henryi	3·1	5·3
Lilium longiflorum	3·3	5·1
Phleum pratense	2·7	3·5
Equisetum arvense	14·1	1·8

Table 3. MOLECULAR FORMULAE OF SOME MODERN SPOROPOLLENINS FROM WALLS WITH THE CELLULOSE INTINE REMOVED, AND OF 'SYNTHETIC' SPOROPOLLENINS FROM CAROTENOIDS AND CAROTENOID ESTERS. MOLECULAR FORMULAE ARE ARBITRARILY RECORDED ON A C_{90} BASIS TO FACILITATE COMPARISON WITH OTHER RECORDED FORMULAE [7]

Material	Molecular formulae
Lycopodium clavatum	$C_{90}H_{144}O_{27}$
Pinus sylvestris	$C_{90}H_{158}O_{44}$
Pinus radiata	$C_{90}H_{149}O_{44}$
Pinus montana	$C_{90}H_{151}O_{33}$
Pinus contorta	$C_{90}H_{150}O_{37}$
Chenopodium album	$C_{90}H_{144}O_{35}$
Cichorium intybus	$C_{90}H_{150}O_{34}$
Chamaenerion angustifolium	$C_{90}H_{145}O_{25}$
Fagus silvatica	$C_{90}H_{144}O_{35}$
Lilium henryi	$C_{90}H_{142}O_{36}$
Lilium longiflorum	$C_{90}H_{144}O_{37}$
Oxidative copolymer from L. henryi carotenoids and carotenoid esters	$C_{90}H_{148}O_{38}$
Oxidative copolymer of β-carotene	$C_{90}H_{130}O_{30}$

which retain the fine detail of the original sculpture, are especially interesting, since the microspores from which they originate may be regarded as templates for the laying down of sporopollenin material during the grain ontogeny (see below). The soluble organic compounds produced during the oxidation [6, 7] were found to consist of a bewildering array of mono- and di-carboxylic acids, both branched (C-methyl) and straight chain, with and without oxy (keto, hydroxy) substituents. There seemed to be no obvious pattern in these products, and one might therefore have been justified in concluding that sporopollenin was derived from the polymerisation of a complex heterogeneous conglomerate of plant chemical oddments all linked into a vast formless structure. However, it seemed to us that there were many reasons for thinking otherwise. First it seemed clear that in spite of minor variations in stoichimetry of the molecular species (see Tables 1 and 3 for data) and architectural variation of exines, the sporopollenins did represent a specific coherent group of biopolymers, since in addition to their common and unique biological function they possessed so many common chemical properties. In addition, it seemed unlikely that nature would use a haphazard mixture of monomeric compounds for a process as vitally important as the provision of a coat for the carrier of genetic material. Indeed all major natural processes —and sporopollenin formation may be so classified—are always marked by the order and simplicity of their biochemistry.

SPOROPOLLENIN OF *LILIUM HENRYI*

It seemed clear, however, that further advances in the subject would require a new approach, and accordingly we decided to seek out directly monomeric precursors of a sporopollenin by following the course of development of the material in a particular plant. Considerable information is available about the development of the exine of pollen grains in the sporogenous tissue of the anthers during the early stages of meiosis [8, 9], and correlation of development with anther or bud size is roughly possible in certain *Lilium* species [10]. Following the advice of Professor Heslop-Harrison, we chose *Lilium henryi*. Over 100 bulbs were grown under greenhouse conditions; at varying intervals anthers were removed, their weights and lengths were measured and portions were examined microscopically [11]. Remaining material was extracted with both aqueous and organic solvents and the extracts were examined chemically. A proportion of the plants (10–15%) was allowed to mature; the pollen was collected and from it nitrogen- and cellu-

lose-free walls were isolated by the usual methods [6]. The sporopollenin so obtained was similar in all respects to material isolated
by us from many other sources.

It soon emerged that apart from simple fats and small amounts
of hydrocarbons the dominant lipid material being formed at quite
early stages of anther growth was a mixture of carotenoids and
carotenoid esters, and this immediately suggested that these were
the long-sought monomers [11]. In retrospect, molecules of the
carotenoid type (Fig. 1) can be seen to be ideally suited for this

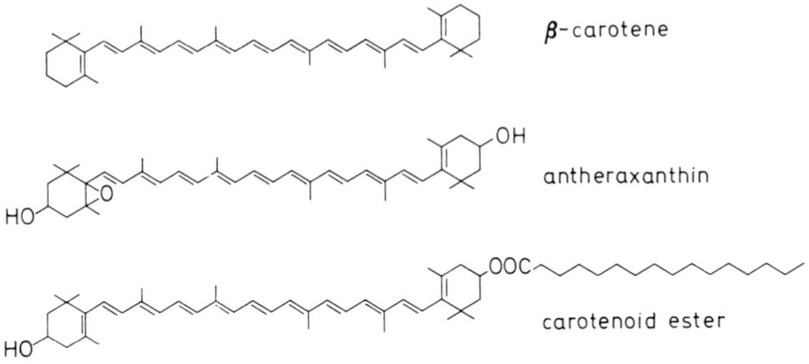

Figure 1

role, since they would be expected to give copolymers with oxygen
which could still retain a high degree of chemical unsaturation and
at the same time degrade with ozone to just that type of mixture of
mono- and di-, straight and branched chain carboxylic acids which
is observed. At the same time our results immediately provided an
explanation of the apparent lack of function of plant carotenoids
which has long puzzled biologists and has even given rise to suggestions about the inefficiency of natural selection in this context [12].

In order to confirm the monomer role of carotenoids in sporopollenin formation, we have examined the copolymerisation of a
number of carotenoids [7, 11], including β-carotene and the purified
carotenoid–carotenoid ester mixture isolated from *Lilium henryi*,
with oxygen in the presence of ionic and free-radical catalysts. Only
ionic catalysts were successful, and in methylene chloride solution
at low temperature (−60°C) in the presence of boron trifluoride
the *Lilium* carotenoids readily produced an insoluble precipitate
of an oxygen-containing chemically unsaturated polymer in excellent yield, with properties virtually indistinguishable from those of
the related sporopollenin. A similar sporopollenin was obtained
from β-carotene. The criteria used for comparison of synthetic

Table 4. COMPARISON OF THE FATTY ACIDS PRODUCED FROM OZONISATION OF *Lilium henryi* POLLEN EXINE, *Lycopodium clavatum* SPORE EXINE AND SOME 'SYNTHETIC' SPOROPOLLENIN OXIDATIVE COPOLYMERS FROM CAROTENOIDS AND CAROTENOID ESTERS

Material	Molecular formula	% Acids produced by ozonisation		
		Branched chain mono-acids	Straight chain mono-acids	Dicar-boxylic acids
Lilium henryi pollen exine	$C_{90}H_{142}O_{36}$	29·3	38·4	32·9
L. henryi carotenoid/ carotenoid ester copolymer	$C_{90}H_{148}O_{38}$	28·3	33·6	36·4
L. henryi carotenoid copolymer	$C_{90}H_{110}O_{33}$	64·1	3·4	33·4
β-carotene co-polymer	$C_{90}H_{130}O_{30}$	61·3	1·9	35·0
Vitamin A-palmitate copolymer	$C_{90}H_{150}O_{13}$	28·3	47·6	25·8
Lycopodium clavatum spore exine	$C_{90}H_{144}O_{27}$	16·7	45·6	37·7

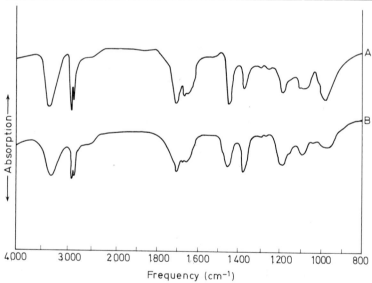

Figure 2. A, Infra-red spectra of sporopollenin from Lilium henryi *pollen exines; B, Infra-red spectra of 'synthetic' sporopollenin from the oxidative copolymerisation of the carotenoids and carotenoid esters from* Lilium henryi.

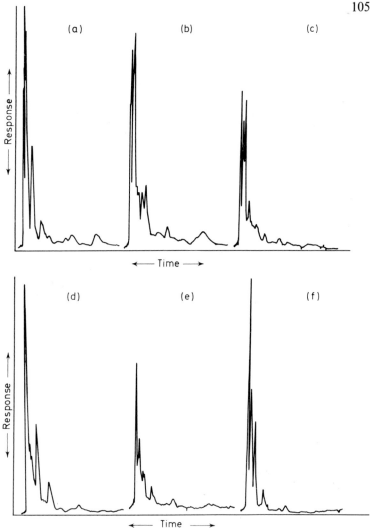

Figure 3. Pyrolysis gas chromatograms of some modern sporopollenins. (a) Lycopodium clavatum *spore exine with cellulose intine; (b)* Lilium henryi *pollen exine with cellulose intine; (c)* Pinus montana *pollen exine with cellulose intine; (d)* Lycopodium clavatum *spore exine with cellulose intine removed; (e)* Lilium henryi *pollen exine with cellulose intine removed; (f)* Pinus montana *pollen exine with cellulose intine removed.*

Column, $\frac{1}{8}$ in × 5 ft stainless steel packed with 5 per cent SE 52 stationary phase on Chromosorb W solid support. Varian-Aerograph 1520 series gas chromatograph with flame ionisation detectors. N_2 carrier-gas flow rate, 20 ml/min. Temperature of the chromatography column, 87°C. Philips chromatography (P.V. 4000 series) pyrolysis head and control unit was used to give pyrolysis temperatures of the materials at 770°C. This temperature (770°C) was attained using the Curie temperature of an Fe (P.V. 4198) 0·5 mm filament wire.

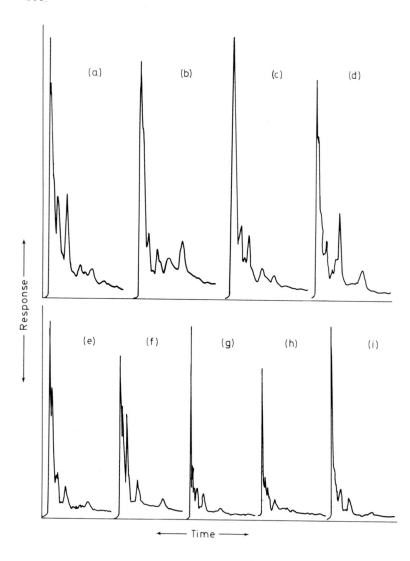

Figure 4. Pyrolysis gas chromatograms of some sporopollenins. (a) Nitella opaca *algal spore with cellulose removed; (b)* Pediastrum duplex *algal spore with cellulose removed; (c)* Mucor mucedo *fungal spore with cellulose removed; (d)* Tasmanites punctatus *fossil spore exines; (e)* Valvisporites auritus *fossil megaspore exines; (f)* Selaginella kraussiana *modern megaspore exines with cellulose removed; (g)* Lycopodium clavatum *spore exines with cellulose removed after heating in a sealed bomb; (h) Synthetic oxidative polymer from* Lilium henryi *carotenoids and carotenoid esters; and (i) Oxidative polymer of β-carotene*

and natural sporopollenins included elemental analyses (Table 3); resistance to acetolysis; infra-red spectra (Fig. 2); examination of oxidation (ozone) products (Table 4) by gas–liquid chromatography and combined gas chromatography–mass spectrometry, where the spectrum of organic acids produced is very similar; comparison

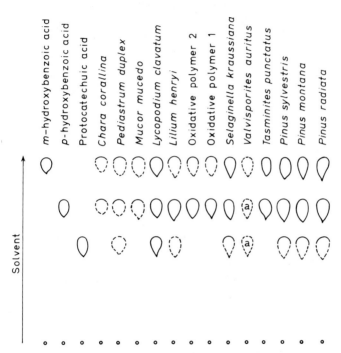

Figure 5. Thin-layer chromatography of the phenolic acids produced from some sporopollenins by potash fusion. Adsorbent, microgranular cellulose powder (Whatman thin-layer chromedia CC41). Solvent, benzene–methanol–glacial acetic acid (90:16:8). Detection, spraying with diazotised p-nitroaniline produces coloured spots corresponding to the phenolic acids.

Solid symbol, major concentration of phenolic acid detected; half-solid symbol, medium concentration of phenolic acid detected; dashed symbol, minor concentration of phenolic acid detected. 'a' indicates that there was insufficient material to allow a complete study of the potash fusion products. The absence of these spots from the thin-layer plate does not necessarily mean that these phenolic acids are absent

of pyrolysis gas–liquid chromatograms (Figs. 3 and 4); and fusion with potassium hydroxide, when in all cases almost identical 'spectra' of phenolic acids (m- and p-hydroxybenzoic and protocatechuic acids) are observed on thin layer chromatograms (Fig. 5).

These last results had previously prompted us to postulate reluctantly that there may be a lignin component in sporopollenin in spite of the absence of all normal lignin tests; we have now been able to withdraw this suggestion [7], since the production of phenolic acids in the above manner is clearly a property of oxidatively polymerised carotenoids. These results confirmed that sporopollenins are copolymers of carotenoids and carotenoid esters with oxygen and that in plant anthers the polymerisation process occurs at an early stage in the development of pollen [3].

We have similarly, in detail, examined the development of sporopollenin in *Lilium longiflorum* and the results closely parallel those derived from *L. henryi*. We have also made a preliminary survey of sporopollenin and carotenoid production in *Ranunculus repens* and in *Geum borisii*. Here the bud size has been used as a datum measurement and the results, though necessarily less accurate, nevertheless amply bear out that the same type of process is operating in these species also.

POST-TETRAD ONTOGENY OF THE POLLEN EXINE

The basic principles of pollen wall ontogeny have been outlined in some detail by Heslop-Harrison [8, 13, 14], Rowley [15] and Echlin and Godwin [16], but we believe that we are now in a position to underpin the biological observations to a limited extent with a chemical basis.

The sequence of steps involved in exine formation may be summarised as follows.

1. The microspores are released from the tetrad. The microspore consists of a sac which is essentially of a polysaccharide nature including certainly a cellulose layer and probably a second hemicellulose or pectin-like layer hydrolysis of which has been found to give a mixture of D-glucose, xylose and other simple sugars [17]. Morphologically the sac is a replica of the pollen grain and may be regarded as a template for sporopollenin deposition.

2. At about the same time as the microspores are released, spherical globules separate from the tapetum. Measurements made from our microphotographs indicate that these are approximately 5 μm in diameter in *Lilium henryi* and initially are transparent in the electron microscope but soon begin to become electron-dense. At the same time a layer of sporopollenin-like material appears on their surface and at this or a slightly later stage they are usually called orbicules or Ubisch bodies. The sporopollenin is then transferred

to the microspore. These observations may be interpreted in chemical terms as follows.

(a) The globules (pro-orbicules) consist of a strong solution of carotenoids and carotenoid esters in a fat-cum-hydrocarbon solvent presumably stabilised by an emulsifier (many anther extracts have high emulsifying properties and *Ranunculus* anther extracts, in particular, form remarkably stable foams [17]) and possibly protected by a polymer (protein?) outer monolayer. A system of this type would be fairly typical of a standard suspension polymerisation.

(b) When the globules leave the tapetum they come into contact with an aqueous phase which contains a specific catalyst (enzyme or enzyme system?) and a source of oxygen (molecular, chemical or both). The catalyst is probably, though by no means certainly, an ionic one, since to date our only success with *in vitro* polymerisation has been with ionic catalysts in bulk solution, whereas all our attempts to produce polymers by free-radical catalysis with and without photochemical illumination have been unsuccessful. The natural catalyst may be quite unique and is certainly of great activity if of the ionic kind, since at room temperature ionic polymerisation processes *in vitro* are normally very slow.

(c) Copolymerisation of the carotenoid mixture with oxygen occurs by what is apparently a fine suspension process, since the orbicules (1–5 μm diameter in *L. henryi* from our microphotographs) are considerably larger than the swollen micelle (0·01 μm) or average latex particle size (0·04–0·08 μm) of an ideal emulsion process [18] and there seems to be no histological evidence for the formation of varying small sizes of latex particles with concomitant decrease in droplet size. In addition, most observers [15] appear to agree that during development of the orbicules a skin of sporopollenin-like material appears on their surfaces and this would correspond reasonably with a suspension polymerisation mechanism in which a skin of polymer forms on the surface of the droplet and is followed by diffusion of monomer particles to the surface and repeated polymerisation thereof. However, until more data are available the mechanism of the polymerisation process must be regarded as unknown. If, however, one assumes that the catalyst is an ionic one, an enzyme catalyst becomes almost essential to introduce that degree of order required for a reasonably rapid polymerisation which is normally introduced during the artificial polymerisation process, by operating at low temperatures ($-60°C$).

(d) The polymerising globules come into contact with the microspore template. Here there is a transfer of sporopollenin in a fully or partly polymerised form as a skin (Rowley [15]) or possibly by alignment of the increasingly viscid globules on the template surface [11]. The transfer is possibly followed by a curing (hardening) process whereby the final details of the exine sculpture are attained. Some final variation of the external architecture may of course follow prior to complete hardening of the exine either by expansion (invasion of the intine) or contraction (elongation of globules to form bacula-type structures) of the developing pollen grain according to the ionic strength, pH and other physical variants of the surrounding medium at the time.

(e) There are invariably present excess polymerising globules, and these will complete the polymer process and give rise ultimately to hard discrete spheroids of sporopollenin. At the same time one might expect to find deposition of some of this excess sporopollenin on other neighbouring plant tissues.

SPOROPOLLENINS FROM ALGAE, FUNGI AND A MEGASPORE

Most of our published work has been concerned with naturally occurring sporopollenins which have been prepared by methods outlined above from a wide variety of modern pollen grain exines and from microspores of *Lycopodium clavatum*. However, we have been equally interested in examining in like manner the various spores produced by more primitive plants, especially algae and fungi and megaspores of higher plants.

Shortage of biological material has limited our experiments, but we have examined spores [19] derived from the algae *Pediastrum duplex* and *Chara corallina*, the megaspore from *Selaginella kraussiana* and the ($\pm$) spore from the fungus *Mucor mucedo*. The formation of the last spore by fusion of the ($+$) and ($-$) strains is accompanied by a parallel enhanced formation of carotenoids which can be readily seen on culture plates. A nitrogen- and cellulose-free wall was readily extracted from each of these spores after treatment by methods used for the isolation of pollen exines. In each case an acetolysis-resistant material was obtained, in yields analogous to those from typical pollen grains, which had all the characteristic properties of sporopollenin (Figs. 4–6) by the criteria of identity mentioned above. There is a strong association between the formation of all these spores and carotenoid production, but the ontogeny is clearly different [20] from that of pollen in the case of the algal and fungal spores at least, since it would appear that sporopollenin production occurs within the developing spore. It is also interesting to note that neither the asexual 'spores' derived from *Mucor mucedo* [17] nor the algal swarmer *Prasinocladus marinus* [17]

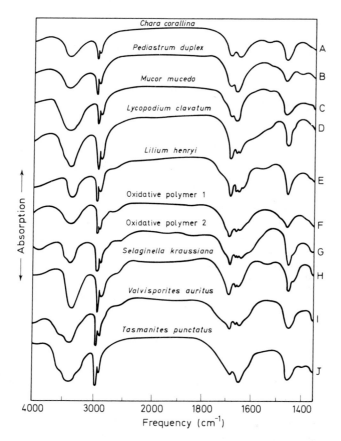

Figure 6. The infra-red spectra of some sporopollenins. (A) Chara corallina *algal spore; (B)* Pediastrum duplex *algal spore; (C)* Mucor mucedo (±) *fungal spore; (D)* Lycopodium clavatum *spore exine; (E)* Lilium henryi *pollen exine; (F) Oxidative polymer from* Lilium henryi *carotenoids and carotenoid esters; (G) Oxidative polymer of β-carotene; (H)* Selaginella kraussiana, *a modern megaspore; (I)* Valvisporites auritus, *a fossil megaspore (250 m. y. old); (J)* Tasmanites punctatus *fossil spore exine (350 m. y. old)*

contained sporopollenin but rather a characteristically polysac-charide-like substance. From our admittedly limited data it would thus appear that there may be a correlation between the sexual process and sporopollenin production in both higher and lower plants.

GEOCHEMISTRY OF SPOROPOLLENIN. A SUMMARY

Pollen and related spore exines are the most ubiquitous of fossils and probably more widely distributed in time and space than any other representative of living matter: this is a direct result of their extraordinary resistance to biological decay and chemical decom-position. The stability and morphological diversity of the exine has led to the evolution of the new science of palynology. It is not always appreciated, however, that many modern pollen exines [3] (e.g. poplar) do not survive as morphologically intact bodies *in vitro* even under what would normally be regarded as favourable conditions of isolation (non-oxidative, low pH, etc.) but undergo physical or physico-chemical abrasion to yield an amorphous insoluble organic material. The chemical structure of these amor-phous sporopollenins is in no way different from those derived from morphologically intact material, since the changes appear to be of a minor nature and presumably a function of the architecture of the exine and its relationship with the intine material. Thinness and, hence, fragility of the exine may be an important factor, although some grass pollen grains (e.g. timothy grass), which contain only small amounts of sporopollenin and consequently possess thin exines, nevertheless readily survive as morphologically intact structures after the vigorous chemical treatment used in their preparation. The point made here is of special importance when the behaviour of exines in sedimentary rock formations is examined.

Structurally amorphous matter of a similar chemical composition should be expected to occur side by side with morphologically intact exines in sediments, since the physico-chemical forces in such an environment may in many ways be equated with the *in vitro* isolation procedures; thus exposure of polysaccharides for thousands or perhaps millions of years (but probably much less) to water even at ambient temperatures is going to result in much the same end-product (monosaccharides) as heating with strong acid for a few days.

From a comparison of limited published analytical data on the insoluble organic matter (kerogen [21]) from various, especially Precambrian, sediments we have recently suggested [22] that the greater part of this material is identical with sporopollenin produced by abrasive changes in many pollen and spore exines of the type mentioned earlier. We have now confirmed these suggestions in a recent communication [19], which records the results of an examination of the insoluble organic matter in samples of the Onverwacht (3.7×10^9 years old) and Fig-tree (3.2×10^9 years old) cherts (Swaziland System); the Nonesuch shale (1.0×10^9 years old, Michigan, U.S.A.); the Green River shale (6.0×10^7 years old, Colorado, U.S.A.); Tasmanin spore coal (3.5×10^8 years old, containing morphologically intact *Tasmanites punctatus*, a fossil planktonic alga, Mersey river, Tasmania); and a carboniferous megaspore, *Valvisporites auritus* (2.5×10^8 years old, Kansas, U.S.A.). All these materials were shown to have a common chemical identity with sporopollenin derived from modern pollen, algal, fungal and megaspore wall exines, and some synthetic analogues prepared by copolymerisation of plant carotenoids with oxygen, the general criteria of identity referred to earlier (Tables 1 and 5; Figs. 4–6) being used.

Table 5. MOLECULAR FORMULAE OF SOME FOSSIL SPOROPOLLENINS

Material	Age (years)	Molecular formula
Tasmanites huronensis	350×10^6	$C_{90}H_{134}O_{17}$
Tasmanites punctatus	350×10^6	$C_{90}H_{132}O_{16}$
Tasmanites erratius [24]	350×10^6	$C_{90}H_{133}O_{11}$
Gleocapsamorpha prisca [24]	350×10^6	$C_{90}H_{331}O_{17}$
Geiseltalpollenin [4]	250×10^6	$C_{90}H_{129}O_{19}S_7N$
Lange sporonin [4]	250×10^6	$C_{90}H_{82}O_{17}N$
Valvisporites auritus	250×10^6	$C_{90}H_{102}O_{16}$

In a separate publication [23] we have also recorded the results of an examination of the insoluble organic matter present in two (Orgueil and Murray) carbonaceous chondrites. Both meteorites contained relatively large amounts (3–4%) of insoluble organic matter and this has proved to be indistinguishable from sporopollenin. It will be appreciated that in these experiments the amounts of material involved are so large as to exclude interference to the results by contamination with odd pollen or spore exines, since the detection of such minute amounts would be beyond the capacity of the techniques used.

REFERENCES

1 BAILEY, I. W., *J. Arnold Arbor.*, **41**, 141 (1960)
2 ZETZSCHE, F. and HUGGLER, K., *Justus Liebigs Annln Chem.*, **461**, 89 (1928)

3 SHAW, G. In: *Phytochemical Phylogeny* (Ed. J. B. HARBORNE), 31 Academic Press, London (1970)

4 ZETZSCHE, F., KALT, P., LIECHTI, J. and ZIEGLER, E., *J. prakt. Chem.*, **148**, 267 (1937)

5 SHAW, G. and YEADON, A., *Grana palynol.*, **5** (2), 247 (1964)

6 SHAW, G. and YEADON, A., *J. chem. Soc.*, **1966** (C), 16

7 BROOKS, J. and SHAW, G., *Nature, Lond.*, **219**, 532 (1968)

8 HESLOP-HARRISON, J., *Science, N.Y.*, **161**, 230 (1968)

9 GODWIN, H., *New Phytol.*, **67**, 667 (1968)

10 HESLOP-HARRISON, J. and MACKENZIE, A., *J. Cell Sci.*, **2**, 387 (1967)

11 BROOKS, J., and SHAW, G., *Grana palynol.*, **8** (2–3), 227 (1968)

12 BURNETT, J. H. In: *Chemistry and Biochemistry of Plant Pigments* (Ed. T. W. GOODWIN), 400, Academic Press, London (1965)

13 HESLOP-HARRISON, J., *Can. J. Bot.*, **46**, 1185 (1968)

14 HESLOP-HARRISON, J., *New Phytol.*, **67**, 779 (1968)

15 ROWLEY, J. W., *Grana palynol.*, **3** (3), 3 (1962); ROWLEY, J. W., *Science, N.Y.*, **137**, 526 (1962); ROWLEY, J. W., *Grana palynol.*, **4** (1), 25 (1963); ROWLEY, J. W. and SOUTHWORTH, D., *Nature, Lond.*, **213**, 703 (1968)

16 ECHLIN, P. and GODWIN, H., *J. Cell Sci.*, **3**, 161 (1968)

17 BROOKS, J. and SHAW, G. (Unpublished results)

18 DUCK, E. W., 'Emulsion Polymerization'. In: *Encyclopedia of Polymer Science and Technology*, Vol. 5, 801, Interscience, New York (1966)

19 BROOKS, J. and SHAW, G., *Grana* (In press)

20 HAWKER, L. E. and GOODAY, M. A., *J. gen. Microbiol.*, **54**, 13 (1968)

21 FORSMAN, J. P. In: *Organic Geochemistry* (Ed. I. A. BREGER), 148, Pergamon, Oxford (1963)

22 BROOKS, J. and SHAW, G., *Nature, Lond.*, **220**, 678 (1968); **227**, 195 (1970)

23 BROOKS, J. and SHAW, G., *Nature, Lond.*, **223**, 754 (1969)

24 EISENACK, A., *Arch. Protistenk.*, **109**, 207 (1966)

Incorporation of Radioactive Precursors into Developing Pollen Walls*

Darlene Southworth, *Department of Botany, University of California, Berkeley, California, U.S.A.*

The pollen wall of *Gerbera jamesonii* (Compositae) has very thick ektexine and endexine layers which can be distinguished in sections by light microscopy. Exine synthesis must involve the transfer of a relatively large quantity of sporopollenin and its precursors into this massive wall. In the study of *Gerbera jamesonii* wall development a number of developmental questions were considered. What are the routes of transfer of substrates to the exine? Is the callose wall a barrier isolating the mother cell and the microspore tetrad? Does exine material pass directly from tapetum to wall or does it first enter the microspore cytoplasm? What is the time of exine synthesis? Does the same compound enter the primexine as well as the ektexine and endexine? Which substrates can serve as precursors for the exine and into which larger molecules are they incorporated? Results with radioactive tracers and autoradiography were used to propose answers to most of these questions. Other studies on the ultrastructure and cytochemistry of developing pollen of *Gerbera jamesonii* are in progress [10, 11]. The tracers used were as follows:

Tracer	Activity	Labelling method
DL-3-Phenylalanine-T(G)	5, 10, 20 μCi/ml	aqueous solution
D-Glucose-6-T	20 μCi/ml	5% glucose solution
Sodium acetate-T	20, 25 μCi/ml	aqueous solution
DL-Mevalonic-5-T Acid (DBED Salt)	20, 25 μCi/ml	aqueous solution

Individual florets were picked from the capitulum, surface-sterilised, washed in sterile distilled water and floated on the tracer

* Abstract.

115

solutions for 5–17 h at room temperature. Fixation was by freeze-substitution in methanol [7] or by formaldehyde [8]. Material was embedded in polyvinylpyrrolidone [3] or epon. Sections cut at 2 μm were coated with Ilford L-4 emulsion; the exposure was for 1–10 weeks. Before emulsion coating, some slides were treated as follows:

(a) Periodic acid–Schiff's reaction for insoluble polysaccharides [7].

(b) Coomassie blue stain for protein [2].

(c) Ethanolamine extraction to remove sporopollenin [1]. Slides were coated with celloidin [7]. Several drops of ethanolamine were placed on the slide, which was placed in a Petri dish at 60°C overnight.

(d) Pronase extraction to remove proteins. Slides were coated with celloidin and the sections were ringed with paraffin. The resulting depression was filled with a solution of 0·1 mg/ml B-grade pronase (Calbiochem) in 1 mM $MgSO_4$ at 37°C overnight.

Phenylalanine did not label tetrads of microspores (Fig. 1). The label appeared in the endexine at the time of its formation and in the ektexine at the time of its thickening (Fig. 2). At slightly later stages, after exine synthesis has been completed, no phenylalanine label was produced. Pronase removed nearly all of the exine label, but left the exine morphologically intact (Fig. 3). Coomassie blue gives a positive stain for protein in the ektexine. Ethanolamine removed the exine, but left a label in and around the cytoplasm (Fig. 4).

Glucose produced a dense label in the callose wall during the pollen mother cell stage, during the furrowing process and after the furrowing was completed (Fig. 5). A light label was found around each microspore during the time of primexine formation (Fig. 6). A light label was formed throughout the exine synthesis period in the ektexine (Figs. 7 and 8). The label in the endexine was nearly absent (Fig. 8). The periodic acid–Schiff's reaction gave a positive test for polysaccharides in the primexine and in the developing ektexine, but gave a negative result in the endexine. The glucose label was light in the microspore cytoplasm at all stages, but within the cytoplasm the label was more concentrated in the peripheral region than in the central region. Stages including intine and starch synthesis were not studied.

Acetate produced a label within the tetrad in the microspore cytoplasm and in the primexine region (Figs. 9 and 10), and also in the developing endexine and ektexine (Figs. 11 and 12). The label became extremely light after exine synthesis was completed.

Mevalonate produced no label in the anther.

117

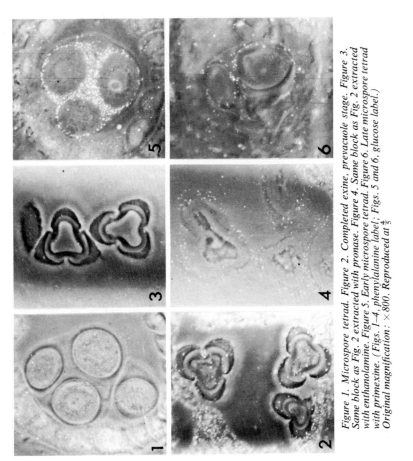

Figure 1. Microspore tetrad. Figure 2. Completed exine, prevacuole stage. Figure 3. Same block as Fig. 2 extracted with pronase. Figure 4. Same block as Fig. 2 extracted with enthanolamine. Figure 5. Early microspore tetrad. Figure 6. Late microspore tetrad with primexine. (Figs. 1–4, phenylalanine label; Figs. 5 and 6, glucose label.) Original magnification: ×800. Reproduced at $\frac{4}{5}$.

118

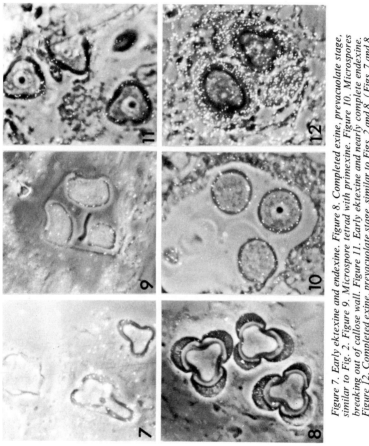

Figure 7. Early ektexine and endexine. Figure 8. Completed exine, prevacuolate stage, similar to Fig. 2. Figure 9. Microspore tetrad with primexine. Figure 10. Microspores breaking out of callose wall. Figure 11. Early ektexine and nearly complete endexine. Figure 12. Completed exine, prevacuolate stage, similar to Figs. 2 and 8. (Figs. 7 and 8, glucose label; Figs. 9–12, acetate label.)
Original magnification: ×800. Reproduced at $\frac{4}{5}$

In 1964 Heslop-Harrison [4] proposed that the callose wall invested the tetrad functions as a 'molecular filter', permitting passage of nutrients but preventing the penetrance of larger molecules which might impair the autonomy of the haploid nucleus. Heslop-Harrison [5] and Heslop-Harrison and Mackenzie [6] observed that radioactive thymidime entered *Lilium* pollen mother cells before formation of the callose wall, but did not penetrate the mother cells or the microspores while they were invested by the callose wall. After release from the tetrad, the microspores did become labelled. Rowley and Dunbar [9] observed that colloidal iron passed readily through the callose wall of the tetrad into the microspores of *Populus* and *Salix*. My results with *Gerbera jamesonii* show that glucose and sodium acetate pass through the callose wall, but phenylalanine apparently does not.

Table 1. TIMES OF INCORPORATION OF RADIOACTIVE COMPOUNDS DURING *Gerbera jamesonii* EXINE DEVELOPMENT

Label	Meiosis	Primexine formation	Endexine formation	Ektexine thickening
Phenylalanine		———————		
Glucose	———————————————————	— — — — —		
Acetate		———————————————		

These experiments, in which the label is incorporated over a long time period, do not allow one to determine whether the exine or the microspore cytoplasm is labelled first. At the light-microscope level it is impossible to distinguish between a label in the innermost endexine and a label in the peripheral cytoplasm with its organelles, including ribosomes, dictyosomes, lipid vesicles, and grey bodies. Chase experiments with a 5 h labelling period followed by a 5 h period on a cold medium show only that the tapetum is labelled before the microspores at all stages studied.

In *Gerbera jamesonii* exine synthesis proceeds with the formation of the primexine within the microspore tetrad. Following callose dissolution the endexine is laid down and finally the ektexine is greatly thickened and expanded. After the exine is completed, a large vacuole forms. Vacuole formation may be taken as the approximate end of exine synthesis [11]. None of the three tracers was incorporated into the wall after vacuole formation. Table 1 shows the correlation of time of incorporation of the tracers with stages of wall formation. Of the three tracers studied, acetate correlates in time most closely with exine synthesis, although phenylalanine, glucose and acetate all are closely involved. Phenylalanine seems to be removed by pronase, although a light label does remain in the wall which could be sporopollenin. Glucose is incorporated into components which are known to be polysaccharide in nature, i.e. the callose wall and the PAS-positive regions of the exine. A small amount of glucose could also be incorporated into the sporopollenin. Acetate is incorporated into those morphological regions which are associated with the exine and sporopollenin. On the basis of these results, I propose the working hypothesis that acetate is incorporated into a (lipid?) fraction of sporopollenin and that small amounts of breakdown products from other precursor molecules, e.g. glucose and phenylalanine, are similarly incorporated.

REFERENCES

1 BAILEY, I. W., 'Some useful techniques in the study and interpretation of pollen morphology', *J. Arnold Arbor.*, **41**, 141–51 (1960)

2 FISHER, D. B., 'Protein staining of ribboned epon sections for light microscopy', *Histochemie*, **16**, 92–6 (1968)

3 FISHER, D. B., JENSEN, W. A. and ASHTON, M. E., 'Histochemical studies of pollen: storage pockets in the endoplasmic reticulum', *Histochemie*, **13**, 169–82 (1968)

4 HESLOP-HARRISON, J., 'Cell walls, cell membranes and protoplasmic connections during meiosis and pollen development'. In: *Pollen Physiology and Fertilization* (Ed. H. F. LINSKENS), 39–47, North-Holland, Amsterdam (1964)

5 HESLOP-HARRISON, J., 'Cytoplasmic continuities during spore formation in flowering plants', *Endeavour*, **25**, 65–72 (1966)

6 HESLOP-HARRISON, J. and MACKENZIE, A., 'Autoradiography of soluble $[2-^{14}C]$ thymidine derivatives during meiosis and microsporogenesis in *Lilium* anthers', *J. Cell Sci.*, **2**, 387–400 (1967)

7 JENSEN, W. A., *Botanical Histochemistry*, Freeman, San Francisco (1962)

8 PETERS, T. and ASHLEY, C. A., 'An artefact in radioautography due to binding of free amino acids to tissues by fixatives', *J. Cell Biol.*, **33**, 53–60 (1967)

9 ROWLEY, J. R. and DUNBAR, A., 'Transfer of colloidal iron from sporophyte to male gametophyte' (Manuscript in preparation)

10 SOUTHWORTH, D., 'Ultrastructure of *Gerbera jamesonii* pollen', *Grana palynol.*, **6**, 324–37 (1966)

11 SOUTHWORTH, D., 'Development of *Gerbera jamesonii* pollen', *XI International Botanical Congress (Abstract)*

Surface Membrane Specialisations and the Pollen Wall*

J. J. Flynn, *Department of Botany, University of Massachusetts, Amherst, Massachusetts, U.S.A.*

After meiosis the four microspores of *Nelumbo* become separated by a callose-containing wall continuous with the special cell wall. The special cell wall appears mostly homogeneous. Except for an irregular region near the periphery which is moderately stained by uranyl acetate and lead hydroxide and which presumably contains remnants of the cell wall which surrounded the sporogenous cell, the special cell wall exhibits little affinity for uranyl or lead stains and only moderate affinity for phosphotungstic acid.

While still enclosed by the special cell wall, each microspore forms a fibrous wall interrupted in places by radially directed posts, the probacules [4]. In cross-section the fibrous wall consists of interlocking fibres of stainable material. The fibres are associated into plaques at the cell surface and the plaques become detached and form a system of anastomosing lamellae. The lamellae lie parallel to the cell surface and the wall is highly reminiscent of the primary cell wall of meristematic cells. The radially directed posts are moderately osmiophilic and temperately stained by uranyl acetate and lead hydroxide. The posts are permeated by a ramifying, non-staining network of voids which at times contains tubular, membranous profiles. The posts are occasionally surrounded by membranes, and in some degree are reminiscent of plasmodesmata. When the microspores are mildly plasmolysed, the posts remain attached to the plasmalemmal surface.

Thiocarbohydrazide (TCH), a multidentate ligand, binds to certain cations found in thin sections of biological material [3]. It also forms stable thiocarbohydrazones when allowed to react with aldehydes. If TCH is attached to either a cation or an aldehyde, it is possible to locate the complex by attaching osmium, phosphotungstic acid or silver proteinate (Pro) to the TCH [8]. Cations are commonly found in cell walls and free aldehydes are found in

* Abstract.

121

lignified walls. Aldehydes may be formed by periodic acid (PA) oxidation of 1,2-glycols such as are found in polysaccharides having 1,4-linkages, or by the deamination of α-amino groups of amino acids by ninhydrin (N). As controls, aldehydes may be blocked by phenylhydrazine or dimedone (5,5-dimethyl-1,3-cyclohexandione), and α-amino groups may be removed by nitrous acid.

When microspores of *Nelumbo* in the tetrad stage are treated with PA–TCH–Pro, the fibrous wall gives a positive response, as do restricted areas on the plasma membrane, small vesicles in the cytoplasm, and distal cisternae of dictyosomes. If similar sections are treated with N–TCH–Pro, the wall fails to stain, the cytoplasm is weakly stained, but the posts are intensely stained except for the voids. Abundant product after N–TCH–Pro indicates the presence of numerous short peptides.

The economy of microspore development demands an unequal balance of payments: more material must be taken into the microspore than passes out in order that the cell volume (which after dissolution of the special cell wall may increase thirtyfold) be filled. The impediment to transfer of material presented by the pollen wall may partially be circumvented by specialisations of the pollen wall and cytoplasm, e.g. wall microtubules of *Nuphar*, tubules between the exine and tapetal surface in *Aegiceras*, and an elaborate plug of polysaccharide material penetrated by divaricating evaginations of the plasma membrane in *Impatiens*. The wall microtubules of *Nuphar* are perpendicular to the cell surface. They permeate the ektexine and small spines, but are absent from the large spines. They are unlike cytoplasmic microtubules: they are smaller (about 100 Å in diameter), have five sub-units [6], are preserved after fixation with acetone or osmium, and give a positive response for 1,2-glycols. The wall microtubules resist pronase, acetolysis and warm pyridine. If microspores are briefly incubated in calcium chloride before fixation, the microtubules decrease in size (to about 70 Å in diameter) and if microspores are incubated in ethylene diaminetetra-acetic acid, they increase in size, to about 140 Å in diameter [2].

The microtubules of *Aegiceras* have a similar diameter to the wall microtubules in *Nuphar*, but they are found on the exine surface and the locular surface of tapetal cells. Some microtubules extend from the exine surface to the tapetal cells, a distance of several microns. The *Aegiceras* microtubules are stained by cationic dyes and complexes (alcian blue and colloidal iron), which is indicative of exposed acidic groups [7]. The well-known fuzzy coats of animal cells also have acidic groups. Fuzzy coats trap materials at the cell surface, where they can be engulfed by the cell via pinocytosis [5].

A plug of polysaccharide is found in the aperture of *Impatiens* after the microspores are released from the special cell wall. Already the microspores are half their final size and their cytoplasm is highly vacuolate. The only noticeable openings in the exine are at the apertures. The fingers of the plasma membrane penetrate the plug until the cell's volume becomes packed with cytoplasm after microspore mitosis. When the intine is laid down, the membranous fingers become sealed off. In the mature pollen grain only a small oncus with membrane-bounded channels remains in the apertural region between the exine and the intine which is continuous with the non-apertural intine [1].

From these fragmentary and scattered observations it seems that the plasmalemmal surface possesses a mosaic of sites, some responsible for polysaccharide deposition, some for probacule formation. The probacules may represent sites, analogous to plasmodesmata, which at times are punctured by tubules or evaginations of the plasma membrane. In *Nuphar* the microtubules have become 'frozen' into the wall. Here, where the microtubules have been most studied, it seems that they may operate as peristaltic pumps to transport materials across the pollen wall. They might operate in such a way that lipophilic materials may pass through the lumen when the tubules are expanded and hydrophilic materials pass between the microtubules and exine when the microtubules are contracted [2]. In *Impatiens* the plug could function as a combined molecular sieve–ion-exchange mechanism where the membranous fingers would scavenge the entrapped material.

This work was supported by NSF grant GB 7077 to Dr. John R. Rowley. The author's present address is: Palynology Laboratory, Nybodagatan 5, S-171 42 Solna, Sweden.

REFERENCES

1 FLYNN, J. J., *Impatiens: Pollen Development, Maturity, and Germination*, Ph.D. Thesis, University of Massachusetts, Amherst (1969)

2 FLYNN, J. J. and ROWLEY, J. R., 'Wall microtubules in pollen grains', *Zeiss Information* (In press)

3 HANKER, J. S., DEB, C., WASSERKRUG, H. L. and SELIGMAN, A. M., 'Staining tissue for light and electron microscopy by bridging metals with multidentate ligands', *Science, N.Y.*, **15**, 1631–4 (1966)

4 HESLOP-HARRISON, J., 'An ultrastructural study of pollen wall ontogeny in *Silene pendula*', *Grana palynol.*, **4**, 7–24 (1963)

5 REVEL, J.-P. and ITO, S., 'The surface components of cells'. In: *The Specificity of Cell Surfaces* (Ed. B. D. DAVIS and L. WARREN), 211–34, Prentice-Hall, Englewood Cliffs, New Jersey (1967)

6 ROWLEY, J. R., 'Fibrils, microtubules and lamellae in pollen grains', *Rev. Palaeobot. Palynol.*, **3**, 213–26 (1967)

7 ROWLEY, J. R., FLYNN, J. J., DUNBAR, A. and NILSSON, S., 'Influence of pinocytosis and membrane specializations on pollen wall form', *Grana palynol.* (In press)

8 THIÉRY, J.-P., 'Mise en évidence des polysaccharides sur coupes fines in microscope électronique', *J. Microscopie*, **6**, 987–1018 (1967)

A Function of the Sporocyte Special Wall*

L. Waterkeyn and A. Bienfait, *Laboratoire de Cytologie et Morphologie Végétales, Louvain, Belgium*

By squeezing a tetrad of microspores of *Ipomoea purpurea* (L.) Roth. still enclosed in the callosic special wall, the spores can be caused to escape from their individual chambers ('logette sporale'). The empty chambers show a distinct structure, the callosic wall possessing an inner relief forming a regular geometric pattern of hexagonal or pentagonal meshes with a central knob. Each released spore has a reticulate wall which stains selectively with toluidine blue, assuming a characteristic metachromatic colour. This wall, the primexine, forms a network of mostly hexagonal meshes. At most of the angles of these polygonal meshes of the primexine there are gaps, and along the sides a row of little pores. In slightly older spores isolated elements appear which fill exactly these gaps and pores; these are the rudiments of spines and the probacula.

The callosic wall of the spore chambers constitutes, in our opinion, an authentic mould for the primexine matrix, and represents thus a first general negative template for the exine patterning. It is effectively in the hollow network of the spore chamber wall that the material of the primexine is cast. Each knob of a polygonal mesh of the spore chamber acts as a centre for a mesh of the primexine. In the same way each of the six (or five) sides of the polygonal meshes of the former becomes a new centre for additional meshes of the latter. From this it is easy to understand why the primexine comes to have four times more meshes than its template.

This double origin of the primexine meshes leaves its traces even on the mature exine. Large spines are regularly absent at some angles of the polygonal network. This is the case with the hexagonal meshes centred on the sides of the large polygons; the distribution of the empty spaces is thus not random but related to the original spore chamber pattern.

* Abstract.

The Tapetal Membranes in Grasses and Ubisch Body Control of Mature Exine Pattern*

U. C. Banerjee and E. S. Barghoorn, *Department of Biology, Harvard University, Cambridge, Massachusetts, U.S.A.*

After the microspores are released from the tetrads into the anther loculus containing the thecal fluid, they go through the following developmental phases: (*a*) rapid increase in size; (*b*) reorientation within the loculus to bring the germ pores close to the wall, and (*c*) accumulation of the characteristic exine material, sporopollenin, on the primexine. At the same time, sporopollenin continues to be deposited on the inner tangential and partly on the radial surfaces of the tapetal cells exposed to the thecal fluid. A mature tapetal membrane of the grasses exhibits the spinulate Ubisch bodies (orbicules), a stranded layer, a backing layer (which becomes fenestrated at maturity), and microrods on the surface facing outwards. The tapetal membrane system obtained from a microsporangium shows a clear imprint of the tapetal cells and encloses the pollen grains like a sac. Four such sacs can be obtained from a mature undehisced anther by acetolysis.

When samples were collected early in phase (*c*), the freshly deposited sporopollenin on the microspores and tapetum was found not to be resistant to acetic anhydride treatment; as the phase advanced, however, resistance to acetolysis was acquired. Single-stage and two-stage replicas of the pollen grain and Ubisch bodies were made for transmission electron microscopy, and direct observations were made by scanning electron microscopy. The transmission electron micrographs show that sporopollenin strands, which were transformed from the cytoplasmic strands by direct deposition of sporopollenin, project outwards from the tips of the pollen grain spinules and make contact with the spinules of the Ubisch bodies. Scanning electron micrographs provide additional

* Abstract.

evidence that the distal (pore) end of the exine remains attached to the tapetal membrane before the anther dehisces. The proximal half of the pollen grain exine situated farthest from the vicinity of the tapetal membranes also exhibits sporopollenin protrusions. Moreover, isolated Ubisch bodies were seen attached to the pollen grain spinules. We interpret the Ubisch bodies as being formed, perhaps independently, by the periplasmodium present in the loculus; their structure is similar to those on the tapetal membranes. The strands observed at the end of phase (c) form a continuous bridge between the pollen grain spinules and the Ubisch bodies of the tapetum. This indicates a possible transfer of substrate from the tapetal cells to the exine.

During the early stages of exine development, the spinules are essentially extensions of the columellae, but at maturity a large number of new spinules, not formed by the primexine, are produced, apparently by the Ubisch bodies. These new spinules are unsupported by columellae. This suggests that the Ubisch bodies play an important role in sporoderm formation in grasses.

Immediately prior to anther dehiscence, some depolymerisation of the mature sporopollenin occurs, possibly controlled by secretion from the centrum of the Ubisch bodies. In the tapetal membrane itself this activates the breakdown of the stranded layer and causes fenestration of the backing layer. Almost at the same time that the depolymerisation system effects breakdown of the sporopollenin strands from the ektexine and Ubisch body spinules, it provides them with their final pointed appearance just before anther dehiscence.

This work was supported by grants from the National Science Foundation, U.S.A., and U.S. Public Health Service.

Pollen and Pollen Tube Metabolism

Pollen Chemistry and Tube Growth*

Robert G. Stanley, *Forest Physiology-Genetics Laboratory, University of Florida, Gainesville, Florida, U.S.A.*

INTRODUCTION

The first detailed chemical analyses of pollen were published in 1829 [1]. Since then progress in elucidating the chemical composition of pollen has primarily resulted from studies on pollen in relation to human allergic responses and bee nutrition. Endogenous chemicals in pollen have been correlated with patterns of inheritance and growth requirements for pollen. Growing pollen *in vitro* also permits the study of chemicals required for growth processes such as nuclear division and cell wall extension.

This paper will present a brief overview of pollen chemistry and growth responses of pollen to specific chemicals. An effort will be made to summarise the frontiers in pollen chemistry and the key problems still to be solved.

Two major groups of pollen have been analysed chemically. One type, wind-borne, anemophilous pollen, occurs in abundance, e.g. pine and palm. This type of pollen is easier to collect than the second, insect-transported, entomophilous type. Bee-transmitted pollens can be collected in traps at the hive entrance, but such pollen loads contain solidifying chemicals added by the bees. To circumvent errors in chemical analyses of bee-collected pollens, these types are also collected by hand. Chemical analyses reported in this review will be limited to hand-collected pollens. Thus good data on hand-collected pollens, both wind-borne and bee-transmitted, are available.

GENERAL ANALYSIS

The gross chemical analyses show that, of major pollen components, mineral and fat contents vary the least (Table 1). At dehiscence time

* A contribution of the *Florida Agricultural Experiment Station Journal*, Series No. 3, 579.

grass pollen, including *Zea mays*, has a water content above 50%; other pollens, e.g. *Typha* or *Pinus*, usually contain about 20% or less water at time of shedding. Compositional data are best compared on a dry weight basis.

Some components, such as carbohydrates and fibre residue, vary more between and within species than other chemical constituents. The carbohydrate content of grass pollen (corn) may be

Table 1. GROSS CHEMICAL ANALYSIS OF POLLEN

Species	Ash	% dry weight Carbo-hydrates	Protein	Lipid	Reference
Zea mays	2·55	36·59	20·32	3·67	[2]
Zea mays	3·46	34·26	28·30	1·48	[3]
Typha latifolia	3·70	17·78	18·90	1·16	[4]
Pinus sabiniana	2·59	13·15	11·36	2·73	[2]
Pinus radiata	2·35	13·92	13·45	1·80	[2]

more than twice that of other angiosperm pollens; carbohydrates in pine and most gymnosperms are considerably lower. Considerable variation in the level of pollen carbohydrates occurs at pollen maturity. Fibre residue of most angiosperm pollens runs between 5 and 7%, but in gymnosperm pollens it averages 15–20% of the dry weight. The high fibre residue in pine pollen may also lower the expressed percentage of carbohydrates and other materials assayed. The wall residue materials, including the wings of pine and some other gymnosperm pollens, present a considerably different structure from that of the multipored pollen grain common in angiosperms. Variations in exine structure may contribute to the survival capacity of pollens from different species and families.

Proteins vary widely with species, usually accounting for 10–30% of the pollen dry weight. Again, the shorter-lived, more rapidly growing pollens, such as *Zea mays* and *Typha*, are generally highest in protein. Lipid contents average 1·5–4% and may be higher in those with oils on the grain surface. Ash content usually is about 2–4%, but may be as high as 7% in some species.

Variation in chemical constituents of pollen is accounted for by (1) species differences and (2) environmental differences during maturation and after dehiscence. During maturation excessively high temperatures tend to reduce the pollen carbohydrate; low light also results in less carbohydrate accumulating in mature pollen, probably as a direct effect of reduced photosynthesis. If the plant nutrient supply, particularly microelements, is below optimum, pollen mineral content may be reduced. Lower content of trace elements in pollen can modify the protein–enzyme levels and thus affect growth and seed set [5].

How pollen is collected and stored can markedly influence some chemical components. For example, if hand-collected pollen is left at room temperature, high metabolic rates induce catabolism of endogenous carbohydrates and organic acids. Thus when pollen is stored under conditions which do not reduce metabolism, its chemical components may be adversely affected and this will be reflected by changes in the analyses.

MINERAL CONTENT

Mineral analyses are usually run on pollen ashed in concentrated acid. Pine, one of the first pollens analysed for minerals [6], contains, as principal elements, potassium, phosphorus, calcium and iron. Studies with *Zea mays* [3] added chlorine and magnesium to the list of pollen constituents. One must be careful in comparing some mineral data, since elements such as chlorine and boron may volatilise in ashing. Date palm, *Phoenix dactylifera*, is usually high

Table 2. MAJOR MINERALS IN POLLEN [2]

Species	Total ash	K	% dry weight P	Ca	Mg
Zea mays	2·55	0·67	0·26	0·10	0·21
Typha latifolia	3·82	0·97	0·49	0·30	0·24
Phoenix dactylifera	6·36	1·14	0·71	1·18	0·38
Pinus sabiniana	2·35	0·88	0·30	0·03	0·11
Pinus radiata	2·59	0·87	0·36	0·04	0·09

in mineral content (Table 2) and this may be related to the sites on which it grows. In such soils a large hardpan accumulates minerals; also, this tree accumulates silicon at very high levels throughout its organs and tissues.

The phosphorus content in pine pollens averages about one-half of that of angiosperm pollens [7]. In a single species pollen samples can sometimes differ by 200% in the level of an element. For example, one variety of corn pollen showed 0·26% while another contained 0·75% dry weight as phosphorus. Corn pollen of the same variety grown in two successive years on the same field had a total ash of 4·90% in both years, but phosphorus values of 0·56 and 0·75% [8].

There is also a considerable difference between levels of minerals in pollen and other plant tissues. Elements such as sulphur and phosphorus, on percentage dry weight, are about 5–10-fold higher in pollen. Others, i.e. potassium, occur at about the same percentage in pollen, leaves and roots.

Trace elements in pollen vary with species and conditions during maturation. Iron and zinc are the most abundant (Table 3). Aluminium, silicon and other non-metabolic elements also occur in

Table 3. TRACE ELEMENTS IN POLLEN, μg/mg DRY WEIGHT [9]

Species	Al	Cu	Fe	Element Mn	Ni	Ti	Zn
Corylus avellana	0·3	1·5	120	37	0	0·3	30
Tulipa suaveolena	49	10	250	36	0	2·0	251
Anemone nemorosa	27	14	286	112	75	1·5	150
Prunus avium	80	18	363	40	0	4·0	140
Prunus spinosa	1·6	20	250	25	0	1·6	80

pollen in variable amounts depending upon the species and environment. In general, plants of the same family accumulate comparable levels of an element.

Boron is a trace element of particular significance in pollen. Since Schmucker [10] showed that boron is involved in growth of many pollens, the element has been widely compared in pollen and floral organs. In most plants the maximum boron level is not found in pollen but rather in the stigmatic tissues. Boron occurs in pollen at about 0·7 μg/mg dry weight, while the stigma often contains 10 times that level. The style may also be high in boron. The level of boron in pollen is influenced by the amount of boron available during development [11]. When floral organs develop in a nutrient solution high in boron, the pollen germinates better, and requires less boron to grow than when little or no boron is available during microsporogenesis.

To understand the role of minerals in pollen, the sites of their localisation are usually sought. Some minerals, i.e. calcium, are generally associated with cell wall formation. Autoradiographs have been employed to localise ^{45}Ca incorporation in germinating pollen [12]. However, it is not easy to analyse pollen for some minerals *in situ* because of the solubility of unbound elements and the imprecision of many particle tracks on film emulsion. An interesting technique employed in our laboratory to gain insight into the localisation of specific elements in pollen uses the X-ray Microprobe Analyzer. Germinated pollen is dried on an aluminium grid and a 100 Å-wide X-ray beam is scanned across the pollen. In Fig. 1 the fluorescence spectrum for calcium was recorded as the X-ray beam moved across the pollen along transect 5. The exine is high in calcium with moderately high levels occurring at the intine and in a few places inside the pollen. The X-ray Microprobe can

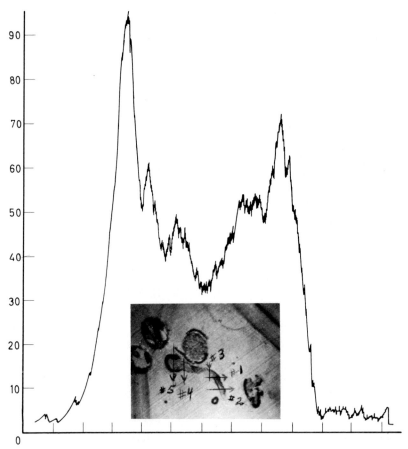

Figure 1. Electron microprobe X-ray analysis of pollen. Scan of Pyrus communis *for calcium, No. 5, from top to bottom*

scan and localise two elements simultaneously, thereby elucidating where specific elements are located and how they are involved in pollen growth. Another method of ascertaining where specific nutrients are involved in pollen growth is to observe the response on addition of the element to the growth media. But observed growth responses can involve secondary interactions with substrates, or problems of diffusion and uptake.

CARBOHYDRATES

Simple sugars are the principal metabolic substrates used by germinating pollen. It is interesting that most grass pollens are shed in the trinucleate stage [13] and do not maintain their viability in storage. These pollens—*Zea mays*, for example—contain 36–40% of their dry weight as carbohydrates (Table 4). Other pollens,

Table 4. POLLEN CARBOHYDRATES [2]

Species		% dry weight Sugars		
	Total	Reducing	Non-reducing	Starch
Zea mays	36·6	6·9	7·3	22·4
Typha latifolia	31·9	0·05	18·9	13·0
Phoenix dactylifera	1·2	1·1	0·1	0·0
Pinus sabiniana	13·2	7·5	3·5	2·2

relatively stable in storage for long time periods, are binucleate and relatively low in total soluble carbohydrates; date palm *(P. dactylifera)* pollen, for example, contains only 1–2% soluble carbohydrate and lacks starch when shed. In contrast, *Typha* pollen starch content is so consistently high that it has occasionally been used as a flour substitute. Starch content of *Zea mays* varies from 12 to 30%, depending on the variety and method of handling. Attempts to correlate the starchy endosperm characteristic of corn and the starch content of the pollen have not been successful [14].

The major component of free sugars in pollen is generally associated with a particular species. In pine pollens over 93% of the free sugar is sucrose, but in angiosperm pollens sucrose represents usually only 20–50% of the free sugar. Other soluble sugars occur in most pollens. In 15 conifer pollens examined, raffinose occurred in all, and stachyose in 10 [15]. Rhamnose occurs in many pollens at maturity, but in *Rosa* it disappears during storage [16]. Arabinose, xylose and galactose also frequently occur as free sugars in pollen. Each of these three sugars also occurs in the hydrolysates of pectin and hemicelluloses of the pollen tube walls. Rare sugars reported in pollen include turanose, nigerose and lactose. These latter sugars are probably fragments from polysaccharides. Several unidentified sugars have been reported in extracts of pine [17] and corn pollen [15]. No one has pursued these leads or determined the significance of these sugars for pollen growth.

Soluble sugars in pollen change markedly with storage and handling conditions. In a long-term experiment in which pine pollen was stored for 15 years at 25% and 10% relative humidity, only pollen stored at 10% relative humidity germinated *in vitro* [18]. After 15 years' storage, polysaccharide content did not differ significantly at the two humidities, but a substantial decrease in glucose and sucrose accompanied the decrease in germination capacity. Pollen collected by bees is higher in reducing sugars than the same pollen collected directly from the plant [2]. The increase in reducing sugars in bee-collected pollen is due to bee secretions and nectar added to the pollen mass by the bees.

Carbohydrates in pollen occur primarily in thè cell walls and as cytoplasmic polysaccharides. Thus little seasonal or species variation occurs in total insoluble polysaccharides compared with the large variation in soluble carbohydrates. Pollens metabolise many sugars other than those which they contain. Examples of this broad capacity are illustrated by experiments with pine pollen germinating in 0·2 M solutions of different sugars (Fig. 2). Pine pollen does not require sugar or boron to germinate *in vitro*, yet oxygen uptake is doubled by exogenous sucrose.

Glucose, galactose and lactose all stimulate pollen respiration. Pine pollen, like other pollens, not only contains many soluble sugars, but has enzymes to metabolise a broad range of sugars absorbed from the external media. Some pollens accumulate starch as excess levels of sugars are absorbed. At dehiscence pine pollen is quite low in starch (Table 4), but on germinating in the presence of sugar, starch is formed in the pollen cytoplasm [19].

Factors such as micronutrients can also affect the ability of pollen to metabolise sugars. Boron is not required for pine pollen to germinate, yet when boron is placed in the media with pine pollen the capacity to metabolise glucose-6-^{14}C is increased by about 60%. Boron stimulation reflects not only the species and substrates available but the type of boron supplied. Butyl borate is superior to boric acid in stimulating glucose metabolism in pollen [20]. Phenyl borate, toxic to pollen at low concentrations, is absorbed more rapidly by pollen than other borate forms. Possibly the dissociated phenyl-moiety inhibits the pollen.

As yet there is no exact recognition of the role of boron in plant metabolism; it is required for meristematic bud development as well as pollen growth. Several hypotheses involving metabolism of sugars and growth of the pollen tube membrane, particularly by incorporation of pectin precursors, have been proposed [21]. Some

138

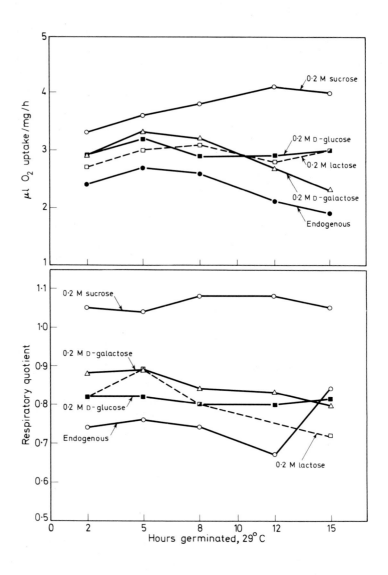

Figure 2. Respiratory patterns of Pinus ponderosa *pollen on different sugars. Each sample contained 50* mg *pollen, 0·3* mg *mysteclin and 2·0* ml *0·002* M CaPO₄ *buffer, pH 5·7. Means of three replicate samples*

evidence also suggests boron acts at the messenger-RNA level. However, we still do not know exactly how boron is involved in pollen or plant growth.

Media for pollen growth *in vitro* containing sugars such as raffinose often yield better growth than media using the simpler disaccharide sucrose. Interestingly, pollen also metabolises lactose (Fig. 2), a fact reported by many workers. Pine pollen tubes form starch when grown on lactose–agar [19]. The ability of pine pollen to metabolise radioactive lactose increased if the pollen was initially grown with galactose [22]. The enzyme that hydrolyses lactose, α-galactosidase, appears to be induced, activated or at least increased in activity in pollen pretreated with galactose. This increase in ability to metabolise lactose does not occur when pine pollen is germinated on lactose alone. This suggests that galactose is the natural substrate and is more readily absorbed than lactose; the latter is merely hydrolysed at the α-galactoside bond. Thus not only can pollens metabolise non-endogenous sugars, but sugars supplied exogenously may affect the levels or types of enzymes formed by germinating pollen.

The cyclitols are an interesting group of compounds that are often related metabolically to sugars. Myoinositol frequently occurs as a free compound in pollen; it also occurs as phosphoinositol. Increases in inositol occur as pollen germinates and the phosphatides are hydrolysed to free inositol. Other cyclitols isolated from pollens include pinitol and sequoyitol. These latter two appear only in trace amounts. In pollen, inositol is not merely an enzyme co-factor as is usually assumed, but can actually be incorporated into pectins during germination [21, 23].

Pentose sugars, ribose and deoxyribose, isolated as free sugars from pollen, are probably hydrolytic products of nucleic acids. Some sugars are also found in association with proteins and pollen lipids.

OTHER FACTORS AFFECTING TUBE GROWTH

Boron is only one factor that can markedly affect pollen growth. Potassium and sodium at concentrations of 0·01 M and barium and magnesium at 0·0015 M inhibit pollen germination [24]. Brink [24] found that calcium at 0·05 M enhances pollen growth. Other researchers have confirmed that calcium enhances pollen growth, and have found that it may interact with potassium, magnesium and possibly hydrogen ions in the germinating medium [25]. Cations and anions affect pollen growth in a lyotrophic-like series. In the cation series silver, lead, magnesium and sodium are most active and calcium, copper and cobalt least active [26]. Cobalt, zinc and other minerals have occasionally been reported, along with calcium and magnesium, as stimulating pollen growth [27]. The anion series decreases in effect on pollen tube growth with boron as most active down to phosphate.

When Ca^{++} was compared to K^+ in its effect on pine pollen growth and metabolism of glucose-1-^{14}C and glucose-6-^{14}C in phosphate buffer, pH 5·9, the 0·002 molar Ca^{++} yielded several hundred times more $^{14}CO_2$ than the K^+ solution [28]. Mascarenhas and Machlis [29] found that calcium in the plant ovary is present in many times the concentration found in pollen, style or stigma. They concluded that a calcium gradient acts as a chemotactant for pollen tube growth. This observation has been difficult to repeat in other laboratories and, in fact, calcium concentration has been found to be lower in the placenta than in the ovary wall [30]. Another approach to determining the nature of pollen tube attractant is to analyse for specific chemical differences in the ovule. In one such study of water-soluble extracts of ovules, induction of tropic pollen tube growth was obtained with each of 4 peptides, 2 amines, 5 aromatic amines and 11 sugars isolated from the extract [31]. As previously indicated, factors such as light or nutrients affect flower tissue development and modify the response of pollen to a particular cation or anion.

Oxygen levels also affect pollen growth and the metabolism of various substrates. This is particularly true of pollen tubes growing down a style where O_2 level can modify use of substrates in the style and the pollen. Factors which affect pollen growth, and which may influence use of specific substrates *in situ*, include: minerals, sugars, CO_2 and O_2, as already discussed; and pH, temperature, osmotic pressure, moisture, growth hormones and irradiation.

CALLOSE AND SPOROPOLLENIN

Callose, a β-1,3-glucose polymer present at high levels in pollen at maturity, declines and is resynthesised during germination. Callose may occur in higher levels than cellulose and pectin [32]. Cellulose, the β-1,4-glucose polymer, usually does not exceed 2% by weight in pollen.

In pine callose occupies considerable cytoplasmic area around the microspore generative cell. While callose is an important and fairly common carbohydrate in plants, we are not exactly sure of its role. In pollen tubes, callose forms a plug as the tube elongates in the style. The plug first forms at the upper end, sealing off the pollen tube from the exine, and thus shortens the cytoplasmic stream. Callose may also act to seal the tube wall if a very small hole occurs in the membranes during growth. At present we assume callose functions in pollen as a storage carbohydrate and also to seal off parts of the tube cytoplasm.

Callose has been localised by two techniques, fluorescent ultra-violet light microscopy [33] and autoradiography [34]. Both methods detect callose plugs in the tubes, in the microspore pore areas, and as a lining inside the tube wall as it grows. Whenever pollen stops growing, callose begins to form at the tube tip. In normal growth callose forms inside the tube wall during growth, starting a little behind the tip. To follow the formation of callose and other wall materials derived from radioactive glucose, we used specific enzymes to hydrolyse pear pollen tubes [34]. This method indicated that the pollen tip is highest in pectins, moderately high in cellulose, but low or lacking in callose. Increased levels of β-1,3-polyglucan occur farther back along the tube wall lining the inside of the primarily cellulose membrane, which has low pectin and hemicellulose. As pollen tubes grow, pectin–hemicellulose is the primary material laid down at the tip with some cellulose, followed by a small amount of callose which forms farther back inside the intine. The chemical composition of the principal exine constituent, sporopollenin, is still not definitely known [but see pp. 100–14—Ed.], although we now recognise that the tapetal cells are involved in its synthesis. Standard techniques used to extract cellulose, 72% sulphuric acid and acetyla-tion, permit us to localise cellulose in the pollen walls; but after removal of all the cellulose, a substantial non-cellulosic residue re-mains in the exine. The amount of pollen wall residue varies from 2–5% in angiosperms to 20–25% in gymnosperms. Pollen wall residue remaining after hydrolysis was first called 'pollenin' by John in 1814. Von Planta, in 1886, called the pollen residue in *Pinus sylvestris* 'cuticula' [6], and other workers called it a modified cellulose [35]. By 1920 the pollen exine was thought to be a cuticle layer over a modified cellulose. Starting in the mid-20s, Zetzsche and his col-leagues analysed lycopodium spores and pollen walls, and termed the spore residues 'sporonine'. Because of the similarity of sporonine to the earlier pollenin of John, they subsequently used the term 'sporopollenin' [36]. They found that sporopollenin had a C/H ratio of 1:1·6. Since exine residue includes oxygen, on further analyses they computed it as having a C_{90} unit, a polymer of iso-prene units. They therefore suggested that sporopollenin had a formula of $C_{90}H_{134}O_{31}$.

Shaw and Yeadon found sporopollenin was composed of fatty

acids with lipid–lignin-like components [37]. They reported a series of mono- and dicarboxylic acids as exine breakdown products with a lignin-like fraction of 10–15%. They also reported about 10% hemicelluloses and 15% cellulose in pine pollen. The lignin-like material might also arise by hydrolysis of carotene polymers [38]. Hydrolysis of the ozonated exine of lily pollen yielded products similar to the ozonated β-carotene products. Since carotenoids are common in many pollens, it was concluded that exine is similar in composition to carotene [38]. However, carotenes have never been isolated from pine pollen. Hopefully, we may obtain information on the exine chemical composition by following the biosynthesis of the exine [39]. In addition to sporopollenin and other wall components, exine often contains other chemicals, i.e. fats, oils or proteins, in *Pollenkitt*, or near the surface [40]. This array of materials may be dissolved off the pollen surface as the tubes grow or when grains land in solution [41].

ORGANIC ACIDS

When carbohydrates are metabolised, organic acids generally result. All the Krebs cycle acids have been found in pollen, but they vary markedly with stage of development and handling. Fatty acids or phenolic acids may be involved in synthesis of more insoluble pollen wall constituents.

The phenolic acids, *p*-hydroxybenzoic, *p*-coumaric, vanillic, protocatechuic, gallic and ferulic, can be extracted and their levels can be measured. Phenolic acid levels in pine pollens have been proposed as an indicator of taxonomic relationships [42]. Except for a few acids such as vanillic, missing in both ponderosa and shortleaf pine pollen, species growing in the western and southern United States, respectively, few distinct patterns are apparent. It is therefore difficult to separate the pine species on the basis of phenolic acids in their pollen. We do not understand the role of these acids in pollen growth; some may lead to lignin-like constituents or be involved in formation of amino acids and other compounds.

Fatty acids are also common in pollen. Large quantities of certain fatty acids, particularly palmitic acid, and linoleic and linolenic acids, exist in many pollens [43]. Pine and other gymnosperm pollens are very high in linolenic acid [44]. There is some correlation between species and the principal fatty acids in the pollen lipids, but percentages found vary widely because of the rapid metabolism and ease of saturation of double bonds in the compounds. The major portion of fatty acids exists in pollen as esters combined with sugars, phosphates and other constituents [45] which have not yet been studied in any great detail.

AMINO ACIDS

All the essential amino acids are present in pollen. Because of the importance of pollen protein in allergic responses and in bee nutrition, most reports on pollen amino acids give hydrolysis data of the total protein. However, the free amino acid pool does not reflect the ratio of amino acids found in the pollen protein [46]. Levels of glutamic acid and other free amino acids decrease markedly during long periods of storage [47]. Fatty acid, or carbohydrates, on metabolism yield acetate and organic acids which can be quickly transaminated to amino acids. Some reported values of amino acid content might reflect variations in pollen metabolism with different handling procedures.

Efforts have been made to apply the data from pollen free amino acid content to understand pollen fertility potential, taxonomic relations and pollen growth potential.

In corn, sterile anthers have a lower proline level and a higher alanine level than viable, fertile anthers [48]. Studies comparing free amino acid levels with apple pollen fertility showed that a higher ratio of proline to histidine occurs in viable pollen from a diploid than in sterile pollen obtained from a triploid [48]. This suggests that compounds such as proline are essential for many pollens to germinate, and that endogenous levels may indicate growth ability. However, analysis of viable or non-viable pollen in terms of free amino acids must be limited to comparisons within a single species or its varieties. Many pollens have only trace amounts of the amino acids leucine or proline in the free state, while histidine may occur at very high levels in other pollens [50]. The exact relationship is still obscure.

Efforts to establish a taxonomic index based on free amino acids in pollen have failed. Seven species of pine pollens studied in this regard [51] yielded no significant differences in free amino acid content. Although pollens of some species were considerably lower in certain free amino acids than others, the differences were of no taxonomic significance.

Even less promising is the concept relating growth potentials to the free amino acid content of pollen. For example, Durzan [52] compared the free amino acid pools in pollen, embryos and dormant seeds of *Picea glauca*. Total soluble nitrogen per gramme of fresh weight was considerably higher in the pollen: i.e. 4000 μg compared to about 100 μg in the embryos or seeds. When the different proportions of amino acids in the soluble nitrogen fraction are compared, one finds that compounds such as arginine and alanine, essential for protein synthesis, occur at relatively low levels in pollen, but at high levels in embryos. On the other hand, proline, often considered to be connected with cell wall synthesis, and found free in levels as high as 1·65% in some pollens [53], occurs in low levels in ungerminated embryos and seeds. It is not easy to equate levels of different free amino acids directly with ability to grow.

When free amino acids were compared in 107 different pollens [50], the maximum number of free amino acids, 19, occurred in Western cottonwood *(Populus sargentia)*. More than one-half of the species tested had above 10 free amino acids. γ-Amino-butyric acid and several exotic amino acids are also found in pollens in the free state, even though these are not found in plant proteins. Amino acids commonly arise from organic acids. Germinating pollen can also incorporate carbon dioxide, $^{14}CO_2$ from the germinating solution, into organic acids and amino acids [54, 55]. The pathway for fixation is primarily phosphoenolpyruvate carboxylase. Other pathways such as ribulose diphosphate may also operate in germinating pollen. Thus, in germinating pollen, sugars are not the only source of organic and amino acids; CO_2 in the surrounding stylar tissues may also provide a carbon source.

PROTEIN AND NUCLEIC ACIDS

This discussion will be limited primarily to quantitative aspects, since a complete discussion of pollen enzymes, isozymes and RNA synthesis occurs elsewhere in this volume [56, 57].

The amount of protein in pollen ranges from about 11 to 30%, with the lower amounts generally found in the species with slower-growing tubes *in situ*, and higher amounts in rapid growing pollens (Table 1). As recently as 1960, studies reported that no net protein synthesis occurred in germinating pollen [58]. Methods in these earlier studies were not as sensitive as current radioisotope tracer work. Protein synthesis during pollen germination has been clearly demonstrated, and the role of functioning polysomes has been indicated [59].

Evidence suggests that dormant messenger RNA (mRNA) and transfer RNA (tRNA) pre-exist in mature microspores and are quickly activated upon germination [60, 61]. These studies were done on pollen germinated *in vitro* and on pollen extracts. A different complement of organelles exists in growing areas of pollen tubes germinated *in vitro* and *in situ* [62]. Caution must therefore be exercised in extrapolating organelle and cytoplasmic patterns from the *in vitro* studies to normal conditions in the style.

In mature microspores DNA occurs at levels of about 0·05% dry weight, while RNA may comprise up to 0·5% of the pollen dry weight. The base composition of pollen RNA and nature of RNA changes during germination *in vitro* have been followed. Base composition of RNA in pine pollen varies with the specific cell organelle [63]. The ratio of purines/pyrimidines in pine pollen microsomal fraction was 1·00, but mitochondrial RNA yielded 1·25, and the soluble fraction 1·28. Guanine was particularly high in the pollen subcellular components. Pseudouridylic acid was found in the nucleotide hydrolysate of *Corylus* pollen [64]. Obviously the general procedure of extracting RNA from pollen for base composition determination [64] can lead to misinterpretation of the changes occurring, or of the type of RNA related to growth. The presence of a given nucleotide, or mole ratios of nucleotides, should be related to specific subcellular organelles as well as stage of development.

Polysomes form as germination proceeds, with only few polysomes found under conditions of inhibited pollen growth [59]. This suggests that the enzyme–protein activities producing pollen tubes are under intimate molecular control of the pollen and/or tissue surrounding the germinating pollen. The lack of new tRNA in growing tubes may in part help explain the interesting observation that 2-thiouracil (2-TU), an antimetabolite which usually inhibits cell growth, actually stimulates pollen tube formation [65]. Added to pollen, 2-TU decreases the level of RNA formed but does not affect the formation of protein [66]. With adequate turnover and low destruction of tRNA in a system of limited growth, protein formation would continue on functional polysomes without new tRNA being required by the pollen germinating *in vitro*. It will be interesting to relate the ability of spermidine and related polyamines to increase nucleic acid synthesis in germinating pollen [67] to the ability to activate or bind preformed mRNA to polysomes. Of course pollen growing *in vitro* seldom if ever attains the tube length it does when growing in the style. Understanding this area of pollen chemistry affords a considerable challenge for future studies.

It may be valuable to speculate on the possible evolutionary significance of the observed lack of new tRNA and limited mRNA formed during pollen growth, albeit *in vitro*, while active polysome and protein formation occurs. One obvious fact is that pollen has a very limited ability to adapt, as evidenced by the narrow range of compatibility and physiological environments in which it can grow. This limited growth potential may be related to inability to form new mRNA and tRNA.

PIGMENTS

The chemistry of pollen pigments has been relatively well investigated, yet comparatively little is known of the physiological role of these pigments. About 80% of several hundred randomly selected species of pollen were yellow in colour [68]. The dominant yellow colour was characterised as due to carotenoids or flavonoids. The carotenoids in pollen are primarily α-carotene and some β-carotene, lycopene, xanthophyll and zeaxanthin; chlorophyll has not been found [69], but anthocyamins are occasionally reported in pollen. Varieties and hybrids of *Petunia* yielded pollen with a range of 38 colour gradations, from yellow-green to grey and blue, and these colours breed true [70].

Free carotenes or their derivatives have never been isolated from pine pollen. While it is possible that carotenes exist in these pollens in some highly modified form, the visible pigments in the exine are primarily flavonoids. Further study of pigments during microsporogenesis might help resolve this question in pines and other species. Study of flavonoid pigments present in pollens of more than 150 species showed that quercetin, kaempferol and isorhamnetin are the most common forms in angiosperm pollens. In gymnosperm pollens isorhamnetin does not occur, quercetin and kaempferol occur occasionally, and considerable amounts of naringenin are found. Naringenin is only rarely found in angiosperm pollens [71].

The nature of pollen pigments and their site of localisation affect the ease with which the pigments are removed. Carotenes are generally removed with the lipid fraction. After removal of carotenes in some species the flavonoid–glucosides can be isolated [72]. In many pollens with surface oils, carotenes are in the oils, but carotenes may also be found in the exine and in the cytoplasm. Some pigments, in particular flavones, are water-soluble and readily diffuse from the pollen.

Several different functions have been ascribed to the pigments in pollen. Carotenoids in orchid pollen tubes appear to stimulate the sexual process [73]. The pigments, according to this hypothesis, are related to sexual compatibility. Carotene is normally accumulated in several varieties of plants during microsporogenesis, but in sterile lines carotenes are not accumulated [74]. Correlations have been made between levels of carotene, the ability of pollen to set fruit and the viability of resulting seeds. Pollen tube growth *in vitro* was stimulated on addition of carotene to the growth media. This suggests that carotenoids can be involved in metabolism, possibly as enzyme co-factors leading to increased tube growth. Flavonoids added *in vitro* also stimulate pollen tube growth [75].

Other suggestions have related flavonols and their glucosides to the incompatibility reaction between pollen and style in *Forsythia* and similar species. The hypothesis proposed that pollen flavonoids are the inhibiting substances, with specific hydrolytic enzymes being present in compatible styles. This idea has not been supported by further experiments [76].

Several roles have been proposed for pollen pigments. One suggestion is that pigments protect the genetic content of the pollen; another that the pigment assures transmission by the proper vector. Pigments, according to this hypothesis, screen potentially harmful light radiation. Wind-borne pollen is supposedly higher in ultraviolet screening pigments than is insect-transmitted pollen [77]. The wind-transmitted pollens would be expected to be higher in carotenes of flavonols and contain different types from insect-transmitted pollens. This hypothesis breaks down, since some pollens collected by bees are actually very high in carotenes. In fact, pollen carotenes have often been suggested as fulfilling the role of vitamin A precursor in bee nutrition.

VITAMINS, HORMONES AND STEROIDS

Pollen vitamins assayed as essential in insect and mammalian nutrition often behave chemically as enzyme co-factors. Some vitamins, i.e. biotin, occur at very low levels in all species (Table 5);

Table 5. POLLEN VITAMIN CONTENT [7, 8]

Vitamin	Pinus montana	μg/g *dry weight* Alnus incana	Zea mays
B$_2$ (riboflavin)	5·6	12·1	5·7
B$_3$ (nicotinic acid)	79·8	82·3	40·7
B$_5$ (pantothenic acid)	7·8	5·0	14·2
B$_6$ (pyridoxine)	3·1	6·8	5·9
C (ascorbic acid)	73·7	—	58·5
H (biotin)	0·62	0·69	0·52

others, e.g. nicotinic acid, occur at much higher levels in pollen. Free folic acid was found at levels of 0·42–2·20 μg/g dry weight in pollen of 10 species assayed [78]. The vitamin level essential for pollen growth is contained in the mature microspore or in tissues through which the pollen germinates. However, decreases in pollen vitamin content can occur during storage [79]. Riboflavin and pantothenic acid were most markedly affected. Such losses may cause reduced growth of pollen tubes or low viability when older, stored pollens are germinated.

Vitamin-like components, ascorbic acid and inositol, the latter already discussed, often occur in pollen in mg/g dry weight levels. Inositol is particularly high in corn and grass pollens; ascorbic acid occurs at relatively high levels in pine and palm pollen.

Pollen hormones and growth substances will be discussed in two separate categories. First, the pollen components already recognised as growth substances in plants, the auxins, kinins and gibberellins; second, those substances considered as, or related to, hormones in man which also occur in pollen, namely the steroids.

Early workers recognised that pollen supplied some substances or induced formation of a chemical in the pollinated flowers which inhibited formation of the peduncle abscission layer. Orchid pollinia were one of the first sources of auxin discovered [80]. Pollen studies are usually run on the homogenised extracts of whole pollen and not on the diffusates during germination. When chromatographic separation procedures are used, whole pollen extracts yield R_fs and indicator colours corresponding to indole acetic acid, auxin inhibitors and gibberellins [81]. Which substances function during pollen germination, or are released by the female tissues and affect pollen growth in the style, is difficult to discern from such studies.

In other studies auxin and gibberellins are added to artificial pollen germination media, and growth stimulation or inhibition is observed. It is not valid to assume that a similar effect occurs in the style unless the growth factor is found in the tissues in which the pollen grows. Many chemicals can stimulate tube growth when added to the pollen media and yet never be found in the living system, e.g. 2-thiouracil and edetic acid (EDTA).

After the discovery of the steroid nature of animal sex hormones, plant scientists started to search for similar substances in pollen. Curative and sex-restorative powers have often been attributed to pollen and have led to its inclusion in some human diets. Initially when rats were used as test materials, and more recently by chemical analysis, oestrone was found in *Phoenix dactylifera*, date palm pollen [82]. Androgenic activity has seldom been found in pollen, but cholesterol, sitosterol and several other related steroids are widely distributed in pollens [83]. These steroid components often occur in association with fatty acids in pollen [45].

The physiological role of steroids in pollen can only be speculated upon at present. The isolation in 1967 of the first steroid sex-hormone in lower plants [84] encouraged our laboratory to reinvestigate the effect of steroids on pollen germination. Pear pollen stimulation was obtained at low moderate concentrations of cholesterol (Table 6) but not with other steroids tested in alcohol or buffer. While our

Table 6. STEROID EFFECT ON POLLEN GERMINATION
Pyrus communis var. W. Nellis

Cholesterol (mg/ml)	$\bar{X}$ tube length (relative units)	% increase
0	32·4	
1	35·0	8
2·5	41·5	28
5·0	37·5	17

Pollen germinated 1 h, 30°C, 0·002 M $CaPO_4$, pH 6·0; H_3BO_3, 50 μg/ml; cholesterol crystals added to the solution.

observations suggested an interaction or metabolism of this common steroid with germinating pear pollen, the more detailed studies with *Chrysanthemum* pollen [85] were much more positive and definitive in their results.

The role in pollen of such ubiquitous components as steroids may not be apparent until we have additional data concerning endogenous levels in the female tissues through which pollen grows and the possible mechanism of steroid involvement in pollen metabolism. Different inherent levels of fatty acids and steroids [45] may explain the variations in response of different pollens to exogenous sterols. It is hoped that speculations in this area will be backed with good isotopic experiments.

Isotopes can help provide an understanding of where specific chemicals are involved in pollen growth and interaction with the female tissues. Radioactive phosphorus was used in 1945 to resolve whether or not cytoplasm from more than one pollen grain is incorporated in maturing seeds. It was found that each aspen seed contained ^{32}P contributed by the cytoplasm of eight pollen grains [86]. These studies offer a tool for interpreting the results of multiple pollinations and possibly understanding cytoplasmic inheritance [28, 87]. The 'Dauermodifikation' and phenotypic expressions of multiple pollinations may be characterised as extranuclear cytoplasmic influences contributed by pollen to the developing progeny. It is a challenging area of research, which closely associates pollen chemistry with plant breeding.

SUMMARY OF POLLEN CHEMISTRY AND GROWTH

An overview of pollen growth suggests that the tube nucleus and probably the generative nucleus are involved in RNA synthesis and protein formation in the elongating pollen tube. Observations on pollen growing *in vitro* show that only a limited amount of new mRNA, and possibly no new tRNA, is required during germination and growth. Enzymes produced and activated during germination metabolise internal and external substrates. This metabolism supplies components and energy for pollen tube growth.

The area with cytoplasm inside the tube is delimited by a callose plug which crowds the cytoplasm forward as the tube extends. In binucleate pollens the division of the generative nucleus precedes bursting of the tube.

Past studies of pollen chemistry and growth provide the following general conclusions:

(1) Metabolic pathways in pollen are those common to most non-green tissues.

(2) The over-all composition and balance of pollen chemical constituents vary with the species, plant nutrient level, environment during development and subsequent handling.

(3) Enzymes or simple chemical constituents rapidly diffuse out of pollen even before the tube is apparent.

(4) Chemicals diffusing out of the pollen, or on the pollen surface, can interact with the female tissues and give rise to incompatible reactions; if a compatible pollination occurs, the pollen metabolises available substrates and grows to the ovule.

(5) Tube growth can be modified by chemicals in the female tissue; external chemicals can be incorporated as building moieties, growth stimulators or inhibitors.

(6) Tube extension occurs by the addition of pectins and hemicelluloses at the end of the pollen tubes via addition of vesiculated membrane-like components at the tube tip; cellulose is probably added after the initial tube membrane is formed. Interestingly, germinating pollen with few dictysomes may produce little or no pectin in the walls [88]. A callose layer generally forms behind the tip and lines the tube membrane. Callose also forms as a plug at the tip when growth ceases.

(7) Decreased pollen viability after dehiscence is generally related to enzyme activities metabolising endogenous substrate in the pollen.

A considerable body of knowledge of constitutive and exogenous chemicals affecting pollen growth exists. Yet there are many components that have not been determined and many growth responses to which we are still seeking answers. A few problems, in particular those which are probably most amenable to investigation at this time, are worth summarising.

(1) Why do pollens grow at different rates in genetically different tissues? For instance, in apples pollination with compatible pollen on to a style 7 mm long results in pollen tubes that grow at 0·35 mm/h, while tubes in incompatible pollinations grow at 0·05 mm/h. Pollen will grow through a style 55 mm long of *Oenothera organensis* at a rate of 6·50 mm/h. This involves many chemical variables, but conditions are known which reverse certain incompatibility reactions; these may be useful tools in exploring this aspect of pollen chemistry.

(2) Why cannot the rates of pollen growth in the style be duplicated *in vitro*? A sustained growth rate of 6·5 mm/h, over 100 μm/min, is never approached in present culture techniques. Apparently, some factor(s) are missing from our most complete media.

(3) Mixed, multiple or successive pollinations can yield interesting patterns of seed set and occasionally apparent genetic modifications. It would be meaningful and of possible significance in plant breeding to study the range, potential and response characteristic of this mechanism.

(4) Where are the complex chemicals of pollen, e.g. pigments, sterols, located and how are they functioning in pollen? Which pollen constituents interact with the female tissue?

(5) What is the role of boron in germinating pollen? Suggestions have been made that boron influences plant metabolism by affecting RNA, or sugar–carbohydrate metabolism, or wall synthesis. Experiments with pollen may help elucidate the general problem of the role of boron in plants.

(6) What chemotactic or attracting substance(s) induce pollen to grow to the ovule?

(7) Cytoplasmic streaming is a problem in cell metabolism for which germinating pollen is a good research tool. In pollen and other cells cyclosis may be related to movement of organelles towards the tip, part of the mechanism to distribute substrates and products to different parts of the cell.

(8) The chemical or information trigger that activates the generative cell to divide is one example of the universal problem of cell division for which pollen may be helpful in providing an answer. What causes the different nuclear division time sequence in grasses compared with other angiosperm pollens?

(9) The function of the vegetative (tube) nucleus in pollen is still not definitely known. The DNA of this nucleus has a different histone pattern from that of the DNA of the generative cell [89] and this nucleus generally disintegrates before the tube bursts. The tube nucleus is involved in RNA–protein metabolism; and while it most often moves to the front of the tube during tube extension, there are cases where it does not and the tube still grows normally.

(10) What is the role of the different isozymes that occur during pollen growth? This is also a problem in other living systems. However, the terminal growth pattern of pollen and its limited functional life and adaptive range make the study of pollen isozymes a particularly interesting problem.

Many other general problems can be approached through a study of pollen chemistry and growth. The most important point of this brief review has been to distinguish what we know about pollen from the large area still unknown. Pollen is a good research tool to elucidate certain basic patterns and mechanisms of plant growth. The challenge is to relate the chemical knowledge we have to the problems we must still solve.

REFERENCES

1 BRACONNOT, H., 'Recherches chimiques sur le pollen du Typha latifolia Linn., famille des Typhacées', *Annls Chim. Phys.* Ser. 2, **49**, 91–105 (1829)
2 TODD, F. E. and BRETHERICK, O., 'The composition of pollens', *J. econ. Ent.*, **35**, 312–17 (1942)
3 ANDERSON, R. J. and KULP, W. L., 'Analysis and composition of corn pollen. Preliminary report', *J. biol. Chem.*, **50**, 433–53 (1922)
4 WATANABE, T., MOTOMURA, Y. and ASO, K., 'Studies on honey and pollen. III. On the sugar composition in the pollen of *Typha latifolia* Linne.', *Tohoku J. agric. Res.*, **12**, 173–8 (1961)
5 LATZKO, E., 'Genetische Stoffwechselregulation und ihre Beeinflussung durch mineralische Nahrelemente', *Bayer. landw. Jb.*, **41**, 259–76 (1964)
6 PLANTA, A. VON, 'Über die Chemische zusammensetzung des Bluthenstaubes der germeinen keifer *(Pinus sylvestris)*', *Landwn Vers Stuen*, **32**, 215–30 (1886)
7 TOGASAWA, Y., KATSUMATA, T. and OTA, T., 'Biochemical studies on pollen. Part VI. Inorganic components and phosphorous compounds of pollen', *J. agric. Chem. Soc. Japan*, **41**, 178–83 (1967)
8 NIELSEN, N., GRÖMMER, J. and LUNDÉN, R., 'Investigations on the chemical composition of pollen from some plants', *Acta chem. scand.*, **9**, 1100–6 (1955)
9 DEDIC, G. A. and KOCH, O. G., 'Zur kenntnis des Spurenelementgehaltes von Pollen', *Phyton B. Aires*, **9**, 65–7 (1957)
10 SCHMUCKER, TH., 'Bors als physiologisch entscheidene des Element', *Naturwissenschaften*, **20**, 839 (1932)
11 VISSER, T., 'Germination and storage of pollen', *Mededel Landbow.*, **55**, 1–68 (1955)
12 STEFFENSEN, D. and BERGERON, J. A., 'Autoradiographs of pollen tube nuclei with calcium-45', *J. biophy. biochem. Cytol.*, **6**, 339–42 (1949)

13 BREWBAKER, J. L., 'Biology of the angiosperm pollen grain', *Indian J. Genet. Pl. Breed.*, **19**, 121–33 (1959)

14 ZUBER, M. S., DEATHERAGE, W. L., MACMASTERS, M. M. and FERGASON, V. L., 'Lack of correlation between the amylose content of pollen and endosperm in maize', *Agron. J.*, **52**, 411–12 (1960)

15 UENO, S., 'Sugars in pollens of cone plants', *Kagaku, Tokyo*, **24**, 90–1 (1954)

16 ZOLOTOVICH, G., SECENSKA, M. and DECHEVA, R., 'Changes in the composition of the sugars and in enzyme activity on storage of rose pollen', *C.r. Acad. bulg. Sci.*, **17**, 295–8 (1963)

17 CHIRA, E. and BERTA, F., 'One of the causes of the cross incompatibility of species of genus *Pinus*', [In Czech., German summ.] *Biológia, Bratisl.*, **20**, 600–9 (1965)

18 STANLEY, R. G. and POOSTCHI, I., 'Endogenous carbohydrates, organic acids, and pine pollen viability', *Silvae Genetica*, **11**, 1–3 (1962)

19 HELLMERS, H. and MACHLIS, L., 'Exogenous substrate utilization and fermentation by the pollen of *Pinus ponderosa*', *Pl. Physiol., Lancaster*, **31**, 284–9 (1956)

20 STANLEY, R. G. and LICHTENBERG, E., 'The effect of various boron compounds on *in vitro* germination of pollen', *Physiologia Pl.*, **16**, 337–46 (1963)

21 STANLEY, R. G. and LOEWUS, F. L., 'Boron and myo-inositol in pollen pectin biosynthesis'. In: *Pollen Physiology and Fertilization* (Ed. H. F. LINSKENS), 128–36, North-Holland, Amsterdam (1964)

22 STANLEY, R. G., 'Methods and concepts applied to the study of flowering in pine'. In: K. V. THIMANN (Ed.), *Physiology of Forest Trees*, 583–99, Ronald Press, New York (1958)

23 KROH, M. and LOEWUS, F., 'Biosynthesis of pectic substances in germinating pollen: labeling with myoinositol-2-^{14}C', *Science, N.Y.*, **160**, 1352–3 (1968)

24 BRINK, R. A., 'The effect of electrolytes on pollen tube development', *J. gen. Physiol.*, **6**, 677–82 (1924)

25 KWACK, B. Y. and MACDONALD, T., 'Role of calcium in pollen growth as expressed by various water-soluble substances', *Bot. Mag., Tokyo*, **78**, 164–70 (1965)

26 ZERLING, V. V., 'Influence of major and minor elements on pollen germination in plants', *Dokl. Akad. Nauk SSSR for. Lang. Edn.* **32**, 439–42 (1941)

27 JOHRI, B. M. and VASIL, I. K., 'Physiology of pollen', *Bot. Rev.*, **27**, 325–81 (1961)

28 STANLEY, R. G., 'Factors affecting germination of the pollen grain', *Internatl. Union For. Res. Org. Proc. Sec, 22, III*, 39–59 (1967)

29 MASCARENHAS, J. P. and MACHLIS, L., 'Chemotropic response of the pollen of *Antirrhinum majus* to calcium', *Pl. Physiol., Lancaster*, **39**, 70–7 (1964)

30 GLENK, H. O., WAGNER, W. and SCHIMMER, O., this volume, 255

31 SCHILDKNECHT, H. and BENONI, H., 'Versuche zur Aufklarung des Pollenschlauchchemotropismus von Narcissen', *Z. Naturforsch.*, **18b**, 656–61 (1963)

32 MARTENS, P. and WATERKEYN, L., 'Structure du pollen "aile" chez les conifères', *Cellule*, **62**, 173–222 (1962)

33 LINSKENS, H. F. and ESSER, K. L., 'Über eine spezifische Anfärbung der Pollenschläuche im Griffel und die Zahl der Kallosepfropfen nach Selbstung und Fremdung', *Naturwissenschaften*, **44**, 16–17 (1957)

34 ROGGEN, H. A. P. and STANLEY, R. G., 'Pollen cell wall hydrolytic enzymes', *Planta*, **84**, 295–303 (1969)

35 KRESLING, K., 'Beiträge zur chemie des Blütenstaubes von *Pinus silvestris*', *Arch. Pharm.*, **229**, 389–425 (1891)

36 ZETZSCHE, F. and HUGGLER, K., 'Untersuchungen über die Membran der Sporen und Pollen. I. *Lycopodium clavatum* L.', *Justus Liebigs Annln Chem.*, **461**, 89–108 (1928)

37 SHAW, G. and YEADON, A., 'Chemical studies on the constitution of some pollen and spore membranes', *J. chem. Soc. (C)*, **1968**, 16–22

38 BROOKS, J. and SHAW, G., 'Chemical structure of the exine of pollen walls and a new function for carotenoids in nature', *Nature, Lond.*, **219**, 532–3 (1968)
39 SOUTHWORTH, D., this volume 115
40 KNOX, R. B. and HESLOP-HARRISON, J., 'Cytochemical localization of enzymes in the wall of the pollen grain', *Nature, Lond.*, **223**, 92–4 (1969)
41 STANLEY, R. G. and SEARCH, R. W., this volume, 174
42 STROHL, M. J. and SIEKL, M. K., 'Polyphenols of pine pollens', *Phytochemistry*, **4**, 383–99 (1965)
43 STANDIFER, L. N., 'Fatty acids in dandelion pollen gathered by honey bees, *Apis mellifera*', *Ann. ent. Soc. Am.*, **59**, 1005–7 (1966)
44 CHING, T. M. and CHING, K. K., 'Fatty acids in pollen of some coniferous species', *Science, N.Y.*, **138**, 890–1 (1962)
45 HOEBERICHTS, J. A. and LINSKENS, H. F., 'Lipids in ungerminated pollen of *Petunia*', *Acta bot. neerl.*, **17**, 433–6 (1968)
46 DJURBABIĆ, B., VIDAKOVIĆ, M. and KOLBACH, D., 'Effect of irradiation on the properties of pollen in austrian and scotch pines', *Experientia*, **23**, 296–8 (1967)
47 KATSUMATA, T., TOGASAWA, Y. and OBATA, Y., 'Biochemical studies on pollen. III. Amino acids of pollen of *Pinus densiflora*' [In Japanese], *J. agric. Chem. Soc. Japan*, **37**, 439–43 (1963)
48 KHOO, V. and STINSON, H. T. JR., 'Free amino acid differences between cytoplasmic male sterile and normal fertile anthers', *Proc. natn. Acad. Sci. U.S.A.*, **43**, 603–7 (1957)
49 TUPÝ, J., 'Free amino acids in apple pollen from the point of view of its fertility', *Biologia Pl.*, **5**, 154–60 (1963)
50 BIEBERDORF, F. W., GROSS, A. L. and WEICHLEIN, R., 'Free amino acid content of pollen', *Ann. Allergy*, **19**, 867–76 (1961)
51 VAN BUIJTENEN, J. P., 'Identification of pine pollens by chromatography of their free amino acids', M.S. thesis, University of California, Berkeley (1952)
52 DURZAN, D. J., 'The nitrogen metabolism of *Picea glauca* (Moench) Voss. and *Pinus banksiana* L. with special reference to nutrition and environment', Ph.D. thesis, Cornell University, Ithaca, N.Y. (1964)
53 BATHURST, N. O., 'The amino acids of grass pollen', *J. Expl Bot.*, **5**, 253–6 (1954)
54 STANLEY, R. G., YOUNG, L. and GRAHAM, J., 'Carbon dioxide fixation in germinating pine pollen', *Nature, Lond.*, **182**, 738–9 (1958)
55 GOSS, J. A. and PANCHAL, Y. C., 'Nonphotosynthetic fixation of $C^{14}O_2$ by *Ornithogalum caudatum* pollen', *BioScience*, **15**, 38–9 (1965)
56 BREWBAKER, J. L., this volume, 156
57 MASCARHENAS, J., this volume, 201
58 POZSÁR, B. I., 'The nitrogen metabolism of the pollen tube and its function in fertilization', *Acta Bot.*, **6**, No. 3/4, 389–95 (1960)
59 MASCARHENAS, J. P. and BELL, E., 'Protein synthesis during germination of pollen: studies on polysome formation', *Biochem. biophys. Acta*, **179**, 199–203 (1969)
60 LINSKENS, H. F., this volume, 232
61 MASCARHENAS, J. P., this volume, 230
62 ROSEN, W., 'Ultrastructure and physiology of pollen', *A. Rev. Pl. Physiol.*, **19**, 435–62 (1968)
63 STANLEY, R. G. and YEE, A., 'Ribonucleic acid composition in cellular fractions of ponderosa pine pollen', *Nature, Lond.*, **210**, 181–3 (1966)
64 MACIEJEWSKA-POTAPCZYK, W., URBANEK, H., KULEC, I. and PACYK, H., 'Nucleic acids and proteins in *Corylus avellana* pollen', *Zesz. nauk. Uniw. łódz. Ser. 2*, **30**, 63–8 (1968)
65 TUPÝ, J., STANLEY, R. G. and LINSKENS, H. F., 'Stimulation of pollen germination *in vitro* by thiouracil and other anti-metabolites of nucleic acid bases', *Acta bot. neerl.*, **14**, 148–54 (1965)

66 TUPÝ, J., 'Synthesis of protein and RNA in pollen tubes stimulated with 2-thiouracil', *Biologia Pl.*, **8**, 398–410 (1966)

67 LINSKENS, H. F., KOCHUTY, A. S. L. and SO, A., 'Regulation der Nucleinsaüren Synthese durch Polyamine in keimenden Pollen von *Petunia*', *Planta*, **82**, 111–22 (1968)

68 MÖBIUS, M., 'Über die Färbung der Anthern und des Pollens', *Ber. dt. bot. Ges.*, **41**, 12–16 (1923)

69 LUNDÉN, R., 'A short introduction to the literature on pollen chemistry', *Svensk kem. Tidskr.*, **66**, 201–13 (1954)

70 FERGUSON, M. C., 'A cytological and a genetical study of *Petunia*. V. The inheritance of color in pollen', *Genetics, N.Y.*, **19**, 394–411 (1934)

71 WIERMANN, R., 'Untersuchungen zum Phenylpropanstoffwechsel des Pollens. I. Übersicht über die bei Gymnospermen und Angiospermen isolierten flavonoiden Verbindungen', *Ber. dt. bot. Ges.*, **81**, 3–16 (1968)

72 TAPPI, G. and MENZIANI, E., 'The flavonolic pigments of the pollen of *Lilium candidum*', *Gazz. chim. ital.*, **85**, 694–702 (1955)

73 TSINGER, N. V., PODDUBNAYA-ARNOL'DI, V. A., 'Physiological role of carotenoids in generative organs of higher plants', *Dokl. Akad. Nauk SSSR*, **110**, 157–9 (1954)

74 SAMORODOVA-BIANKI, G. B., 'Content of carotenoids and their dynamics in fertile and sterile anthers of some plant species', *Dokl. Akad. Nauk SSSR*, **109**, 873–5 (1956)

75 MINAEVA, V. G. and GORBALEVA, G. N., 'Effects of flavonoids on pollen germination and growth of pollen tubes', *Polez. Rast. Prir. Flory Sib.*, 231–5 (1967) [from *Chem. Abstr.*, **69**, 103905c (1968)]

76 LINSKENS, H. F., 'Pollen', *Handbk. Pl. Physiol.*, **18**, 368–406 (1967)

77 WERFT, R., 'Über die Lebensdauer der Pollenkörner in der freien Luft', *Biol. Zbl.*, **70**, 354–67 (1951)

78 NIELSEN, N. and HOLMSTROM, B., 'Occurrence of folic acid, folic acid conjugates and folic acid conjugases in pollen', *Acta chem. scand.*, **11**, 101–4 (1957)

79 NIELSEN, N., 'Vitamin content of pollen after storage', *Acta chem. scand.*, **10**, 332–3 1965)

80 LAIBACH, F., 'Pollenhormon und Wuchstoff', *Ber. dt. bot. ges.*, **50**, 383–90 (1932)

81 MICHALSKI, L., 'Growth regulators in the pollen of pine', *Acta Soc. Bot. Pol.*, **34**, 475–81 (1967)

82 BENNETT, R. D., KO, S-T. and HEFTMANN, E., 'Isolation of estrone and cholesterol from the date palm *Phoenix dactylifera*', *Phytochemistry*, **5**, 231–5 (1966)

83 STANDIFER, L. N., DEVYS, M. and BARBIER, M., 'Pollen sterols. A mass-spectrographic survey', *Phytochemistry*, **7**, 1361–5 (1968)

84 MCMORRIS, T. C. and BARKSDALE, A. W., 'Isolation of a sex hormone from the water mould *Achyla bisexualis*', *Nature, Lond.*, **215**, 320–1 (1967)

85 MATSUBARA, S., this volume, 186

86 EKLUNDH-EHRENBERG, C., EULER, H-V. and HEVESY, G., 'Number of pollen grains identified in the fruit of the aspen', *Ark. Kemi Miner. Geol.*, **B23**, No. 5, 1–5 (1946)

87 POLYAKOV, I. M., 'New data on use of radioactive isotopes in studying fertilization of plants'. In: H. F. LINSKENS (Ed.), *Pollen Physiology and Fertilization*, 194–9, North-Holland, Amsterdam (1964)

88 CRANG, R. E. and MILES, G. B., 'An electron microscope study of germinating *Lychnis alba* pollen', *Am. J. Bot.*, **56**, 398–405 (1969)

89 JALOUZOT, R., 'Differenciation nucleaire et cytoplasmique du grain de pollen de *Lilium candidum*', *Expl Cell Res.*, **55**, 1–8 (1969)

Pollen Enzymes and Isoenzymes

J. L. Brewbaker, *Department of Horticulture, University of Hawaii, Honolulu, Hawaii*

Pollen enzymes have been the subject of two major reviews, that of Paton [1] in 1921 and a recent compilation by Makinen and Macdonald [2]. During the intervening half-century and despite major strides in protein and enzyme chemistry, comparatively few published studies have appeared on the enzymes of pollen. Many questions of pollen biology (e.g. storage longevity, viability *in vitro*, incompatibility phenomena) appear to await more intensive pollen enzymatic research (cf. that of Dickinson and colleagues [3–7] on carbon chemistry of pollen tubes). It is hoped that a more thorough review of available information on this subject may encourage such research.

Table 1 summarises those enzymes which have been reported to be active in the pollen grains of higher plants. Enzyme classification numbers and trivial names follow those of Barman [8]. The enzymes are considered individually here in the order in which they appear in the table, i.e. alphabetically by trivial name within the six classes, dehydrogenases, oxidases, transferases, hydrolases, lyases and ligases.

DEHYDROGENASES

Dehydrogenase (DH) activity has been reported for pollen of many plant species [9–11]. Activity in pollen is high relative to other tissues—for example, styles [12]. The tetrazolium dye tests commonly used for the examination of seed viability also reveal great variation in pollen DH activity, with male- and semi-sterile plants and haploids, for example, producing many non-stainable grains [13]. MTT, a tetrazolium bromide, gave best staining results for pollen in viability tests [14].

The identification of specific dehydrogenase activities has been

156

Table 1. ENZYMES REPORTED TO BE ACTIVE IN POLLEN GRAINS

Enzyme group	E.C. No.	Enzyme trivial name
Dehydrogenases	1.1.1.49	Glucose 6-phosphate dehydrogenase
	1.4.1.3	Glutamate dehydrogenase
	1.1.1.41	Isocitrate dehydrogenase
	1.1.1.27	Lactate dehydrogenase
	1.6.4.3	Lipoamine dehydrogenase
	1.1.1.37	Malate dehydrogenase
	1.1.1.44	6-Phosphogluconate dehydrogenase
	1.3.99.1	Succinate dehydrogenase
	1.2.1.9	Triosephosphate dehydrogenase
	1.1.1.22	UDPG dehydrogenase
Oxidases	1.4.3.2	Amino acid oxidase
	1.11.1.6	Catalase
	1.9.3.1	Cyrochrome oxidase
	1.11.1.7	Peroxidase
Transferases	2.7.7.6	ADPG pyrophosphorylase
	2.6.2.1	Alanine aminotransferase
	2.6.1.1	Aspartate aminotransferase (GOT)
	2.1.3.2	Aspartate carbomoyl transferase
	2.7.1.1	Hexokinase
	2.7.4.6	Nucleoside diphosphate kinase
	2.7.5.1	Phosphoglucomutase
	2.4.1.1	Phosphorylase
	2.7.7.16	Ribonuclease (RNase)
	2.7.7.9	UDPG pyrophosphorylase
Hydrolases	3.1.3.2	Acid phosphatase
	3.1.3.1	Alkaline phosphatase
	3.5.1.14	Aminoacylase
	3.2.1.1	Amylase
	3.2.1.4	Cellulase (β-1,4-glucanase)
	3.1.1.(1,2,4)	Esterases
	3.2.1.26	β-Fructofuranosidase (invertase)
	3.2.1.20	α-Glucosidase
	3.4.1.1	Leucine aminopeptidase
	3.2.1.15	Polugalacturonase (pectinase)
	3.4.4.(4,5)	Protease (trypsin+ chymotrypsin)
	3.2.1.28	Trehalase
Lyases	4.1.3.7	Citrate synthase
	4.1.2.7	Ketose 1-phosphate aldolase
Ligases	6.4.1.(1,2)	Carboxylases

made for at least ten substrates (Table 1). *Glucose 6-phosphate DH* and *isocitrate DH* activities were reported for lily pollen by Dickinson and Davies [7]. *Glutamate dehydrogenase* (GDH) activity has been recorded for pollen of petunia [15], apple [16] and lily [9, 17]. Two GDH bands were resolved on electrophoresed gels (zymograms) of lily pollen [16]. Single GDH isozymes distinguished the pollen and stylar extracts of petunia [15]. Lactate DH activity has been reported for aspen pollen [18] and for lily pollen [19]. *Lipoamine DH* activity has also been recorded in several species [9, 20]. *Malate DH* activity has been observed in pollen of apple and petunia [16, 19], with eight isozymes distinguished in petunia pollen.

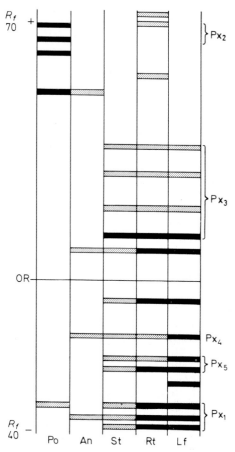

Figure 1. *Maize peroxidase isoenzyme polymorphism for pollen (Po), anther (An), style (St), root (Rt) and leaf (Lf) tissues. Braces to right embrace isoenzymes governed by alleles of five controlling loci, based on starch gels of Hamill and Brewbaker [28]*

6-*Phosphogluconate DH* activity was observed in lily pollen [17]. *Succinate DH* activity has been reported for several plant species [9, 20, 21], and five isozyme bands were distinguished in zymograms from apple pollen [16]. *Triosephosphate DH* (or 'glyceraldehyde phosphate DH') was reported by Okunuki [9] in several species, and *UDPG–DH* activity has been noted in petunia pollen [15].

The dehydrogenase activity of pollen tubes appears to be concentrated in spherosomes, while that of the pollen grains is concentrated in the pollen wall in plants such as *Amaryllis* and *Paeonia* [11]. Lily pollen clearly lacks the alcohol dehydrogenase activity required to oxidise pentaerythritol [6]. The pollen dehydrogenases studied by Poddubnaya-Arnoldi *et al.* [10] are probably not to be interpreted as alcohol dehydrogenases [2].

OXIDASES

Studies of pollen oxidases have been restricted largely to the cytochrome oxidases, catalases and peroxidases. *Aminoacid oxidase* activity has been reported for apple pollen [20]. *Catalase* activity appears to be high in pollen of most species [1, 10, 12] with possible exceptions such as *Tephrosia* sp. [12]. A single zone of activity appears on catalase zymograms of *Oenothera* [2] and maize, in which the activity diminishes rapidly with ageing of pollen [23, 24]. *Catechol oxidase* (or 'polyphenol oxidase') activity is not evident in the natural colour changes of pollen or their extracts. Paton [1] and Poddubnaya-Arnoldi *et al.* [10] could discern no activity in 83 species. The report of a low catechol oxidase activity in maize pollen by Istatkov, Secenska and Edreva [25] thus requires confirmation. *Cytochrome oxidase* activity has been observed in pollen of all species examined [9–12, 20, 26]. Activity was associated with the particulate fractions of corn pollen [26], and appeared to Tsinger and Petrovskaya [11] to be localised in the pollen wall.

Peroxidase activity has been reported for pollen from several hundred plant species [10, 12, 24, 27]. Activity is low relative to other plant tissues [10] and isoenzymic polymorphism (Fig. 1) is similarly low compared to that of vegetative tissues [19, 28]. King's suggestion [27] that the assay of peroxidase activity could serve as an index of pollen viability has not been generally confirmed. Loss of pollen viability in corn, for example, does not involve total loss of peroxidase activity but appears to be accompanied [28] by the disappearance from zymograms of the Px_2 peroxidase bands (Fig. 1). Most plant species show from one to three well-defined zymogram bands, while some species (e.g. tomato, *Vinca*) show little or no activity on starch gels [12, 24].

TRANSFERASES

The aminotransferase (AT) or transaminase activity of pollen grains has been studied in comparatively few species. *Aspartate AT* ('glutamate oxalacetate transaminase') is represented [29] by at least four major regions of activity on zymograms of maize pollen (Fig. 2). Hybrid bands formed by allelic heterozygotes for one of

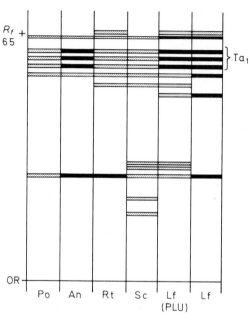

Figure 2. *Maize aspartate aminotransferase zymogram; abbreviations as on Fig. 1 (after Macdonald and Brewbaker [30])*

the controlling loci, Ta_1 (Fig. 2), were interpreted by Macdonald and Brewbaker [30] to suggest a dimeric structure of the active transaminase protein. Aspartate AT activity was high in pollen tubes, while a very low *Alanine AT* activity was also observed in tubes [15]. The comparative activities of *Aspartate–carbamoyl transferases* in pollen grains and pollen tubes were found by Roggen [15] to be similar, and no increase of activity was associated with germination. Dickinson and Davies [7] have provided evidence for *hexokinase* and *phosphoglucomutase* activity in lily pollen, with little or no increase in activity noted during germination. Concurrent studies of lily *nucleoside diphosphate kinases* revealed high activity of GTP kinases, but limited or negligible activity of CTP and UTP kinases.

Phosphorylases have been verified in pollen grains of many species, and a major increase in activity is associated with pollen germination and tube elongation [3, 12, 31–4]. High phosphorylase activity is also recovered in pollen diffusates [12]. Dickinson and Davies [7] have demonstrated high activities of both *ADPG pyrophosphorylase* (related to starch synthesis) and *UDPG pyrophosphorylase* (related to callose and cellulose synthesis) in lily pollen extracts. ADPG–PP was activated most effectively by 3-phosphoglycerate. No activity was observed for CDPG, GDPG and TDPG pyrophosphorylases in the lily material [7].

Ribonuclease was reported to be active and highly localised to the intine of pollen from nine species studied by Knox and Heslop-Harrison [35, 38].

HYDROLASES

Acid phosphatase (AP) has been studied more widely in pollen than any other enzyme, reflecting in part the ease and reliability of enzyme identification. Several hundred species have been analysed and AP activity has been shown to be comparatively high in all [1, 10, 35–8]. Activity is higher in pollen than in most plant tissues, comparing favourably with that of root hairs [37]. The comparatively rare reports of species lacking AP activity [31] would appear to bear confirmation. AP activity reportedly diminished during pollen germination [15] and was particularly low at the tips of pollen tubes [37]. Knox and Heslop-Harrison [35, 38] have observed that AP activity of at least 60 species is associated with the intine, and especially concentrated in pores of the pollen wall. Gorska-Brylass [36] reported that AP activity in pollen tubes was associated with spherosomes and not with mitochondrial fractions. Several diffuse AP activity of at least 60 species is associated with the intine, and [16] and lily [19]. Pollen AP is highly heat-stable [39], and its activity diminishes as pollen ages [37]. A very low AP activity characterises stylar tissues.

The evidence concerning *alkaline phosphatase* activity in pollen is inconsistent. No activity characterises most species [2, 15, 37], although the level of activity was evidently high enough in some pollen to exclude artefacts [12, 36, 39]. Unlike the acid phosphatases, alkaline phosphatase is diffused generally throughout the protoplast [35, 36] and is not associated with spherosomes or the pollen wall. The presence of *aminoacylase* activity and its activation by cobalt was reported by Umebayashi [22].

ATPase activity appears to have been confirmed only for microsporocytes, and the activity diminishes greatly during later stages of microspore maturation [21]. Activity in mature pollen grains and tubes bears confirmation.

Amylase ('diastase') activity was reported for pollen [40] as early as 1886 and confirmed later for many species by Green [41]. Green further noted the increased diastatic activity of pollen and pollen extracts during germination [42]. Early literature on pollen amylases and the pattern of starch grain appearance and digestion was summarised by Tischler [43]. Pollens from about 100 species have been tested, all showing comparatively high amylase activity relative to other tissues [1, 12, 37, 42]. The rapid increase of amylase activity (about fivefold) and disappearance of starch grains during germination and prior to tube emergence is widely documented [32, 37, 42, 44], as is the reappearance of starch during tube elongation [6]. Amylases diffuse rapidly from pollen [2, 12, 39] and are represented by a single slow-moving band on zymograms of *Oenothera* [2], maize and other species [24]. Distinction has not been made between the α- and β-amylase activity of these enzymes in pollen. Pollen diffusates from *Oenothera* actually digest a halo in the surrounding starch on gels [39]. Pollen amylases show little loss of activity during short periods of pollen storage [45]. Activity appeared to be concentrated in the intine of three species studied by Knox and Heslop-Harrison [35, 38].

The *cellulase* (β-1,4-glucanase) activity of pollen ranges widely for different species tested [1, 46]. Stanley and his colleagues [46–8] have reported that exogenous cellulase stimulated pollen growth, and that cellulase activities in tissues such as the stigma and ovary greatly exceeded those of pollen. Rapid release and diffusion of cellulase occurs from immersed pollen [44].

Although *DNase* activity is high in developing sporocytes, often continuing to within a few hours of anthesis, it is apparently absent or very low in activity in pollen and pollen tubes [36].

Esterase activity is high among most of the ca. 50 species tested [2, 24, 35, 36, 38]. The enzyme is largely localised in the intine of pollen grains [35, 38] and in spherosomes of the pollen tube [36]. A high degree of band polymorphism has been observed on esterase zymograms [2, 16, 18, 29, 49, 50], although this polymorphism is exceeded by that of vegetative tissues (Fig. 3). Esterase isoenzymes can distinguish pollen and tapetal tissues, as in *Hemerocallis* (Fig. 4), and at least one isozyme observed in *Oenothera* pollen tubes was synthesised *de novo* upon germination [49]. Genetic control of polymorphism involving several maize esterases was elucidated by Macdonald [29], in part through the use of a chemical inhibitor, DDVP (Fig. 3). The maize pollen esterases comprised at least three

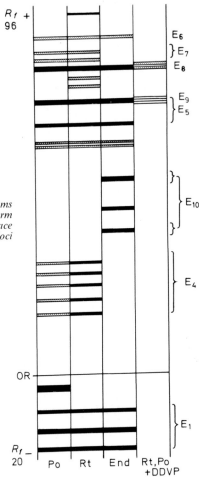

Figure 3. *Maize esterase polymorphisms for pollen (Po), root (Rt) and endosperm (End) tissues. Braces to right embrace isozymes governed by alleles of eight loci (after Macdonald [29])*

groups—carboxylic ester hydrolases, aryl ester hydrolases and acetic ester acetyl esterases [29].

The first recorded enzyme of pollen was *β-fructofuranosidase* (or 'invertase'), originally noted by van Tieghem in 1869 and confirmed by Green [42] for several species. With a few doubtful exceptions, all pollens have high invertase activities [1, 37, 42], and a 4–8-fold increase of activity occurs upon germination [37, 42]. Invertase activity in maize pollen does not diminish even though the grains lose viability [45]. Dickinson [5] reported that the enzyme was localised external to the pollen membrane and probably bound

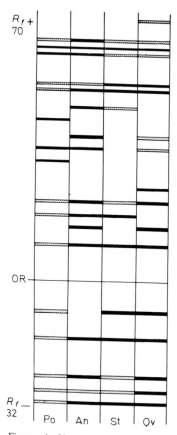

Figure 4. Hemerocallis aurantiaca esterase zymogram [24] of floral tissues, pollen (Po), anther (An), style (St) and ovary (Ov)

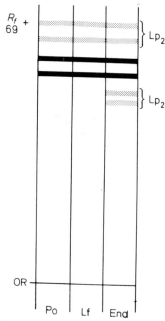

Figure 5. Maize leucine aminopeptidase zymogram for Pollen (Po), leaf (Lf) and endosperm (End) tissues. Braces to right embrace isozymes governed by alleles of two loci [50]

to the intine. The rapid synthesis of starch during pollen germination [6, 32] has been studied especially by Dickinson [6] with an inert pentaerythritol growth medium. The rapid concurrent decrease of sucrose reserves during germination led Dickinson [6] to suggest two major patterns for starch synthesis using pollen sucrose, namely via fructofuranosidase and hexokinase to hexose and hexose phosphate, or via sucrose glucosyltransferases. The regulatory role of ADPG pyrophosphorylases in this conversion has also been implied [7].

Dickinson [5] also observed a high activity of α-*glucosidase* in lily pollen, and concluded that the enzyme was localised primarily within the pollen wall. Gussin *et al.* [51] supported this conclusion, and also identified *trehalase* (trehalose-1-glucohydrolase) as an active enzyme of the pollen grains. The sugar trehalose was not found as a constituent of pollen or pistillate tissues, and the high trehalase activity in pollen was linked to a role in facilitating glucose transport through the pollen wall.

The rapid and extensive changes in callose distribution in developing pollen tubes implies β-*1,3-glucanase* ('callase') activity, although critical evidence of its presence in pollen is wanting. Roggen and Stanley [48] reported a significant response of pear pollen tubes when β-1,3-glucanase was added within a few minutes of the immersion of pollen for germination. In the species tested by Gorska-Brylass [36], β-glucuronidase activity also could not be shown for pollen and pollen tubes.

Leucineaminopeptidase activity characterised the pollen of most species studied by Makinen and Macdonald [2], and two and three bands were distinguished clearly on zymograms (Fig. 5). Genetic polymorphisms were observed in maize pollen for LAP locus, Lp_2.

The report of lipase activity for four of 21 species tested by Paton [1] appears to warrant re-examination. Gorska-Brylass [36] observed no lipase activity in pollen from several species, and, similarly, maize esterases could not be shown by Macdonald [29] to have measurable lipase activity.

Polygalacturonase ('pectinase') activity of pollen appears to be very low, although all species tested show activity [1, 47]. Polygalacturonase (PG) activity is much higher in the ovary than in pollen [47], and little or no increase occurs in pollen during germination and growth. PG rapidly diffuses from pollen [52] and is inferred to be the 'cutinase' described during pollen tube penetration of the stigmatic surface in crucifers [53]. The addition of PG to growth media stimulated pollen-tube growth of pear, and was effective even if added subsequent to tube emergence [48]. PG treatments at higher concentrations greatly distort pollen tube growth and are highly influenced by calcium concentration of growth media [24].

The *protease* and proteinase activities of pollen have not been studied well. A high proteolytic activity has been observed for pollen of several species [1, 31]. *Trehalase* activity of pollen [51] was discussed earlier in this section.

LYASES AND LIGASES

Pollen lyases or synthetases have been the subject of very few studies. Roggen [15] reported low pollen values for activity of *citrate synthease* and for a *ketose 1-phosphate aldolase*. Both enzymes appeared to increase in activity upon germination. Stylar extract activities were comparatively much higher. Activity of ligase, *carboxylase*, was shown by Okunuki [9] to be comparatively high for pollen of five out of seven genera studied. Amino acid synthase activity was implied in studies of pollen tubes [54, 55] but has not been confirmed for pollen grains. RNA synthetase activity has been implied in relation to the rapid appearance of RNA and polyribosomes during pollen germination [54, 56, 57], although methylation of pre-existent RNA has been suggested as an alternative to net synthesis [58]. Steffensen has shown that there is no synthesis of typical 5S, 18–28S, or tRNA molecules by pollen. DNA synthetase is not active in pollen [57]. Enzymes described as 'folic acid conjugases' [58] have been reported for pollen; their classification is obscure.

DISCUSSION

Comparatively few enzymes of plant tissues have been sought without success in pollen or pollen tubes. Pollen grains probably do not normally have catechol oxidases (polyphenol oxidase), or the enzymes associated with plastids and plant pigments (although some pollens are pigmented). Other enzymes sought unsuccessfully in pollen include maltase [31], lipase [36], β-glucuronidase [36], arylsulphatase, *p*-diphenyl oxidase ('laccase') [1], several pyrophosphorylases [7] and 'zymase' [1]. The activities of alkaline phosphatase, ATPase, DNA synthetase, and β-1,3-glucanase in pollen grains are evidently very low or absent. The inability of germinating pollen to synthesise DNA, histone proteins and most forms of RNA [59] indicates the absence or suppression of appropriate enzymes in this tissue.

In the great majority of studies pollen grains thus appear similar to other plant tissues, at least in respect of the absence or presence of particular enzyme activities. These data support other evidence of high physiological competence of the pollen grain. The lack of totipotency of the angiosperm pollen grain, like its other physiological peculiarities, probably does not relate directly to the absence of any major class of enzyme.

Isoenzymic analyses by means of gel electrophoresis are providing evidence, however, for considerable specialisation of the pollen grain. Zymograms of pollen are rarely identical to those of surrounding tapetal or pistillate tissues. Pollen grains commonly exhibit few of the isoenzymes that are seen in the seed or vegetative tissues. As noted in this review, zymograms of dehydrogenases, peroxidases and esterases have revealed bands apparently unique to the pollen grain and tube. The isoenzymic evidence can be expected to contribute increasingly to our understanding of the unique physiological properties of pollen.

In general, pollen enzymes are freely and rapidly diffusible from the pollen grain. A possible role is suggested for pollen enzymes in triggering their own germination process through action on the stigmatic fluid, especially in the trinucleate pollen types [12].

Increased enzymatic activity during germination has been indicated for many enzymes (e.g. amylase, β-fructofuranosidase, phosphorylase, transaminase), but little evidence is available on the nature of this increase (net synthesis vs. activation or release). The fact that many inhibitors of protein synthesis (methyl tryptophane, fluorophenylalanine, puromycin, chloramphenicol) fail to inhibit pollen germination [24] implies that germination can proceed without net enzyme synthesis, at least in the binucleate pollen types which have been studied.

It is probable that most enzymes of the pollen grain are synthesised subsequent to meiosis, under control of the haploid genome. Hybrid enzymes would therefore not be expected in the pollen from diploid plants. In polyploids, however, the possible role of hybrid dimers in self-incompatibility phenomena ('dominance' and 'competition') of pollen has been suggested by Lewis [61] and others. Genetic studies of isoenzymic loci in maize [24, 28, 29] indicate a comparatively low frequency (three of 23 loci) of gene loci at which such hybrid enzymes (polymers) occur.

Among the questions on which pollen enzyme studies might have a bearing is that of the loss of viability of pollen, through either natural or induced causes. Pollen slowly loses certain types of enzymatic activity during storage (e.g. peroxidases, phosphatases) but these cannot account for the rapid loss of viability of, for example, corn pollen [45, 62]. In general, the rapid loss of viability of trinucleate pollen in storage [60] is no better understood in the light of present knowledge of pollen enzymes than is the unique inability of most trinucleate pollens to germinate in culture media. The exponential killing of pollen by UV [63] appeared to imply DNA dimerisation phenomena mediated through blocks to enzyme synthesis, but this conclusion cannot be reconciled with the evidence (discussed above) on enzyme synthesis during pollen germination. The presence of enzymes capable of photoreactivating UV damage to pollen has been indicated by Fujii [64].

REFERENCES

1 PATON, J. B., 'Pollen and pollen enzymes', *Am. J. Bot.*, **8**, 471–501 (1921)
2 MAKINEN, Y. and MACDONALD, T., 'Isoenzyme polymorphism in flowering plants II. Pollen enzymes and isoenzymes', *Physiologia Pl.*, **21**, 477–86 (1968)
3 DICKINSON, D. G., 'Germination of lily pollen: respiration and tube growth', *Science, N.Y.*, **150**, 1818–19 (1965)
4 DICKINSON, D. B., 'Inhibition of pollen respiration by oligomycin', *Nature, Lond.*, **210**, 1362–3 (1966)
5 DICKINSON, D. B., 'Permeability and respiratory properties of germinating pollen', *Physiologia Pl.*, **20**, 118–27 (1967)
6 DICKINSON, D. B., 'Rapid starch synthesis associated with increased respiration in germinating lily pollen', *Pl. Physiol., Lancaster*, **43**, 1–8 (1968)
7 DICKINSON, D. B. and DAVIES, M. D., this volume, 190
8 BARMAN, T. E., *Enzyme Handbook*, Vols. 1 and 2, Springer-Verlag, Berlin (1969)
9 OKUNUKI, R., 'Über den gaswechsel der pollen II, III, IV', *Acta phytochim., Tokyo*, **11**, 27–80, 249–60 (1939)
10 PODDUBNAYA-ARNOLDI, V. A., TSINGER, N. V., PETROVSKAYA, T. P. and POLUNINA, N. N., 'Histochemical study of pollen and pollen tubes in the angiosperms', *Rec. Adv. Bot.*, **1**, 682–5 (1960)
11 TSINGER, N. V. and PETROVSKAYA, T. P., 'The pollen grain wall—a living, physiologically active substance', *Dokl. Akad. Nauk SSSR*, **138**, 466–9 (1961)
12 MARTIN, F. W., 'Some enzymes of the pollen and stigma of the sweet potato', *Phyton, B. Aires*, **25**, 97–102 (1968)
13 SARVELLA, P., 'Vital-stain testing of pollen viability in cotton', *J. Hered.*, **55**, 154–8 (1964)
14 NORTON, J. D., 'Testing of plum pollen viability with tetrazolium salts', *Proc. Am. Soc. hort. Sci.*, **89**, 132–4 (1966)
15 ROGGEN, H. P. J. R., 'Changes in enzyme activities during the progamephase in *Petunia hybrida*', *Planta*, **16**, 1–31 (1967)
16 VEIDENBERG, A. E. and SAFONOV, V. I., 'The composition of the enzymatic complex of pollen in several species and varieties of apple tree', *Dokl. Akad. Nauk SSSR*, **180**, 1242–5 (1968)
17 DESBOROUGH, S. and PELOQUIN, S., 'Disc-electrophoresis of protein and enzymes from styles, pollen and pollen tubes of self-incompatible cultivars of *Lilium longiflorum*', *Theor. appl. Genet.*, **38**, 327–31 (1968)
18 EULER, H. VON, 'Biochemische Untersuchungen an diploiden und triploiden Espen aus normalen und rontgenbestrahlten Pollen', *Ark. Kemi Miner. Geol. 26A*, **30**, 1–19 (1949)
19 LINSKENS, H. F., 'Die Anderung des Protein- und Enzym-musters während der Pollenmeiose und Pollenentwicklung. Physiologische Untersuchungen zur Reifeteilung', *Planta*, **69**, 79–91 (1966)
20 EULER, H. VON and VON EULER, J., 'Über die katalatischen Wirkungen Pflanzlicher Zellen', *Ark. Kemi Miner. Geol. 26A*, **22**, 1–22 (1948)
21 PALUMBO, R. F. 'A cytochemical investigation of enzyme activities in the developing pollen of *Tradescantia paludosa* and *Lilium longiflorum* var. Croft', *Diss. Abstr.*, **13**, 966–7 (1953)
22 UMEBAYASHI, M., 'Aminoacylase in pollen and its activation by cobaltous ion', *Pl. Cell Physiol, Tokyo*, **9**, 583–6 (1968)
23 BECKMAN, L., SCANDALIOS, J. G. and BREWBAKER, J. L., 'Catalase hybrid enzymes in maize', *Science, N.Y.*, **146**, 1174–5 (1964)
24 BREWBAKER, J. L. (Unpublished data)
25 ISTATKOV, S., SECENSKA, S. and EDREVA, E., 'Comparative biochemical studies on

the pollen and stigma of self-pollinated lines and hybrids of corn', *C.r. Acad. bulg. Sci.*, **17**, 73–6 (1964)

26 WALDEN, D. B., 'Preliminary studies on longevity of corn pollen and related physiological factors', *Diss. Abstr.*, **20**, 3488–9 (1960)

27 KING, J. R., 'The peroxidase reaction as an indicator of pollen viability', *Stain Technol.*, **35**, 225–7 (1960)

28 HAMILL, D. E. and BREWBAKER, J. L., 'Isoenzyme polymorphism in flowering plants IV. The peroxidase isoenzymes of maize', *Physiologia Pl.*, **22**, 945–58 (1969)

29 MACDONALD, T., 'Isoenzymic studies of esterase, 1-aspartate, 2-oxoglutarate aminotransferase, and carbohydrase in *Zea mays*', Ph.D. Thesis, University of Hawaii (1969)

30 MACDONALD, T. and BREWBAKER, J. L., 'Isoenzyme polymorphism in flowering plants V. Genetic control and dimeric nature of two transaminase isoenzymes in maize' (Unpublished)

31 BELLARTZ, S., 'Das Pollenschlauchwachstum nach arteigener und artfremder Bestaubung einiger Solanaceen', *Planta*, **47**, 588–612 (1956)

32 IWANAMI, Y., 'Physiological studies on pollen', *J. Yokohama munic. Univ., Ser. C.*, **116**, 1–137 (1959)

33 KATSUMATA, T. and TOGASAWA, Y., 'Biochemical studies on pollen. IX, X, XI, The enzymes related to the metabolism of carbohydrates in pollen', *Nippon Nogei Kagaku Kaishi*, **42**, 1–17 (1968)

34 OKUNUKI, R., 'Über den Gaswechsel der Pollen V', *Acta phytochim., Tokyo*, **13**, 93–8 (1942)

35 KNOX, R. B. and HESLOP-HARRISON, J., 'Cytochemical localization of enzymes in the wall of the pollen grain', *Nature, Lond.*, **223**, 92–4 (1969)

36 GORSKA-BRYLASS, A., 'Hydrolases in pollen grains and pollen tubes', *Acta Soc. Bot. Pol.*, **34**, 589–604 (1965)

37 HAECKEL, A., 'Beiträg zur Kenntnis der Pollenfermente', *Planta*, **39**, 431–59 (1951)

38 KNOX, R. B. and HESLOP-HARRISON, J., 'Pollen wall proteins: localization and enzymic activity', *J. Cell Sci.*, **6**, 1–27 (1970); this volume, 171

39 LEWIS, D., BURRAGE, S. and WALLS, D., 'Immunological reactions of single pollen grains, electrophoresis and enzymology of pollen protein exudates', *J. exp. Bot.*, **18**, 371–8 (1967)

40 STRASBURGER, E., 'Über fremdartige Bestaubung', *Jb. wiss. Bot.*, **17**, 50–98 (1886)

41 GREEN, J. R., 'On the occurrence of diastase in pollen', *Ann. Bot.*, **5**, 511–12 (1891)

42 GREEN, J. R., 'Researches on the germination of the pollen grain and the nutrition of the pollen tube', *Phil. Trans. R. Soc. B.*, **185**, 385–409 (1894)

43 TISCHLER, G., 'Untersuchungen über der Starkegehalt des Pollens tropischer Gewachse', *Jb. wiss Bot.*, **47**, 219–42 (1910)

44 STANLEY, R. G. and LINSKENS, H. F., 'Protein diffusion from germinating pollen', *Physiologia Pl.*, **18**, 37–43 (1965)

45 KNOWLTON, H. E., 'Studies in pollen with special reference to longevity', *Mem. Cornell Univ. agric. Exp. Sta.*, **52**, 745–93 (1921)

46 STANLEY, R. G., this volume, 131

47 KONAR, R. N. and STANLEY, R. G., 'Wall-softening enzymes in the gynoecium and pollen of *Hemerocallis fulva*', *Planta*, **84**, 304–10 (1969)

48 ROGGEN, H. P. J. R. and STANLEY, R. G., 'Cell-wall-hydrolyzing enzymes in wall formation as measured by pollen tube extension', *Planta*, **84**, 295–303 (1969)

49 MAKINEN, Y. and BREWBAKER, J. L., 'Isoenzyme polymorphism in flowering plants I. Diffusion of enzymes out of intact pollen grains', *Physiologia Pl.*, **20**, 477–82 (1967)

50 SCANDALIOS, J. G., 'Tissue-specific isozyme variations in maize', *J. Hered.*, **55**, 281–5 (1964)

51 GUSSIN, A. E. S., MCCORMACK, J. H., WAUNG, L. Y. L. and GLUCKIN, D. S., 'Trehalase: a new pollen enzyme', *Pl. Physiol., Lancaster*, **44**, 1163–8 (1969)

52 STANLEY, R. G. and THOMAS, A., 'Pollen enzymes and growth', *Proc. Ass. sth. agric. Wkrs*, **64**, 265 (1967)

53 LINSKENS, H. F. and HEINEN, W., 'Cutinase-nachweis in pollen', *Z. Bot.*, **50**, 338–47 (1962)

54 TANO, S. and TAKAHASHI, H., 'Nucleic acid synthesis in growing pollen tubes', *J. Biochem.*, **56**, 578–80 (1964)

55 ROSEN, W. G., 'Ultrastructure and physiology of pollen', *Amer. Rev. Pl. Physiol.*, **19**, 435–62 (1968)

56 MASCARENHAS, J. P., 'Pollen tube growth and ribonucleic acid synthesis by vegetative and generative nuclei of *Tradescantia*', *Am. J. Bot.*, **53**, 563–9 (1966)

57 MASCARENHAS, J. P. and BELL, E., 'Protein synthesis during germination of pollen and studies on polyribosome formation', *Biochim. biophys. Acta*, **179**, 199–203 (1969)

58 NIELSEN, N. and HOLMSTROM, B., 'On the occurrence of folic acid, folic acid conjugates and folic acid conjugases in pollen', *Acta chem. scand.*, **11**, 101–4 (1957)

59 STEFFENSEN, D. M., this volume, 223

60 BREWBAKER, J. L., 'The distribution and phylogenetic significance of binucleate and trinucleate pollen grains in the angiosperms', *Am. J. Bot.*, **54**, 1069–83 (1967)

61 LEWIS, D., 'A protein dimer hypothesis on incompatibility', *Proc. Int. Conf. Genet., London, 1963*, **3**, 657–63 (1963)

62 GOSS, J. A., 'Development, physiology, and biochemistry of corn and wheat pollen', *Bot. Rev.*, **34**, 333–58 (1968)

63 BREWBAKER, J. L., ESPIRITU, L. and MAJUMDER, S. K., 'Comparative effects of x-ray and U.V. irradiations on pollen germination and growth', *Radiat. Bot.*, **5**, 493–500 (1965)

64 FUJII, T., 'Photoreactivation of mutations induced by ultraviolet radiation of maize pollen', *Radiat. Bot.*, **9**, 115–24 (1969)

Pollen Wall Enzymes: Taxonomic Distribution and Physical Localisation*

R. B. Knox, *Botany Department, Australian National University, Canberra, Australia*
J. Heslop-Harrison, *Institute of Plant Development, University of Wisconsin, Madison, Wisconsin, U.S.A*

Using cytochemical methods, Tsinger and Petrovskaya-Baranova [1] demonstrated the presence of proteins with enzymic properties in certain strata of the pollen walls of *Paeonia* and *Amaryllis*. We have confirmed these observations in a survey of more than 70 flowering-plant species, including all major pollen-structural types, and for most we have been able to specify with some precision the localisation of the wall-held enzymes.

The wall proteins are rapidly lost when the pollen grain is moistened, and so are not readily detectable by standard procedures using fixed material. Observations were accordingly made on freeze-sectioned material [2]. Proteins were localised by properties other than enzymic activity, UV microscopy and naphthol yellow S staining being used. For many species it was evident that a substantial proportion of the total protein of the pollen grain occurred in the walls, in three sites: (1) most prominently in the cellulosic intine, particularly at the apertures or pores; (2) in cavities in the exine, especially in species of Compositae; and (3) in the superficial material derived from the tapetum. For the localisation of enzyme activity, direct cytochemical procedures were used for acid phosphatase, ribonuclease and esterase, and substrate-film methods for amylase and protease [2]. In the mature, dormant pollen most of the detectable hydrolase activity was found to be associated with the walls, with very little in the protoplast of the vegetative cell. The pollen of *Crocus vernus* is typical of the non-aperturate class, with a thick intine and relatively thin exine. The principal site of acid phosphatase, esterase and ribonuclease activity was found to

* Abstract.

171

be the central zone of the intine over the whole wall. Slight activity was observed just within the plasmalemma, and also in the superficial *Pollenkitt*. The monoporate class is typified by grass pollen: activity here is concentrated in the thick boss of intine underlying the single pore, with essentially no activity elsewhere in the dormant grain. Exceptionally low activity was observed in the poral intine of the cultivars of *Zea mays* examined in comparison with other grasses. Polyporate pollens, such as those of *Malvaviscus arboreus*, *Hibiscus rosa-sinensis*, *Cobaea scandens* and *Silene* spp., showed activity in the intine at each pore, and in a thin layer in the interporal intine also. *Cobaea* showed intense enzyme activity in the *Pollenkitt*, especially for esterase. There was some variation among the dozen or so monocolpate pollens examined. The colpus of *Gladiolus gandavensis* is not conspicuously differentiated by exine features, but the thickened underlying intine is the site of intense hydrolase activity. In species of Liliaceae and Amaryllidaceae the thin intines show much less activity, although in the large grains of *Lilium* spp. the hydrolases are readily detectable in the colpial zone. Similar variation exists among tricolpate and triporate species, common horticultural plants such as *Primula obconica*, *Antirrhinum majus*, *Petunia hybrida* and *Nicotiana tabacum* showing relatively low activity, and composites, such as *Cosmos bipinnatus*, revealing intense activity. In all cases activity is concentrated in the apertural intine. Among wind-pollinated species the two ragweeds, *Ambrosia trifida* and *A. artimisiifolia*, show high activity even in stored pollen, while a species with thick intine, *Carex stricta*, has relatively little.

Among gymnosperms three species of *Pinus* and one of *Abies* show acid phophatase and ribonuclease activity in the intine, again concentrated in the germinal slit region. Enzyme activity was detected throughout the intine of one pteridophyte, *Equisetum palustre*, but the spores of several ferns examined showed no detectable wall enzymes. Similarly, the spore walls of six species of bryophytes showed no associated activity for any enzyme tested.

The development of the intine has been followed in several species, and the period of protein incorporation has been established. In all cases the spores as released from the meiotic tetrads are without enzyme activity in the wall. Intine growth reaches its maximum shortly before or during the vacuolate phase, and incorporation of proteins begins at this time. Enzyme activity is detectable almost from the beginning of cellulose deposition, particularly at the sites of pores and colpi. During the period of synthesis intense enzyme activity is often detectable in the peripheral regions of the protoplast. Stratified endoplasmic reticulum underlies the plasmalemma in the regions of incorporation, and plates or leaflets of proteinaceous material are included between strata of cellulose. This material, which is presumed to carry the wall enzymes, remains visible in the mature intine [3]. At maturity fertile grains show no enzyme activity in the peripheral regions of the cytoplasm, but activity often persists in the protoplast of sterile grains.

We have found that pollen leachates show strong immunogenic activity in rabbits [4], and immunofluorescence methods have been used to establish the source of the antigens. Partly purified globulin fractions from antisera against *Gladiolus* and *Ambrosia* pollen leachates were labelled with rhodamine B isothiocyanate and fluorescein isothiocyanate and used against fresh freeze-sectioned pollen. Localisation was also sought by an indirect method in which the precipitin formed by the reaction of unlabelled antiserum with antigens in the sections was detected with labelled goat anti-rabbit globulin. With both methods good localisation was obtained: the intine, the site of the rapidly leachable enzymic proteins, is also the principal source of much of the antigenic material. However, in both *Gladiolus* and *Ambrosia*, antigens are also present in the superficial *Pollenkitt*.

REFERENCES

1 TSINGER, N. V. and PETROVSKAYA-BARANOVA, T. P., 'The pollen grain wall—a living, physiologically active structure', *Dokl. Akad. Nauk SSSR*, **138**, 466–9 (1961)
2 KNOX, R. B. and HESLOP-HARRISON, J., 'Cytochemical localization of enzymes in the wall of the pollen grain', *Nature, Lond.*, **223**, 92–4 (1969)
3 HESLOP-HARRISON, J., 'Ultrastructural aspects of differentiation in sporogenous tissue', *Symp. Soc. exp. Biol.*, **17**, 315–40 (1963)
4 KNOX, R. B., HESLOP-HARRISON, J. and REED, C., 'Localization by immunofluorescence of antigens associated with the pollen grain wall', *Nature, Lond.*, **225**, 1066–8 (1970)

Pollen Protein Diffusates*†

R. G. Stanley and R. W. Search, *Forest Physiology-Genetics Laboratory, University of Florida, Gainesville, Florida, U.S.A.*

Proteins are among the constituents which rapidly diffuse from germinating pollen [1]. Isozymes of amylase and other enzymes have been characterised in the initial 5 min diffusate of *Oenothera* pollen [2]. In the present experiments variations in total ultra-violet (UV) absorbing materials diffusing from different pollens within seconds after they are placed in solution were studied. Variations in the diffusates, it is suggested, can lead to different reactions to pollen in humans and plants.

Levels and spectra of materials that diffuse from pollens in 2–15 s elutions vary with the pollen species and pH of the eluting solution. Pine (*Pinus elliottii*) and palm (*Caryota* spp.) pollen yielded the lowest amounts of materials in comparison with 20 other pollen species eluted 15 s with double-distilled water. The UV absorption spectra can be classified as those with low or with high levels of 260–270 nm absorbing materials. The spectra can be further characterised as being with or without a secondary absorption peak at 255 nm. The materials solubilised from Russian thistle (*Salsola pestifer*) and English plantain (*Plantago lanceolata*) yielded spectra with primary (265 nm) and secondary (255 nm) absorbing peaks that were among the highest of any pollens tested.

Pear pollen was studied in greater detail. Germination was not significantly reduced after one 15 s elution; however, successive elutions did reduce germination by over 50%. By the fourth 15 s water elution the pollen was essentially non-viable (Table 1).

The amount of materials lost by the pollen is a function of the elution time and the pH of the eluting solution. That different pollen species release proteins at a different rate is also reflected by the total loss of material from the pollens. Pollens with little protein

* Abstract.

† A contribution of the *Florida Agricultural Experiment Station Journal*, Series No. 3, 581.

or pigment detectable in the elution solution, e.g. pine and palm, also showed minimum material losses, as dry weight decreases, on elution. With increased time, however, proteins and other components are removed even from those pollens (Table 2).

Pollens were eluted in 0·01 M barbitol buffer over pH range 5·0–9·0. Buffers at pH 6·0–7·0 removed less material from those

Table 1. EFFECT OF SUCCESSIVE ELUTIONS ON POLLEN GERMINATION, *Pyrus communis*

	Number of successive 15 s elutions				
	0	*1*	*2*	*3*	*4*
% germination after 1·5 h	54	49	22	23	9

pollens in the high, rapid loss category. This pH also corresponded to the pH optimum for germination *in vitro*. Higher or lower pH buffers removed significantly greater amounts of material from these pollens than did the pH 7·0 buffers.

It may be meaningful to compare amounts eluteable with both the rates of germination and nature of the pollen surface. Some pollens are known to have a cuticle-like layer on the exine, and others have oils and loose fibrous strands on their surfaces.

Materials removed from the pollen by rapid water or buffer extraction included the enzymes cellulase and pectinase (small amounts). Enzymes diffusing from the pollen varied with the species and method of extraction. Sodium chloride solution slightly increased the amount of enzyme released by the pollens. However,

Table 2. INFLUENCE OF ELUTION TIME ON POLLEN WEIGHT

Time eluted with water (s)	*% decrease in dry weight*	
	Pyrus communis	Pinus elliottii
5	15	1
15	24	3
60	28	10
120	35	15

the ease of solubilisation and the fact that different pollens release different initial amounts of cellulase and other enzymes suggest that some enzymes are surface-localised, or very near or in the exine, and that rapid changes in surface properties can occur during pollen activation.

A scanning electron microscope was used to compare the surfaces of eluted and non-eluted pine, pear and ragweed pollens. Substantial differences were noted in the pear pollen surface, while no visible changes occurred on the pine surface; questionable differences on ragweed pollen surfaces were obtained at a low ($\times$ 2000) magnification.

These results suggest that pollens release protein and other chemical moieties to the solubilising solutions of their environment at different rates and in different quantities, depending primarily upon the species and, to a lesser degree, upon the extracting solution. The growth reactions of pollen, the rate and type of allergic reaction of people to pollen, and the compatibility response of plants will presumably be influenced by such differences.

REFERENCES

1 STANLEY, R. G. and LINSKENS, H. F., 'Protein diffusion from germinating pollen', *Physiologia Pl.*, **18**, 47–53 (1965)
2 MÄKINEN, Y. and BREWBAKER, J., 'Isoenzyme polymorphism in flowering plants. I. Diffusion of enzymes out of intact pollen grains', *Physiologia Pl.*, **20**, 477–82 (1967)

Pollen Tube Growth and Fine Structure

Walter G. Rosen, *Department of Biology, State University of New York, Buffalo, New York, U.S.A.*

INTRODUCTION

The experimental analysis of pollen tube growth is generally thought to have begun with the work of van Tieghem [32]. It was just 100 years ago that he published his findings that oxygen is required for pollen germination and that oxygen consumption is accompanied by carbon dioxide production and the disappearance of starch during the growth of the pollen tube. He is thought to have been the first to add carbohydrate (cane sugar) to a pollen culture medium. This paper is dedicated to van Tieghem, in recognition of the hundredth anniversary of his pioneering contributions to our knowledge of pollen germination and tube growth.

POLLEN GERMINATION

When we examine the cytoplasm of the mature, non-germinated pollen grain, we find, generally, an absence of vacuoles but an abundance of familiar organelles. There is a high density of quiescent dictyosomes and of mitochondria, rough and smooth endoplasmic reticulum and free ribosomes. *Lilium longiflorum* pollen cytoplasm possesses numerous lipid droplets but lacks starch [26]. The changes which precede and accompany germination are rapid and dramatic. Within minutes after the pollen has been placed in a medium which permits germination but before the emergence of the tube, starch is demonstrable, both by cytochemical methods at the light-optical level and by electron microscopy [26], as well as by chemical analyses [11].

At the fine-structural level Larson [20] found that in the species

177

he examined germination is characterised by the conversion of the Golgi apparatus from a quiescent to an active, vesicle-producing form, and by the formation of vacuoles.

Dickinson [9, 10] analysed the metabolism of lily pollen growing *in vitro* and reported three distinct phases of respiration, corresponding to different metabolic processes. These are (1) an initial high rate of respiration, which lasts about 30 min, (2) a 50–60 min period of lower rate, then (3) another period of high respiration, which is accompanied by tube growth. The first period of high respiration coincides with rapid starch formation. From fine-structure studies on pollen growing in the pistil, I propose that there is probably a fourth stage of respiration which reflects a switch by the pollen tube from endogenous to exogenous substrates and which cannot occur *in vitro*. The reason for this suggestion will become apparent later.

Water uptake, and the activation or synthesis of enzymes, are no doubt the basic initiating factors of the germination process. It has been shown by a number of investigators that germination and pollen tube growth are independent of RNA synthesis [8, 21, 23]. Protein synthesis is initiated during the 'activation phase', before the pollen tube begins to form [23]. Thus it appears that messenger RNA is present in the quiescent grain. There is an increase in polysomes and a decrease in free ribosomes during the activation phase [21, 23], and both protein and RNA synthesis can be demonstrated during growth of the tubes in culture. Echlin [12] noted that elaborate, spiral-shaped polysomes are found in developing pollen mother cells of *Ipomoea*. Crang and Miles reported their appearance at the time of germination in *Lychnis alba* [4], and Jensen found them in the young zygote [16].

In Easter Lily pollen tubes we find both free ribosomes and ribosomes on regions of the endoplasmic reticulum [26]. Cytochemically, however, the greatest concentration of RNA appears in the growing zone at the tip of the tube, a region lacking ribosomes but packed with smooth membranes which are removed by RNase [6]. By its location it would seem that this RNA is active in the process of tube growth, and I hope that workers in this field, who until now have concentrated on ribosome activity, will give this non-ribosomal RNA some attention.

FINE STRUCTURE: *IN VITRO*

Pollen tubes are among that small class of 'cells' in which growth is exclusively restricted to the tips. Easter Lily pollen growing on the surface of solid medium can be readily demonstrated by carbon-marking experiments to possess a growth zone which is restricted to the tipmost 3–5 μm [6]. Cytochemical analysis reveals this zone to be singularly rich in RNA, protein and PAS-positive material. In the electron microscope we find that this growth zone is characterised by the presence of numerous vesicles and an elaborate network of smooth membranes. The vesicles appear to rise from the ends of the dictyosome cisternae. They coalesce with one another and ultimately contribute their membranes and contents to the compartmented cap which covers the growth zone at the tip. Cytochemistry at the electron microscope level indicates that the cap and the vesicles contain pectin, and that the RNA mentioned earlier resides in the smooth membranes. Sassen [31], also, found that *Petunia* pollen tubes are characterised by the accumulation at their tips of vesicles derived from the dictyosomes, and a similar situation was reported by Larson [20] in four additional species, two with hollow and two with solid pistils. It is interesting to note that vesicles are also prominent in the growing region at the tips of root hairs [1].

In the region behind the tip the tube contains those organelles which were encountered in the grain before germination, as well as numerous amyloplasts. In contrast to the wall of the grain, the tube wall is thin, and in sectioned lily material appears to consist of two layers [26]. Tubes growing in the pistil have much more complex walls (Rosen, unpublished).

A few studies have been made of the arrangement of cellulose micelles in the tube wall. While the patterns vary from species to species as reported by different investigators, it is interesting to note that the tip region can generally be distinguished from the more mature regions. At the tip the micelles are either shorter and randomly oriented [24] or, in the case of Lily, absent or hidden by an amorphous material [5, 31]. Crang and Miles [4] failed to detect pectins in the walls of *Lychnis alba* pollen growing *in vitro* and concluded that cellulose is the primary wall component.

The near-absence of dictyosomes from the tube cytoplasm in this species led these workers to stress the possibility of wall growth by means other than vesicle accumulation at the tube tip. They suggest the possibility of wall growth by apposition in addition to apical growth. The cytoplasm of pollen tubes growing *in vitro* apparently lacks microtubules [4, 27].

Electron microscopy clearly establishes that the generative cell is surrounded by its own distinct wall and that its cytoplasm is different from that of the grain proper. Generative cell cytoplasm, studied especially by Bopp-Hassenkamp [2] and by Larson [20], contains a reduced number of organelles, which are also less well developed than those of the grain proper. Also, generative cell cytoplasms possess little if any storage material.

Changes in the generative cell envelope and the possible role of the envelope and associated structures as a secretion system have recently been described [13].

I will not refer further to the generative cell and the sperm cells which arise from it either before or after germination except to mention that further fine-structural analysis of these cells promises to yield important information regarding the roles of these cells in inheritance. For example, Lombardo and Gerola [22] report that in *Pelargonium zonale*, where leaf variegation can be transmitted by the pollen, proplastids are found in the generative cell cytoplasm, while in *Mirabilis japonica*, where variegation is transmitted only through the female line, the generative cell cytoplasm lacks proplastids. On the other hand, Jensen and Fisher [17] found that in cotton none of the sperm cytoplasm enters the egg. Thus the role of the cytoplasm, male or female, in this aspect of heredity can hardly be considered to have been settled by electron microscopy, at least as of this writing.

FINE STRUCTURE: *IN SITU*

Only a few reports are available on the fine structure of pollen tubes growing in the pistil, and not all of these are concerned with the organisation of the tube tips. Further, we have no evidence that growth of the tube in the pistil (that is, elongation) is restricted to the tip, or to the same portion of the tip as *in vitro*. The situation in lily pollen will indicate why this question is important.

The tip of a lily pollen tube, following growth in a compatible pistil (that is, following cross-pollination with another cultivar of the species), looks much different from the tip of a tube which grew *in vitro* [28]. The tube in the compatible pistil has at its tip a series of deep, irregular embayments, and the entire appearance is suggestive of material moving into the tube from outside the vicinity of the tip rather than the reverse, which is seen in the cultured tubes, where the vesicles seem to move outwards from the cytoplasm towards the cap compartments. This difference in appearance is suggestive of functional differences. It appears as though growth *in vitro* is largely autotrophic, with new growth at the tip deriving from stored material which is transformed and transferred to the wall via the vesicles. In the compatible pistal it appears as though the tube is taking up material from the stylar canal, growing by a largely heterotrophic mode. These inferences receive some support from the appearance of the tips of tubes growing in incompatible (selfed) pistils [28]. Here we find that the tip of the tube is covered by a compartmented cap, rather like the tip of the tube growing *in vitro*. Since tubes in incompatible pistils cease growth after completing only about half of the journey to the ovary, one might surmise that growth begins autotrophically, via vesicles and a compartmented cap, whether in a compatible or an incompatible pistil, or *in vitro*. In the incompatible pistil and *in vitro*, growth ceases when stored reserves are exhausted. In compatible pistils, on the other hand, the tube is able to switch, during the growth, from autotrophic to heterotrophic growth, and this is manifested in a switch from compartmented cap to embayments. Growth via embayments would reflect a respiratory phase not detectable by Dickinson [9, 10] in his studies *in vitro*. It must be stressed that this model is far from substantiated. We have searched diligently for signs of the transitional fine structure between compartments and embayments; and though we have seen configurations which are suggestive of transitional stages, we frankly have not seen anything fully convincing. The pistil is long; tube growth is rapid; and if the transition is also rapid, it is possible that we may sample at many different times without catching it.

We have also sought to test the theory that the embayments of compatible tubes function in a manner akin to pinocytosis by employing radioactive and electron-opaque markers. Results of these experiments, described elsewhere in this volume [30], have thus far been inconclusive.

The embayments at the tips of compatible pollen tubes bear a striking resemblance to the wall ingrowths which characterise the cells which Gunning and Pate [15] refer to as 'transfer cells'. These are cells which are specialised in relation to short-distance transport of solutes, either into or out of cells, the ingrowths serving to increase the amount of cell surface across which transport can occur. Whether the embayments at the tips of compatible tubes reflect an active process of ingestion (i.e. pinocytosis) of secretory product from the pistil, or whether they are stable structures similar to the wall ingrowths of transfer cells, remains to be determined [29].

Because of the obvious difficulties which arise as a consequence of having the tubes grow in close association with the transmitting tissue of the solid style, much less is known about the fine structure and physiology of pollen growing *in situ* in solid-styled plants. Since no method has been devised for dissecting out the pollen tubes from the pistil, encounters with tube tips in thin-sectioned material occur only by chance, and the tip therefore cannot be recognised with certainty. Thus we have no clear idea of its organisation. Kroh [19] noted that *Petunia* pollen tubes growing in the pistil have much more elaborate and irregular lateral walls than have tubes in culture, and she suggested that this may facilitate uptake of materials from the pistil. Jensen and Fisher [18] noted that cotton pollen tubes appear to grow through a thick, pectinaceous layer of the wall of the transmitting cells and not through the middle lamella as others have reported (e.g. Ref. 3). Crang [3] has stressed the possibility that at least a part of *Lychnis alba* tube growth in the pistil is by apposition.

POLLEN TUBE GROWTH *IN VITRO*

A problem which has bedevilled students of pollen growth has been that of achieving, *in vitro*, growth which equals that which must be accomplished in the pistil if fertilisation is to occur. A few species are capable of growth in culture which equals that in the pistil but most fall far short, regardless of the culture conditions employed. Lily pollen tubes, for example, will grow to a maximum length of only about a centimetre *in vitro*, while the lily pistil is 10 times that length. Obviously the pistil provides the pollen with an environment which we have not yet succeeded in duplicating. The missing factors may be physical or nutritional, or both. Clues as to what these factors may be are likely to come from an analysis of the interactions between pistil and pollen during tube growth through the pistil.

A discussion of the chemical factors provided by the pistil which may account for the superior growth *in situ* as compared to *in vitro* will be found in another chapter of this volume [30].

A fruitful approach to the development of improved pollen tube growth media would be to go directly to the pistil and to extract from it, and identify, those soluble components which promote pollen tube growth on a minimal medium. This would be easiest in species such as the Easter Lily in which the pollen grows through a canal, where it is bathed in the secretory product of the cells which line the canal. The secretory product is readily available for collection and analysis. It is reasonable to propose that this tactic would lead to the identification of hitherto unknown compounds which regulate the growth of pollen tubes in nature, acting as growth promoters, chemotropic agents, mediators of the incompatibility reaction, or perhaps some combination of these.

We must also, however, be attentive to the physical environment. Growth requirements, physical as well as chemical, may change significantly during growth. Indeed, the general failure to achieve growth in culture which approaches growth in the pistil may come from a failure to recognise changing growth and osmotic requirements at different stages of tube development. Although we employ a constant medium in culture experiments, the environment of the tube may be changing as it grows through the pistil and perhaps we should be designing culture media accordingly.

In neither the solid nor the hollow pistils do the pollen tubes encounter an environment which resembles that of the aqueous solutions which are so frequently employed in culture studies. While growth of pollen within or on the surface of media solidified with agar may come somewhat closer to approximating the natural environment, it is probably nevertheless a very poor approximation. In my laboratory we have made efforts to improve the physical characteristics of the *in vitro* environment for lily pollen. We have tried to replace agar with other gels, such as pectin, gelatin, etc. We have varied the concentration, and, hence, the viscosity, of a variety of gels. We have compared different sources of agar, and agar subjected to a variety of purification procedures. We have compared the growth of pollen on the surface with pollen suspended in the gels. Aside from stating that pollen grows better on agar of bacteriological grade if the agar is first washed thoroughly with water [14], we cannot report any useful insights from these efforts. I feel, however, that some careful and systematic thinking and experimentation will yield dividends in this area. We ought to measure the viscosity of stigmatic exudate, as well as its osmolarity, and duplicate these values in culture media. We ought to question the tacit assumption that these values remain constant as the tube grows into new areas in the pistil; rather we should measure these values, or at least test the effects of modifying these values upwards and downwards in the culture medium.

Another approach derives from the axiom that structure and function are ultimately inseparable. In the case of the function of pollen tube growth stimulants, we might reasonably expect alterations in structure, resulting from addition of stimulatory substances, to give us indications as to their function. Clues to the actions of growth inhibitors are sometimes revealed through alterations of fine structure, as, for example, the well-known action of colchicine on dictyosome function [7]. It seems to me to be equally plausible to seek clues to the action of growth stimulants by searching for fine-structural changes which their presence might induce.

Many compounds, particularly certain amino acids and hormones, have been reported to stimulate the growth of the pollen tubes of one or more species *in vitro*, and for each compound it might be well to consider its role in relation both to its occurrence in the pistil in a form available to the pollen and to its effects on fine structure.

Finally, there is the question of the control of *direction* of growth of the pollen tube. It is generally accepted that the arrival of the tube at the embryo sac is accomplished not merely by a passive guidance mechanism involving growth of the tube along the path of least resistance. The pistil produces chemotropically active substances which are easily demonstrated *in vitro* and which presumably are active during at least a part of the journey of the tube through the pistil. The points at which chemotropic guidance might be expected to be particularly critical would be at the start and at the end of the journey—that is, immediately after germination on the stigma surface and at the time of entry into the embryo sac [33].

Reasoning by analogy from the geotropic and phototropic reactions of other plant parts, one might expect the chemotropic response to result from greater growth of the tube on the side away from the source of the stimulus, resulting in reorientation of growth towards the source of the active material [25]. Thus examination of the fine structure of the tips of tubes which are fixed while in the process of responding (by changing direction of growth) to the chemotropic factor might give an indication of the mechanism of action of the factor.

REFERENCES

1 BONNETT, H. T. JR. and NEWCOMB, E.H., 'Coated vesicles and other cytoplasmic components of growing root hairs of radish', *Protoplasma*, **67**, 59–75 (1966)
2 BOPP-HASSENKAMP, G., 'Elektronenmikroskopishe Untersuchungen an pollen-schauchen zweier liliaceen', *Z. Naturforsch.*, **15**, 91–4 (1960)
3 CRANG, R. E., 'A fine structural study of *in vivo* pollen tube penetration in *Lychnis alba*', *Trans. Am. microsc. Soc.*, **85**, 564–70 (1966)
4 CRANG, R. E. and MILES, P. G., 'An electron microscope study of germinating *Lychnis alba* pollen', *Am. J. Bot.*, **56**, 398–405 (1969)
5 DASHEK, W. V., 'The Lily pollen tube: Aspects of chemistry and nutrition in relation to fine structure', Doctoral thesis, Marquette University, Milwaukee, Wis. (1966)
6 DASHEK, W. V. and ROSEN, W. G., 'Electron microscopical localization of chemical components in the growth zone of lily pollen tubes', *Protoplasma*, **61**, 192–204 (1966)
7 DAUWALDER, M. and WHALEY, W. G., 'A new stage of cell plate formation and some observations on puromycin and colchicine treatments', *J. Cell Biol.*, **27**, 24A (1965)
8 DEXHEIMER, J., 'Sur la synthèse d'acide ribonucléique par les tubes polliniques en croissance', *C.r. hebd. Séanc. Acad. Sci., Paris*, **267**, 2126 (1968)
9 DICKINSON, D. B., 'Germination of lily pollen: Respiration and tube growth', *Science, N.Y.*, **150**, 1818–19 (1965)
10 DICKINSON, D. B., 'Permeability and respiratory properties of germinating pollen', *Physiologia Pl.*, **20**, 118–27 (1967)

11 DICKINSON, D. B., 'Rapid starch synthesis associated with increased respiration in germinating lily pollen', *Plant Physiol.*, **43**, 1–8 (1968)

12 ECHLIN, P., 'An apparent helical arrangement of ribosomes in developing pollen mother cells of *Ipomoea purpurea* (L.) Roth.', *J. Cell Biol.*, **34**, 150–3 (1965)

13 GIMÉNEZ-MARTIN, G., RISUEÑO, M. C. and LÓPEZ-SÁEZ, J. F., 'Generative cell envelope in pollen grains as a secretion system, a postulate', *Protoplasma*, **67**, 223–35 (1969)

14 GOLAS, R. M., 'Studies on the physiology of the pollen of *Lilium longiflorum*', M.S. thesis, Marquette University, Milwaukee, Wis. (1960)

15 GUNNING, B. E. S. and PATE, J. S., '"Transfer Cells"—Plant cells with wall ingrowths, specialized in relation to short distance transport of solutes—Their occurrence, structure, and distribution', *Protoplasma*, **68**, 107–33 (1969)

16 JENSEN, W. A., 'Cotton Embryogenesis—Polysome formation in the zygote', *J. Cell Biol.*, **36**, 403–6 (1968)

17 JENSEN, W. A. and FISHER, D. B., 'Cotton Embryogenesis: The entrance and discharge of the pollen tube in the embryo sac', *Planta*, **78**, 158–83 (1968)

18 JENSEN, W. A. and FISHER, D. B., 'Cotton Embryogenesis: The tissues of the stigma and style and their relation to the pollen tube', *Planta*, **84**, 97–121 (1969)

19 KROH, M., 'Fine structure of *Petunia* pollen germinated *in vivo*', *Rev. Paleobot. Palynol.*, **3**, 197–203 (1967)

20 LARSON, D. A., 'Fine-structural changes in the cytoplasm of germinating pollen', *Am. J. Bot.*, **52**, 139–59 (1965)

21 LINSKENS, H. F., 'Isolation of ribosomes from pollen', *Planta*, **73**, 194–200 (1967)

22 LOMBARDO, G. and GEROLA, F. M., 'Cytoplasmic inheritance and ultrastructure of the male generative cell of higher plants', *Planta*, **82**, 105–10 (1968)

23 MASCARENHAS, J. P. and BELL, E., 'Protein synthesis during germination of pollen: Studies on polyribosome formation', *Biochim. biophys. Acta.*, **179**, 199–203 (1969)

24 O'KELLEY, J. C. and CARR, P. H., 'An electron micrographic study of the cell walls of elongating cotton fibers, root hairs, and pollen tubes', *Am. J. Bot.*, **41**, 261–4 (1954)

25 ROSEN, W. G., 'Studies on pollen-tube chemotropism', *Am. J. Bot.*, **48**, 889–95 (1961)

26 ROSEN, W. G., GAWLIK, S. R., DASHEK, W. V. and SIEGESMUND, K. A., 'Fine structure and cytochemistry of *Lilium* pollen tubes', *Am. J. Bot.*, **51**, 60–71 (1964)

27 ROSEN, W. G. and GAWLIK, S. R., 'Fine structure of Lily pollen tubes following various fixation and staining procedures', *Protoplasma*, **61**, 181–91 (1966)

28 ROSEN, W. G. and GAWLIK, S. R., 'Relation of lily pollen tube fine structure to pistil compatibility and mode of nutrition', *Electron Microsc.*, **2**, 313 (Proc. Int. Conf. Electron Microsc., Kyoto, Maruzen, Tokyo, 1966)

29 ROSEN, W. G. and THOMAS, H. R., 'Secretory cells of lily pistils. I. Fine structure and function', *Am. J. Bot.*, **57**, 1108–14 (1970)

30 ROSEN, W. G., this volume, 239

31 SASSEN, M. M. A., 'Fine structure of *Petunia* pollen grain and pollen tube', *Acta bot. neerl.*, **13**, 175–81 (1964)

32 VAN TIEGHEM, P., 'Végétation libre du pollen et de l'ovule et sur la fécondation directe des plantes', *Annls Sci. nat., Bot.*, **12**, 312–28 (1869)

33 WELK, M. SR., MILLINGTON, W. F. and ROSEN, W. G., 'Chemotropic activity and the pathway of the pollen tube in lily', *Am. J. Bot.*, **52**, 774–80 (1965)

Effects of Steroids on Pollen Germination*

S. Matsubara, *College of Agriculture, Kyoto University, Kyoto, Japan*

INTRODUCTION

The germination of chrysanthemum pollen *in vitro* is promoted by adding dissected floral organs, such as stigma, style, ovary, filament or petal, irrespective of kind of plant [1]. Water, ether or methanol extracts of some other plant organs, such as young fruits of tomato and onion bulbs, are also effective. The active material in methanolic extracts of onion bulbs has been partially fractionated, and it has become clear that more than one germination-promoting substance (GPS) is present [2]. Identification of these is in progress. Because the onion bulb GPSs were detected in the neutral fraction soluble in ethyl acetate, there seemed a possibility that steroids might be concerned, and for this reason tests of the effectiveness of several such compounds on the germination of chrysanthemum pollen have been made. Tests have also been made on the effectiveness of the enzyme hyaluronidase, because of the possibility that it might aid pollen germination by softening the cell wall of the germ pore.

Chrysanthemum leucanthemum pollen was used for the germination tests because pollen of this species was the most sensitive in the genus to GPSs. The basal medium contained 25% sucrose and 10 p.p.m. boric acid, and incubation was for 2 h at 20°C. The steroids (Table 1) were tested at 1, 10, 50 or 100 p.p.m. in the basal medium. Effectiveness was compared against onion bulb scale GPS B-2 fraction [2], dried down and dissolved in the basal medium at 100 p.p.m. Tests were also made on various phytosterols (Table 2), at concentrations of 1, 10 and 100 p.p.m. in the basal medium.

As shown in Table 1, the germination-promoting activity of testosterone at the concentrations of 10 and 100 p.p.m. was higher than that of the GPS. 4-Androstene-3,17-dione and androstenolone

* Abstract.

186

Table 1. EFFECTS OF SEVERAL STEROID HORMONES ON GERMINATION OF CHRYSANTHEMUM
POLLEN

Substance added to medium	Concentration (p.p.m.)	Percentage of germination of Chrysanthemum leucanthemum pollen
Androsterone	100	11·3
	10	0·0
	1	0·0
Androstenolone	50	31·6
	10	17·8
	1	0·7
4-Androstene-3,17-dione	100	38·4
	10	16·2
	1	0·0
Testosterone	100	51·8
	10	39·3
	1	0·7
Progesterone	50	2·8
	10	0·7
	1	0·0
Estradiol	100	0·0
	10	0·0
	1	0·0
Promoting substance (GPS)	100	36·0
Control (basal medium)		0·0

were also active, but their activities were lower than those of
testosterone and the GPS. Activity of androsterone was the lowest.
Estrogens were not effective at any concentration.

The results with phytosterols are shown in Table 2. Stigmasterol
and diosgenin were effective on pollen germination, but their
activities were lower than the activity of testosterone. Saponin
was not effective at any concentration.

For the biological activity of steroid hormones, a keto group at
the 3-position and a double bond (Δ^4) and methyl groups at the
10- and 18-positions are effective, and a saturated A ring decreases
activity. Active steroids, testosterone and 4-androstene-3,17-dione,
are both androstenes with a double bond (Δ^4) and a keto group at
the 3-position. Another active steroid, androsterone, is an androstene with a saturated A ring and no keto group at the 3-position,

Table 2. EFFECTS OF PHYTOSTEROLS ON GERMINATION OF CHRYSANTHEMUM POLLEN

Substance added to medium	Concentration (p.p.m.)	Percentage of germination of Chrysanthemum leucanthemum pollen
Saponin	100	0·0
	10	0·0
	1	0·0
Diosgenin	100	30·3
	10	28·7
	1	0·0
Stigmasterol	100	22·5
	10	21·6
	1	0·0
Testosterone	100	45·8
	10	30·1
	1	0·0
Control (basal medium)		0·0

and its activity on pollen germination is the least. Androstenolone was found to be as active as 4-androstene-3,17-dione, and this compound is also another androstene with a Δ^5 double bond, whose structure is similar to that of stigmasterol or diosgenin. It is of interest that the promoting activities vary depending on the structural differences among androgens.

However, progesterone has a Δ^4 double bond and a keto group at the 3-position, but it is a C_{21} steroid with additional 2-carbons, and it was not effective. Estradiol, a C_{18} steroid, was not effective. It is noteworthy that pollen germination was promoted by androgens (C_{19} steroids) but not by estrogens (C_{18} or C_{21} steroids). Active phytosterols on pollen germination, such as diosgenin and

Table 3. R_f VALUES OF THE GPS AND STEROIDS IN VARIOUS SOLVENTS

Compounds	$R_f \times 100$ with solvent	
	chloroform	80:20 chloroform:acetone
Testosterone	23	78
4-Androstene-3,17-dione	28	82
Diosgenin	26	78
Stigmasterol	39	80
GPS	0	0

stigmasterol, have methyl groups at the 10- and 18-positions and a Δ^5 double bond. From the present study it is not clear whether estradiol was not effective or whether it had an inhibiting effect on pollen germination.

Chromatographic R_f values of the active steroids and the GPS on silica gel plates are compared in Table 3 for two solvent systems. There is no indication of the identity of the GPS with any of the steroids.

The action of hyaluronidase was tested at concentrations of 5, 50 and 500 units per millilitre of the basal medium; and to examine whether hyaluronidase and testosterone interact a further test was made with the enzyme at 100 units per millilitre and testosterone

Table 4. EFFECTS OF TESTOSTERONE AND HYALURONIDASE ON GERMINATION OF CHRY-
SANTHEMUM POLLEN

Substance added to medium	Concentration	Percentage of germination of Chrysanthemum leucanthemum pollen
Testosterone	50 p.p.m.	41·8
Hyaluronidase	500 units/ml	0
	50 units/ml	0
	5 units /ml	3·3
Testosterone + hyaluronidase	50 p.p.m. + 100 units/ml	61·9
Control (basal medium)		9·5

at 50 p.p.m. The result is shown in Table 4. Although hyaluronidase alone was not effective, it interacted synergistically with testosterone. It is not known whether the chrysanthemum pollen wall, especially that of the germ pores, contains materials attacked by the enzyme. However, it is possible that hyaluronidase may increase the permeability of the germ pores and as a result aid the penetration of testosterone.

REFERENCES

1 TSUKAMOTO, Y. and MATSUBARA, S., *Pl. Cell Physiol.*, *Tokyo*, **9**, 237 (1968)
2 MATSUBARA, S. and TSUKAMOTO, Y., *Pl. Cell Physiol.*, *Tokyo*, **9**, 565 (1968)

Metabolism of Germinating Lily Pollen: Pollen Enzymes*

David B. Dickinson and Michael D. Davies, *University of Illinois, Urbana, Illinois, U.S.A.*

This study was conducted to learn whether enzymes that produce carbon skeletons and energy for polysaccharide biosynthesis occur in mature lily pollen or appear during germination. The enzymes studied represent key steps in the pathways by which hexoses yield reduced pyridine nucleotides, high-energy phosphates and sugar nucleotides. The latter would be precursors of cell wall polysaccharides and starch.

Pollen of *Lilium longiflorum* cv. Ace was incubated in pentaerythritol + 3 mM KPO_4 [2]. Pollen was centrifuged from the pentaerythritol (500 g, 15 min, 0°C) and 95% or more of the cells were ruptured in isolation medium (100 mg pollen + 1 ml of 10 mM HEPES, 0·5 mM $MgCl_2$, 0·1 mM EDTA, 1 mM dithiothreitol) in an ice-cold mortar. Cellular debris was removed (38 000 g, 15 min, 0°C), and the supernatant was the source of enzyme.

Optimum concentrations of H^+, substrate, co-factor and (in one case) allosteric activator were established for each enzyme. Proportionality between enzyme activity and volume of supernatant used was demonstrated for all enzymes. Protein was determined [10] after precipitation with 10% trichloroacetic acid, and pollen counts are expressed as the mean ± standard error. Spectrophotometric enzyme assays were done at 22°C and were based on absorbance changes of pyridine nucleotides at 340 mμ. Total reaction mixture was 0·6 ml, and HEPES buffer (50 mM) was used in all assays. Assay conditions were as follows. Hexokinase [6]: 5 mM glucose, 2·5 mM ATP, 2·5 mM $MgCl_2$, 0·5 mM NADP and 85–255 μg/ml enzyme; pH optimum was 7·5. Glucose-6-phosphate dehydrogenase [13]: 2·5 mM glucose-6-phosphate, 10 mM $MgCl_2$, 0·5 mM NADP and 0·425–1·28 mg/ml enzyme; pH optimum was 8·0. Phosphoglucomutase [6]: 8 mM glucose-1-phosphate, 0·5 mM NADP

* Abstract.

190

and 34–102 μg/ml enzyme; pH optimum was 7·5. Isocitric dehydrogenase [1]: 0·3 mM isocitrate, 0·5 mM NADP, 5 mM MgCl$_2$ and 28·3–56·6 μg/ml enzyme; pH optimum was 8·0. Malic dehydrogenase [9]: 0·1 mM oxalacetate, 0·5 mM NADH and 0·23–0·7 μg/ml enzyme; pH optimum was 7·5. Nucleoside diphosphate kinase [12]: 10 mM glucose, 2 mM MgCl$_2$, 0·5 mM NADP, 1 mM ADP, 2 mM GTP and 5·2–10·5 μg/ml enzyme; pH optimum was 7·5.

Pyrophosphorylases were assayed by adsorption of radioactive product to DEAE–cellulose paper and counting in a Packard scintillation counter [4, 5, 11]. Identity of products was confirmed by thin-layer chromatography and location of radioactive spots with a Packard Radiochromatogram Scanner. All isotopic assays

Table 1. ACTIVITY OF ENZYMES ISOLATED FROM MATURE LILY POLLEN

Enzyme	Enzyme activity (milliunits/mg pollen)	
	Not germinated	Incubated 2 h
Nucleoside diphosphate kinase*	86·0	—
Hexokinase†	5·1	5·5
Glucose-6-phosphate dehydrogenase†	1·8	0·7
Phosphoglucomutase§	5·5	5·7
Isocitric dehydrogenase†	10·2	6·8
Malic dehydrogenase†	1227·0	1105·0
CDP-glucose pyrophosphorylase‡	<0·1	<0·1
GDP-glucose pyrophosphorylase‡	<0·1	<0·1
TDP-glucose pyrophosphorylase‡	~0·2	~0·3
UDP-glucose pyrophosphorylase‡	131·0	109·0
ADP-glucose pyrophosphorylase§		
No activator	2·2	1·5
+5 mM 3-phosphoglycerate	17·4	10·3

* 211 μg protein/mg pollen, a determination done on a pollen homogenate filtered through muslin. Subsequent determinations done on supernatant fluid following high-speed centrifugation (Exp. 1-10-69).

† 5645±48 pollen grains/mg. 73% germination at 2 h. Protein (μg/mg pollen): not germination, 170; incubated 2 h, 150 (Exp. 7-10-69).

‡ 33% germination at 2 h. Protein (μg/mg pollen): not germinated, 131; incubated 2 h, 101 (Exp. 15-124, 1-90).

§ 2912±150 pollen grains/mg. 16% germination at 2 h. Protein content (μg/mg pollen): not germinated, 68; incubated 2 hr, 50 (Exp. 16-1, 1-143).

were done at 30°C in a reaction volume of 0·1 ml which contained 50 mM HEPES.

Assay conditions were as follows: UDP-glucose pyrophosphorylase: 1 mM UTP, 1 mM ^{14}C-glucose-1-phosphate, 10 mM MgCl$_2$, 300 μg/ml BSA and 1·1–4·4 μg/ml enzyme. ADP-glucose pyrophosphorylase: 2 mM ATP, 1 mM ^{14}C-glucose-1-phosphate, 10 mM MgCl$_2$, 5 mM 3-phosphoglycerate, 300 μg/ml BSA and

16–64 μg/ml enzyme. CDP-, TDP- and GDP-glucose pyrophosphorylase: assayed as for ADP-glucose pyrophosphorylase with the appropriate nucleotide triphosphate and 65–262 μg/ml enzyme.

The enzyme data in Table 1 show that all were present in abundance except for CDP-, GDP- and TDP-glucose pyrophosphorylase. No enzyme increased during the 2 h incubation of pollen in culture medium, but glucose-6-phosphate dehydrogenase and ADP-glucose pyrophosphorylase decreased about 50%. The 20–30% decrease in soluble protein may be due in part to incomplete recovery of germinated pollen. No ADP-glucose pyrophosphorylase was detected in the 2 h culture medium after removal of pollen. About 2% of UDP-glucose pyrophosphorylase was present in the medium.

The results indicate that no general synthesis or activation of enzymes occurs during germination and that soluble cytoplasmic enzymes remain within the cells.

Starch biosynthesis may be regulated at the ADP-glucose pyrophosphorylase step [5]. In Table 2 the *in vitro* activity of this enzyme

Table 2. COMPARISON OF *in vivo* RATE OF STARCH ACCUMULATION AND *in vitro* ACTIVITY OF PYROPHOSPHORYLASES

Reaction or enzyme	Product of reaction	Rate of product formation ($\mu\mu$mol/min/pollen grain at 30°C)
Starch accumulation*	glucosyl units	1·5
ADP-glucose pyrophosphorylase†	ADP-glucose	
No activator		0·8
+5 mM 3-phosphoglycerate		6·0
UDP-glucose pyrophosphorylase†	UDP-glucose	121·0

* Calculated from Table II and Fig. 2 of Ref. 3.
† Exp. 16-1.

is compared with the *in vivo* rate of starch accumulation reported earlier [3]. With optimal activator (3-phosphoglycerate), ADP-glucose is produced four times more rapidly *in vitro* than would be necessary to form starch *in vivo*. Without activator the *in vitro* rate of ADP-glucose synthesis is about one-half of that of starch accumulation. It is possible that *in vivo* rates of starch synthesis are regulated by the intracellular level of allosteric activator. However, UDP-glucose pyrophosphorylase is present in an apparent 80-fold excess, and it is not yet known whether pollen starch synthetase reacts with UDP-glucose. In either case, the enzyme catalysing this step of starch synthesis operates at considerably less than maximal velocity in the living cells.

UDP-glucose pyrophosphorylase probably catalyses a vital step in pollen wall synthesis because UDP-glucose is thought to be the precursor for callose, cellulose, pectin and at least part of the hemicellulose [7, 8]. The high level of this enzyme in pollen may be related to rapid cell wall synthesis during tube growth. As yet, there is no information concerning pollen enzymes which transfer sugar residues from nucleotides to the growing polysaccharide chains.

This work was supported by National Science Foundation Grant No. GB 8764.

REFERENCES

1 DAVIES, D. D. and ELLIS, R. J., 'Enzymes of the Krebs cycles'. In: H. F. LINSKENS, B. D. SANWAL and M. V. TRACEY (Eds.), *Modern Methods of Plant Analysis*, Vol. VII, 620–1, Springer-Verlag, Berlin (1964)

2 DICKINSON, D. B., 'Permeability and respiratory properties of germinating pollen', *Physiologia Pl.*, **20**, 118–27 (1967)

3 DICKINSON, D. B., 'Rapid starch synthesis associated with increased respiration in germinating Lily pollen', *Pl. Physiol.*, *Lancaster*, **43**, 1–8 (1968)

4 DICKINSON, D. B. and PREISS, J., 'ADP-glucose pyrophosphorylase from maize endosperm', *Archs Biochem. Biophys.*, **130**, 119–28 (1969)

5 GHOSH, H. P. and PREISS, J., 'Adenosine diphosphate glucose pyrophosphorylase. A regulatory enzyme in the biosynthesis of starch in spinach leaf chloroplasts', *J. Biol. Chem.*, **241**, 4491–4504 (1966)

6 GIBBS, M. and TURNER, J. F., 'Enzymes of glycolysis'. In: H. F. LINSKENS, B. D. SANWAL and M. V. TRACEY (Eds.), *Modern Methods of Plant Analysis*, Vol. VII, 525–6, Springer-Verlag, Berlin (1964)

7 HASSID, W. Z., 'Transformation of sugars in plants', *A. Rev. Pl. Physiol.*, **18**, 253–80 (1967)

8 HASSID, W. Z., 'Biosynthesis of oligosaccharides and polysaccharides in plants', *Science, N.Y.*, **165**, 137–44 (1969)

9 HIATT, A. J. and EVANS, H. J., 'Salt influence on malic dehydrogenase', *Pl. Physiol.*, *Lancaster*, **35**, 662–72 (1960)

10 LOWRY, O. H., ROSEBROUGH, N. J., FARR, A. L. and RANDALL, R. I., 'Protein measurement with the folin phenol reagent', *J. Biol. Chem.*, **193**, 265–75 (1951)

11 NEWSHOLME, E. A., ROBINSON, J. and TAYLOR, K., 'A radiochemical enzymatic activity assay for glycerol kinase and hexokinase', *Biochim. biophys. Acta*, **132**, 338–46 (1967)

12 NORMAN, A. W., WEDDING, R. T. and KAY BLACK, M., 'Detection of phosphohistidine in nucleoside diphosphokinase isolated from Jerusalem artichoke mitochondria', *Biochem. biophys. Res. Commun.*, **20**, 703–9 (1965)

13 WAYGOOD, E. R. and ROHRINGER, R., 'Enzymes of the pentose phosphate cycle'. In: H. F. LINSKENS, B. D. SANWAL and M. V. TRACEY (Eds.), *Modern Methods of Plant Analysis*, Vol. VII, 522–4, Springer-Verlag, Berlin (1964)

The Significance of a Wall-bound, Hydroxyproline-containing Glyco-peptide in Lily Pollen Tube Elongation*

W. V. Dashek† and H. I. Harwood,‡ *Department of Biology, Boston University* and W. G. Rosen, *Department of Biology, State University of New York, Buffalo, New York, U.S.A.*

Plant cell walls contain a structural glycoprotein, 'extensin', which has been proposed as a regulator of cell extension [3]. 'Extensin' is rich in hydroxyproline (hypro) which arises from the hydroxylation of peptide-bound proline. Regulation of pollen tube extension by endogenous substances, especially proteins, has received little attention. Although the presence of enzymes in pollen tube walls is well documented, the possibility that tube walls also contain structural proteins has not been widely considered.

Previously we reported the incorporation of ^{3}H-proline into lily pollen tube walls [1]. Because this and other labelling patterns resembled certain of those for 'extensin', we suggested that this glycoprotein was a component of the tube wall. Thus one, perhaps the major, metabolic role of the abundant free proline found in pollen cytoplasm could be its conversion to wall-bound hypro.

The present study was carried out to determine: (*a*) whether hypro is indeed a constituent of the lily pollen tube wall; (*b*) whether the tube wall contains hypro-glycopeptides ('extensin' fragments) and (*c*) whether hypro-glycopeptides regulate tube elongation.

Is hypro a constituent of the lily pollen tube wall? We attempted to answer this question by determining hypro levels colorimetrically in a purified wall fraction and in a cytoplasmic fraction precipitated with cold trichloroacetic acid (TCA) according to the method of

---

* Abstract.

† Present address: Biology Department, Virginia Commonwealth University, Richmond, Virginia.

‡ Present address: Biology Department, Brandeis University, Waltham, Mass.

194

Holleman and Key [2]. Whereas the ungerminated grain lacks wall-bound hypro, its cytoplasm contains 0·05% hypro on a dry weight basis. Following 3 h of germination the amount of hypro in the wall increased to 0·08% while the level of cytoplasmic hypro decreased from 0·05% to 0·02%. These results suggested that hypro is a constituent of the tube wall. To confirm this by another procedure, we exposed germinating pollen to ^{14}C-proline (the precursor of hypro) and then followed changes in specific activity for both wall-bound and cytoplasmic (TCA-precipitable) hypro during an 8 h time course. The specific activities of wall-bound and cytoplasmic hypro differed with respect to both their levels and their labelling patterns. At peak labelling times the specific activity of

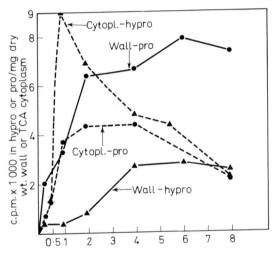

Figure 1. Changes in specific activity of wall-bound and cytoplasmic hydroxyproline and proline during an 8 h time period. 20 mg fresh weight (fr. wt.) lots of Lilium longiflorum *pollen were germinated in Petri dishes containing 10 ml Dickinson's medium [6] and 0·5 μCi ^{14}C-proline (specific activity 205 mC/mmol; walls and cytoplasm hydrolysed with 6N HCl for 18 h at 105°C; hydrolysates evaporated to dryness; residue dissolved in H$_2$O and spotted on Whatman paper; proline and hypro separated by electrophoresis; spots containing labelled proline and hypro added to scintillation fluid and counted*

cytoplasmic hypro was three times that of the specific activity of wall-bound hypro (Fig. 1). As for labelling patterns, the specific activity of cytoplasmic hypro increased rapidly for 1 h and then declined during the next 7 h. In contrast, the specific activity of wall-bound hypro did not rise markedly until 1 h after the beginning of the experiment. Between 1 and 4 h there was a linear increase in specific activity of wall-bound hypro, which then remained constant from 4 to 8 h.

The results from both the colorimetric assays and from labelling experiments are consistent with the view that peptide-linked hypro is indeed present in the lily pollen tube wall. Since a marked increase in wall-bound hypro did not occur until the onset of tube elongation, the increase between 1 and 4 h appears to be exclusively associated with tube wall synthesis. Also, since the decline in the specific activity of cytoplasmic hypro was paralleled by a rise in the specific activity of wall-bound hypro, it was concluded that cytoplasmic hypro in peptide linkage (TCA-precipitable) is a precursor of wall-bound hypro.

Does the lily pollen tube wall contain hypro-glycopeptides? To answer this question we applied Lamport's procedures [5] with

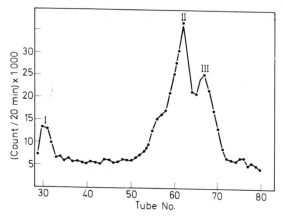

Figure 2. Sephadex G-25 elution profile of a cellulase digest of pollen tube walls labelled with ^{3}H-proline. $\frac{1}{2}$ g fr. wt. pollen germinated for 6 h in 100 μCi uniformly labelled ^{3}H-proline (specific activity 403 mCi/mmol); refluxed walls treated with cellulase (10 mg/mg dry wt. of wall) at 37°C and pH 4·5; digest centrifuged, 10 000 g, 20 min; supernatant evaporated to 1 ml; gel-filtered on Sephadex G-25 column (exclusion volume ~ 30 ml)

minor modifications for isolating hypro-glycopeptides. Briefly, these methods consisted of exposing germinating pollen to ^{3}H-proline for 6 h and then rupturing the tubes with attached grains by sonication (6 min at 40–60 mA). Next, walls were separated from the cytoplasm by centrifugation (2 min at 500 g). Following purification with five salt (1M NaCl) washes and three water washes, walls were refluxed in water for 18 h and then treated for 18 h with cellulase, an enzyme known to release hypro-glycopeptides [5]. The cellulase-released material (CRM) was then gel-filtered on Sephadex columns which were eluted with 0·1N acetic acid.

Filtration on G-25 (exclusion = >5000 molecular weight) revealed three labelled peaks (Fig. 2). Peak I eluted in the exclusion volume and was rich in labelled hypro. Peak III was retarded and contained labelled free hypro. Peak II was also retarded and, unlike the other two peaks, lacked labelled hypro and was not

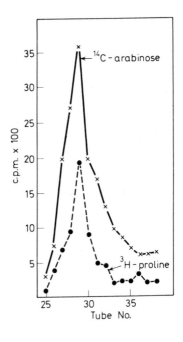

Figure 3. Sephadex G-75 elution profile of cellulase digest of pollen tube walls labelled with ^{3}H-proline and ^{14}C-arabinose. $\frac{1}{2}$ g fr. wt. of pollen simultaneously exposed 6 h to ^{3}H-proline (100 μCi and L-1-^{14}C arabinose (10 μCi; specific activity 10 mCi/mmol); wall-digest gel-filtered on a Sephadex G-75 column (exclusion volume ~28 ml)

consistently found. The finding that a peak containing ^{14}C-hypro eluted in the exclusion volume indicated that hypro was in a component(s) possessing a molecular weight of 5000 or more, a weight characteristic of some peptides.

To determine whether hypro was a constituent of a glycopeptide(s), we carried out double labelling experiments with ^{3}H-proline and ^{14}C-arabinose, the predominant sugar in the hypro-glycopeptides isolated from tomato cell walls [5]. Gel filtration of CRM on Sephadex G-75 revealed a coincidence of label derived from proline and arabinose (Fig. 3). The label derived from proline was in the form of hypro. These results suggested that hypro-glycopeptides are present in lily pollen tube walls. This suggestion was further supported by our findings that some proteases (pronase, pepsin) also release such glycopeptides.

Do hypro-glycopeptides regulate pollen tube elongation? Lamport [4] isolated hypro-*o*-arabinosides from tomato cell wall glycopeptides. He proposed that breakage of hypro-*o*-arabinoside

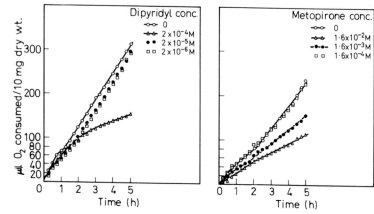

Figure 4. Respiration of pollen tubes exposed to hydroxylation inhibitors. 10 mg fr. wt. lots of pollen germinated in medium containing inhibitors, as indicated

linkages leads to increased wall plasticity, which allows cell extension to proceed.

To determine whether blockage of linkage formation promotes pollen tube elongation, we exposed tubes to compounds which

Table 1. EFFECTS OF HYDROXYLATION INHIBITORS ON ELONGATION OF LILY POLLEN TUBES

Dipyridyl		Hydroxyquinoline	
Molar concentration	Tube length (mm)	Molar concentration	Tube length (mm)
0	1·0	0	1·5
2×10^{-4}	<0·1	2×10^{-4}	<0·1
2×10^{-5}	0·5	2×10^{-5}	0·3
2×10^{-6}	1·3	2×10^{-6}	0·5

Phenanthroline		Metopirone	
Molar concentration	Tube length (mm)	Molar concentration	Tube length (mm)
0	1·4	0	1·3
2×10^{-4}	<0·1	$1·6 \times 10^{-2}$	<0·1
2×10^{-5}	0·4	$1·6 \times 10^{-3}$	0·3
2×10^{-6}	1·4	$1·6 \times 10^{-4}$	2·6

could inhibit proline hydroxylation, and thus might be expected to yield walls deficient in both hypro and the linkage. Three chelating agents, $\alpha\alpha'$-dipyridyl (DP), 8-hydroxyquinoline (HQ), and o-phenanthroline (Phen), and an inhibitor of steroid hydroxylations, metopirone (MP), were used. The chelating agents, at all concentrations tested, either inhibited elongation or were without effect

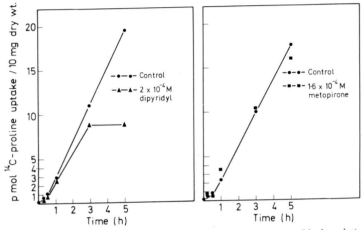

Figure 5. Uptake of ^{14}C-proline by pollen tubes in the presence of hydroxylation inhibitors. 20 mg fr. wt. lots of pollen sown in medium containing ^{14}C-proline; pollen collected by filtration under light suction on to millipore filter discs; adhering pollen exhaustively washed with medium lacking label; discs with pollen placed in scintillation fluid and counted

(Table 1). In contrast, MP at 1.6×10^{-4}M stimulated elongation twofold.

To check the specificity of these compounds as inhibitors of proline hydroxylation, we examined their effects on respiration, uptake of proline and arabinose, and protein synthesis. Of the four

Table 2. EFFECTS OF HYDROXYLATION INHIBITORS ON PROTEIN LEVELS IN ELONGATING LILY POLLEN TUBES

Inhibitors	μg protein/10 mg dry wt.	
	3 h*	5 h*
None	388	419
2×10^{-4}M dipyridyl	159	193
1.6×10^{-4}M metopirone	378	407

* Culture time.

compounds, only MP promoted tube elongation without depressing any of the above parameters (Figs. 4 and 5; Table 2). Because the findings with DP, HQ and Phen were similar, Figs. 4 and 5 and Table 2 contrast the results of only one of the growth inhibitory compounds, DP, with a growth promoting compound, MP. These results suggest that MP can be used as a specific tool for evaluating the role of the hypro-*o*-arabinose linkage in regulating tube elongation. Experiments are currently in progress to find out whether MP does indeed affect linkage formation.

This work was supported by grant GB-5402 from the National Science Foundation and by funds from the Graduate School, Boston University.

REFERENCES

1 DASHEK, W. V. and ROSEN, W. G., 'Electron microscopical localization of chemical components in the growth zone of lily pollen tubes', *Protoplasma*, **61**, 192–204 (1966)
2 HOLLEMAN J. M. and KEY, J. L., 'Inactive and protein precursor pools of amino acids in the soybean hypocotyl', *Pl. Physiol., Lancaster*, **42**, 29–36 (1967)
3 LAMPORT, D. T. A., 'Oxygen fixation into hydroxyproline of plant cell wall protein', *J. Biol. Chem.*, **238**, 1438–40 (1963)
4 LAMPORT, D. T. A., 'The hydroxyproline-*o*-glycosidic linkage of the plant cell wall glycoprotein extensin', *Nature, Lond.*, **216**, 1322–4 (1967)
5 LAMPORT, D. T. A., 'The isolation and partial characterization of hydroxyproline-rich glycopeptides obtained by enzymic degradation of primary cell walls', *Biochemistry, N.Y.*, **8**, 1155–69 (1969)
6 DICKINSON, D. B., 'Germination of lily pollen: respiration and tube growth, *Science, N.Y.*, **150**, 1818–19 (1965)

RNA and Protein Synthesis During Pollen Development and Tube Growth

J. P. Mascarenhas, *Department of Biological Sciences, State University of New York at Albany, New York, U.S.A.*

INTRODUCTION

The male gametophyte of flowering plants is developmentally a very simple system as compared to the sporophyte. Although it apparently contains the genetic information for a normal plant, all it does is put out a tube, which grows in length for several millimetres. During the course of this growth the generative nucleus in the tube, which is already in late interphase or early prophase of mitosis, undergoes a division. In the pollen of many species this division has already occurred in the pollen grain before anthesis. It seems likely that the only synthetic processes occurring in the pollen tube are those concerning tube wall polysaccharide synthesis, and the synthesis of membranes. All other normal sporophyte developmental processes are repressed.

The variable or differential gene activity theory is currently believed to be the most likely to account for cellular differentiation. This theory states that cells become specialised because of the expression of a select group of genes in each special cell type. The molecular relationship between the structure of the various proteins found in the cell and the chromosomal DNA is now well known. It is the functioning of these proteins that determines the characteristics of a particular cell type. The differentiation of cells thus depends ultimately on the select transcription of genetic information. The answers to questions concerning the development of the pollen tube must accordingly be looked for in the patterns of synthesis of RNA and protein.

No attempt is made in this review to cover the very extensive literature on nucleic acid and protein synthesis during the meiotic stages of pollen development, since several recent reviews on the subject are available [55, 75]. The treatment will be restricted to events in pollen development starting from just before microspore mitosis and ending with pollen tube growth preceding fertilisation. The pollen tube will be considered first, because the synthetic events in pollen grain development are more meaningful if one is familiar with the macromolecular events in pollen tube development.

RNA SYNTHESIS DURING POLLEN GERMINATION AND TUBE GROWTH

When pollen grains are placed in a medium containing a labelled RNA precursor such as uridine, it is found that RNA is synthesised by the germinating pollen grain [13, 41, 78]. ^{3}H-uridine was incorporated into RNA of both the vegetative and generative nuclei of *Tradescantia paludosa* within 15 min of transfer of pollen to the growth medium, as studied by autoradiography [41]. Label incorporation was also found in the cytoplasm. In the presence of actinomycin D, which inhibits DNA-dependent RNA synthesis [16, 18], the incorporation of ^{3}H-uridine into both vegetative and generative nuclei was abolished. No silver grains were observed over the nuclei when the material was treated with ribonuclease. These observations indicated that the incorporation of [^{3}H]-uridine into newly synthesised RNA was being studied.

These results were similar to those of Young and Stanley [78] with pine pollen, although the time sequence was very different. In pine pollen the nucleus of the tube cell and the generative nucleus incorporated labelled RNA precursors after 27 h of growth. In pine pollen growth is slow, and pronounced development of the pollen tube does not take place until about 30 h after germination. The stage of growth after 27 h in pine pollen might thus be comparable to the early stages of growth in *Tradescantia*. There are, however, some differences between *Tradescantia* and pine pollen. In pine the tube cell nucleus was labelled during early phases of germination, and the generative nucleus only at a later time [78]. In *Tradescantia* no such differential labelling of nuclei was observed.

The vegetative nucleus and the generative nucleus thus appear to be biochemically functional at least during the early stages of growth. Since it is not yet possible to block selectively RNA synthesis by one or other of the nuclei, we have no way of knowing what the functions of the nuclei are in pollen tube growth.

When pollen grains of *Tradescantia* are grown in a medium containing actinomycin D sufficient to inhibit all RNA synthesis, it is found that germination and growth of the pollen tube up to a length of about 250–350 μm is not inhibited (Fig 1). The migration of the two 'nuclei into the pollen tube is also not affected. The RNA species synthesised by the vegetative and generative nuclei are not required for germination, early growth of the pollen tube and migration of the nuclei into the tube during growth. However, further tube elongation and also generative nucleus division in the

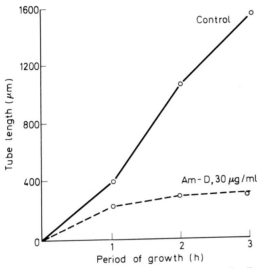

Figure 1. *Effect of actinomycin D on the growth of pollen tubes of* Tradescantia paludosa *(graph plotted from data of Ref. 41)*

tube are inhibited when RNA synthesis is blocked. It thus appears that the messenger RNA, ribosomal RNA and transfer RNA required for early growth of the pollen tube are synthesised before the pollen is released from the anthers [41].

The lack of inhibition by actinomycin D of pollen germination and early tube growth has been confirmed by Dexheimer [13] for pollen of several species, *Iris pseudoacorus, Iris sibirica, Narcissus odoratus, Hippeastrum vittatum, Endymion nutans, Lotus corniculatus, Sarothamnus scoparius, Streptocarpus caulescens, Aeschynanthus pulchrum* and *Lobelia splendens.*

The reports discussed above have all been autoradiographic studies of RNA synthesis. Autoradiography does not allow one to distinguish between the various RNA species.

Tano and Takahashi [66], using chemical extraction procedures and fractionation on a methylated albumin kieselguhr (MAK) column, found small but significant synthesis of RNA by tobacco pollen after 7 h of growth. At the end of this time about 30% of the pollen was found to have germinated and the average length of tubes was about 300 μm. The RNA synthesised had a base ratio high in adenine (A) and uridine (U), and this resembled the DNA base ratio of the plant. It was different from that of the bulk of RNA present in the pollen, which was high in guanosine (G) and cytosine (C). They suggested that the RNA synthesised was not ribosomal and might be messenger RNA.

Steffensen [63], working with lily pollen and using similar techniques, made essentially similar conclusions. The MAK column ^{3}H-uridine profile of the pollen tube RNA did not match the optical density profile of ribosomal RNA, and the base composition of pollen tube RNA was also quite different from that of ribosomal RNA. By conventional cytological criteria no nucleolus was demonstrable in the lily pollen tube nuclei. Steffensen concluded that little, if any, ribosomal RNA was made in the pollen tube in the absence of nucleolar activity, and that what RNA was made was probably largely messenger.

In the MAK column separations used by Tano and Takahashi

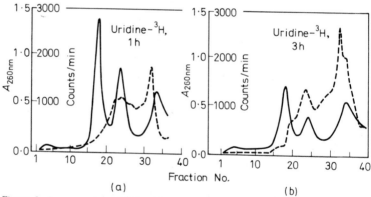

Figure 2. Incorporation of ^{3}H-uridine into phenol-extracted cytoplasmic material layered on 15–30% sucrose gradients in 0·5% (SDS) sodium dodecyl sulphate buffer and centrifuged in Spinco SW 41 rotor at 26 000 rev/min for $14\frac{1}{3}$ h at 24°C. (a) 15 mg pollen grown 1 h with ^{3}H-uridine; (b) 15 mg pollen grown 3 h with ^{3}H-uridine. Solid line, absorbance at 260 nm ($A_{260\ nm}$); dashed line, counts/min. (Data from Ref. 44)

[66] and Steffensen [63] radioactivity overlapped the ribosomal absorbance peaks. The MAK column separations did not entirely eliminate the possibility of synthesis of small amounts of ribosomal RNA. Moreover, the tobacco pollen used by Tano and Takahashi grew to only a fraction of the length it would normally have grown

in the style. With lily pollen, too, one does not obtain *in vitro* more than 10–15% of the normal growth in length [56]. The apparent lack of synthesis of ribosomal RNA could for these reasons be an artefact caused by the inhibited growth of the pollen tubes. It is known that when microbial and mammalian cells are transferred to a poorer medium which stops growth, ribosomal synthesis is also inhibited [10, 48, 76].

Tradescantia paludosa pollen grown in liquid shaking cultures grows at about the same rate as it does in the style. Moreover, generative nucleus division occurs at about the same time *in vitro* as *in situ* [41]. For studies carried out *in vitro* to be comparable to the situation in the style, it is necessary that growth of pollen in the test tube be similar to that *in vivo*.

When *Tradescantia* pollen is grown in the presence of ^{3}H-uridine, and a cytoplasmic extract purified by phenol extraction is analysed by sucrose density gradient sedimentation, the profiles seen in Fig. 2 are obtained [44]. The two ribosomal RNA (rRNA) peaks have

Table 1. BASE COMPOSITION OF DIFFERENT FRACTIONS OF POLLEN AND POLLEN TUBE RNA (FROM REF. 44)

Type of RNA	C	% ^{32}P counts in A	U	G	% G+C	% A+U
Pollen grain RNA synthesised before anther dehiscence						
15–19S (ribosomal)	21·6	24·1	24·9	29·3	51·0	49·0
23–28S (ribosomal)	26·2	24·2	19·4	30·0	56·2	43·8
4–5S	25·2	24·2	23·4	27·1	52·3	47·6
RNA synthesised by the growing pollen tube						
16–20S	16·6	39·0	26·3	18·1	34·7	65·3
22–27S	17·3	38·2	27·9	16·6	33·9	66·1
10–14S	14·7	42·8	24·7	17·8	32·5	67·5
4–7S	27·6	40·2	20·6	11·6	39·2	60·8

sedimentation values of 25S and 16S when compared with *E. coli* rRNA. The results indicate that although an apparent RNA species having a sedimentation value of about 16S is synthesised by growing pollen tubes, no RNA sedimenting with a peak at 25S is made. One would expect to find labelled 16 and 25S rRNA after 3 h if there was a similarity between plant systems and those of bacteria or mammals [17, 52].

To characterise the labelled material further, pollen was grown in ^{32}P medium and the RNA was purified and analysed for its base composition. The ^{32}P profile is different from the profile

206

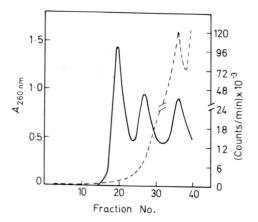

Figure 3. Profile of incorporation of ^{32}P into cytoplasmic, phenol-extracted material from 30 mg pollen grown in ^{32}P for 2 h. Layered on 15–30% sucrose in 0.5% SDS buffer. Spinco SW 41 rotor; 26 000 rev/min for $14\frac{1}{3}$ h at 25°C. Solid line, $A_{260\,nm}$; dashed line, counts/min. (From Ref. 44)

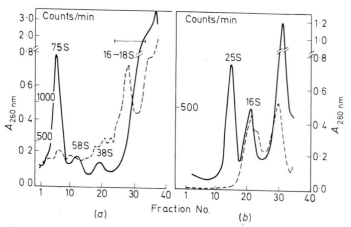

Figure 4. (a) Cytoplasmic extract of 20 mg pollen grown for 1 h in ^{3}H-uridine. DOC and Brij 58 added to 0.5% each, and extract layered on 15–30% sucrose gradient and centrifuged in Spinco SW 25.1 rotor for 17 h at 22 000 rev/min at 4°C. Fractions collected, precipitated with 5% TCA and counted. Solid line, $A_{260\,nm}$; dashed line, counts/min. (b) Material from a gradient identical with that of (a) shown by bracketed line (approximately 8–26S) pooled, made 1% with SDS, marker rRNA added, concentrated by precipitation with 2 volumes of ethanol, dissolved in 0.5% SDS buffer and layered on 15–30% sucrose gradient in same buffer. Spinco SW 41 rotor, 41 000 rev/min for $5\frac{3}{4}$ h at 25°C. Solid line, $A_{260\,nm}$; dashed line, counts/min. (From Ref. 44)

obtained when ^{3}H-uridine is used as the label. Figure 3 shows a cytoplasmic ^{32}P profile from pollen grown in ^{32}P. No distinct peak is seen at 16S although there is a shoulder in this region. As will be discussed later, a fraction of the 16–18S material labelled with a ^{3}H-uridine label is not RNA.

In order to compare the base composition of the RNA made by the pollen tube with that of rRNA, pollen obtained from inflorescence cuttings placed in ^{32}P solutions was used. The data for pollen and pollen tube RNA are presented in Table 1. The base composition of RNA made by the growing pollen tube is entirely different from that of the pollen rRNA, having a very low G + C content and a very high A + U content. The smaller species of RNA ($<$ 7S) from the mature pollen grain, which include transfer and 5S rRNA, also have a different base composition from that of pollen tube RNA.

When a cytoplasmic extract of pollen tubes growing in the presence of ^{3}H-uridine is directly analysed on a sucrose gradient, a peak of radioactivity is seen at about 16–18S (Fig. 4). The 75S, 58S and 38S peaks are the single ribosome, and large and small ribosomal sub-unit peaks, respectively. Fractions from an identical gradient as shown in Fig. 4 (*a*) were pooled and made 1% with SDS to release RNA from proteins. When now analysed on a gradient it is seen (Fig. 4*b*) that the 16–18S material found in the cytoplasm continues to sediment at the same position. Further, phenol extraction, pancreatic ribonuclease, micrococcal nuclease or pronase treatments also fail to change its sedimentation properties measurably, which indicates that this fraction is not RNA and probably is not complexed with proteins, or if it is, that the protein component is of small molecular size [42, 44].

The 16–18S material is also synthesised in the presence of actinomycin D.

If a ^{14}C-sucrose label is used instead of ^{3}H-uridine, there is incorporation of radioactivity into the 16–18S material. With a couble label of ^{3}H-uridine and ^{14}C-sucrose, when the 16–18S material is analysed on a sucrose gradient, the two peaks of radioactivity, i.e. ^{14}C and ^{3}H, are coincident, which indicates that both labels are in the same material (Fig. 5). When purified ^{14}C-sucrose labelled material is hydrolysed and the hydrolysate is analysed by paper chromatography in several solvent systems, label is found in arabinose, galactose and a few other unidentified spots. The 16–18S material is thus a polysaccharide to which a uridine phosphate molecule is attached. This is also true for a fraction of the smaller material that sediments with a peak around 4–6S [42, 44].

In an autoradiographic study, Dexheimer [13] has found cytoplasmic ^{3}H-uracil incorporation in pollen tubes even in the presence of actinomycin D. It is likely that in the pollen species he has worked with a similar uridine containing polysaccharide is produced.

By use of density gradient analyses coupled with MAK column separations and base analyses of fractions, it has been found that

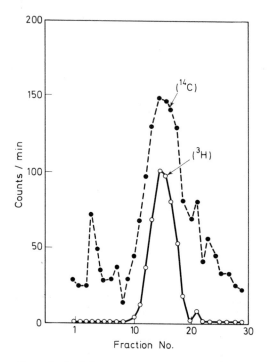

Figure 5. Pooled 10–29S material from a sucrose gradient of a cytoplasmic extract of pollen labelled with ^{14}C-sucrose and ^{2}H-uridine, treated with RNase, phenol-extracted, and layered on a 15–30% sucrose gradient in 0·5% SDS buffer. Spinco SW 41 rotor; 37 000 rev/min for 9¼ h at 25°C. (From Ref. 44)

apparently no transfer RNA is synthesised by pollen tubes of *Tradescantia paludosa* [45].

In summary, it appears likely that the lack of synthesis of ribosomes by pollen tubes is a fairly general phenomenon. As pointed out by Steffensen [63], conventional staining techniques do not show the presence of nucleoli in the pollen tube nuclei of most flowering plants. This is in agreement with current knowledge, which considers the nucleolus as the site of ribosome synthesis [8, 51, 52].

A word of caution must, however, be introduced here. Jensen and co-workers [26, 27] have found from electron microscope studies of cotton pollen that a nucleolus is present in the generative cell nucleus in the pollen tube. After division, each of the two sperm nuclei also contains a single nucleolus. The nucleoli are, however, very small. No nucleolus has been seen in the vegetative nucleus in the cotton pollen tube. It is possible that in some pollens the generative nucleus and the sperm nuclei might synthesise small amounts of ribosomal RNA.

The pine pollen tube cell nucleus often contains several nucleoli in initial germination stages [15]. In the autoradiographic studies of Young and Stanley [78] the nucleoli seemed to be the most heavily labelled structures during initial stages of pollen germination and incorporation of labelled nucleosides. The situation in pine pollen and possibly other gymnosperm pollen might be quite different from that of angiosperm pollen as far as ribosome synthesis is concerned. It could be for this reason that it is much easier to obtain tissue cultures from mature gymnosperm pollen. Haploid callus cultures have been obtained from the pollen of Ginkgo biloba [69], Taxus [70], Torreya nucifera [71] and Ephedra [33]. In angiosperm pollen, at least in those species where the matter has been studied, the genes for ribosomal RNA and transfer RNA appear to be shut off at the time of anther dehiscence. It seems to me that unless a method is found to turn these and possibly other critical genes on again, it will not be possible to obtain callus cultures from such mature pollen grains. There have been a few reports of haploid tissue cultures obtained from the culture of anthers in defined media [20, 21, 47, 49, 65]. Only one of these studies reported the precise stage in pollen development when tissue cultures can be obtained; Nitsch and Nitsch [49] were able to obtain haploid tobacco plantlets only from spores that were fully individualised, uninucleate and devoid of starch grains.

As will be seen later, the ribosomal and transfer RNA genes are not turned off at this early stage during Tradescantia and lily pollen development. It should theoretically be possible to obtain from Tradescantia, lily and probably other pollens, haploid tissue cultures at somewhat later stages than obtained by Nitsch and Nitsch [49].

Although the pollen tube does not synthesise ribosomal or transfer RNA, it does synthesise RNA. This RNA in Tradescantia is polydisperse in size, sedimenting between 2 and 30S, with a peak at about 6–8S. The RNA is low in guanosine and cytosine and high in adenosine and uridine. RNA of similar base composition is synthesised by pollen tubes of tobacco [66] and lily [63]. The RNA could possibly be messenger, but this has not yet been demonstrated.

PROTEIN SYNTHESIS DURING POLLEN GERMINATION AND TUBE GROWTH

In the preceding section we have seen that the pollen grain, when it is shed from the anther, is prepackaged with the messenger, ribosomal, and transfer RNAs needed for germination and early tube growth. Does the pollen grain also contain the necessary proteins for germination and tube growth?

Various workers have demonstrated protein synthesis in pollen tubes: Stanley, Young and Graham [60] in pine pollen; Tupý [72–74] in *Nicotiana alata*; Linskens [37] in lily; Linskens, Kochuyt and So [38] in *Petunia*; and Mascarenhas and Bell [43] in *Tradescantia* pollen.

In pollen tubes of *Nicotiana alata* Tupý [74] showed that after a lag phase of about an hour the incorporation of ^{14}C-leucine into protein was linear over 6 h of growth *in vitro*.

In a sucrose density gradient analysis of ribosomes from lily pollen Linskens [37] showed that after 2 h of germination there was a substantial increase in polysomes and a decrease in monosomes as compared to pollen incubated for 4 min in growth medium. After 4 min, germination had not yet occurred. After 10 h of incubation, pollen tubes of lily stop growing, and at this time there is according to Linskens a reduction of incorporation of labelled amino acids into nascent proteins in the polysomal region. From Figs. 2 and 3 in his paper, however, this does not seem very evident.

Linskens [37] also reports the presence of ribonuclease-insensitive polysomes in ungerminated lily pollen. He suggests that possibly the messenger RNA attached to the ribosomes is protected by some protein moiety against ribonuclease destruction. In this configuration the polysome aggregate is inactive in protein synthesis. During germination and the initiation of protein synthesis this protection is lost.

Tradescantia pollen, like most other pollens, is dehydrated and metabolically inactive [7, 32]. When pollen is placed in a growth medium, it imbibes water very rapidly, and within a few minutes a tube grows out and germination is complete. Cytoplasmic extracts of ungerminated *Tradescantia* pollen, and germinating pollen incubated in the growth medium for different intervals of time, were analysed on sucrose density gradients [43]. The polysome profile in ungerminated pollen shows one main peak at approximately 75S, that of single ribosomes. Mild treatment of the cytoplasmic extract with ribonuclease increased the area under the single ribosome peak, indicating conversion of polysomes to free ribosomes. About 38% of the ribosomes in ungerminated pollen are present in the form of polysomes, i.e. aggregates of ribosomes sensitive to ribonuclease. Most of these are very large and sediment in the pellet after density gradient centrifugation. After 5 min of incubation there is a dramatic reduction in single ribosomes and a corresponding increase in the number of polysomes which are sensitive to ribonuclease. At 5 min 55% of the ribosomes are present in polysomes. After 2 min of incubation the profile obtained is intermediate between that at 5 min and that of ungerminated pollen, showing even at this early time an increase in polysomes (50% ribosomes in polysomes). At 10 min 62% of the ribosomes are in polysomes. When pulse-labelled with ^{14}C-amino acids, the distribution of nascent protein being examined, the highest specific activity is found in the polysomal region. Ribonuclease degrades the polysomes so that all the radioactivity associated with nascent polypeptides is now found on single ribosomes. This confirms that the polysome profile changes reflect changing protein synthesis patterns.

As seen earlier, the mature, ungerminated *Tradescantia* pollen grain contains all the ribosomes that are required for the growth of the pollen tube, since no ribosomes are made during pollen tube growth [44]. About 62% of the ribosomes are present as free ribosomes in the pollen grain, while a significant number, 38%, are present in the form of polysomes. Unlike the polysomal material in ungerminated lily pollen described by Linskens [37], the polysomes in *Tradescantia* pollen are very sensitive to ribonuclease.

The fact that aggregates of polysomes sensitive to ribonuclease are present in *Tradescantia* indicates that a fraction of the ribosomes in the metabolically inactive, ungerminated pollen grain are prepackaged with messenger RNA. The importance of this can be seen in the rapid activation of protein synthesis on imbibition (less than 2 min) and the rapid formation of a germ tube (less than 10 min). In other pollen species it is known that respiration and metabolism of external sucrose occurs long before tube growth is visible [14, 61].

Electron micrographs of ungerminated *Tradescantia* pollen do show aggregates of ribosomes (J. J. Flynn and J. P. Mascarenhas, unpublished). Most of the ribosomes are, however, associated with membranes. This makes it practically impossible to count ribosomes to calculate what percentage of them are in polysomes.

None of the work described above distinguishes between protein synthesis in the vegetative cell cytoplasm and that in the generative cell cytoplasm. In electron micrographs, since the bulk of the ribosomes are seen in the vegetative cell cytoplasm [27, 35, 40], it would

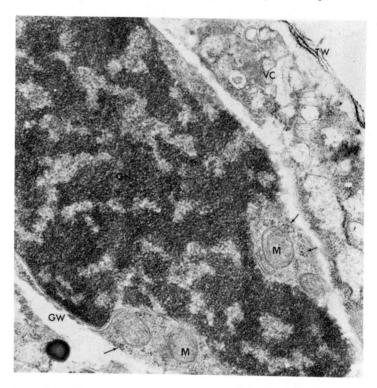

Figure 6. Electron micrograph of the generative cell in the pollen tube of Trade-scantia paludosa *after 1 h of growth* in vitro, *showing mitochondria and poly-ribosomes in the cytoplasm of the generative cell. Magnification: ×35000.*
   GW = wall of generative cell; GN = dense chromatin of generative nucleus; M = mitochondria; VC = vegetative cell cytoplasm; TW = pollen tube wall. Arrows point to ribosomes and polyribosomes in the generative cell cytoplasm. (Picture taken by Dr James J. Flynn in the author's laboratory)

appear that most of the protein synthesis observed is in the vegetative cell. However, the generative cell cytoplasm in several species contains many ribosomes. We have observed ribosomes in the generative cell cytoplasm of *Tradescantia* (Fig. 6), and Jensen and co-workers have reported that both the generative cell cytoplasm and also the cytoplasm of the sperm cells is rich in ribosomes and polysomes [26, 27]. What sorts of proteins are synthesised by the generative cell and what are their functions are interesting questions to which there is no answer at present.

We have seen that just as RNA synthesis occurs very early in pollen germination, so also does protein synthesis. Are the proteins synthesised early in germination required for germination and tube growth? The answer, unlike that for RNA, appears to be in the affirmative.

When pollen grains of *Tradescantia* are grown in the presence of cycloheximide, a potent inhibitor of protein synthesis in higher organisms [4, 22, 34], it is seen that germination is greatly reduced and pollen tube growth is inhibited (Fig. 7). The fact that some

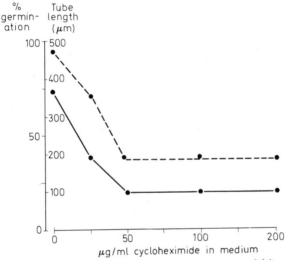

Figure 7. *Effect of cycloheximide on the germination (solid line) and tube growth (dashed line) of* Tradescantia paludosa *pollen. Observations made after 35* min *of incubation. Pollen grown in a liquid medium on a shaker* [41]

grains germinate and a small amount of tube growth occurs is probably a function of how rapidly the cycloheximide penetrates the pollen grain, since in *Tradescantia* germination is a very rapid process. Similar results have been obtained with puromycin, although puromycin is not as effective an inhibitor as cycloheximide.

Dexheimer [11] has observed that chloramphenicol, an inhibitor of bacterial, mitochondrial and chloroplast protein synthesis [2, 50, 59], arrests the growth of pollen tubes of *Lobelia*. The presence of chloramphenicol and other inhibitors causes very drastic changes in the appearance of the endoplasmic reticulum of the pollen tube [11, 12]. It thus appears that the proteins synthesised in the initial stages of pollen germination are required for germination and early tube growth.

There are numerous reports in the literature of enzymes of various sorts that are present in pollen grains and pollen tubes [14, 19, 28–30, 39, 61, 68]. To my knowledge there is no proof yet for the *de novo* synthesis of any enzyme during pollen germination and tube growth. We do not even know whether certain enzymes are present in an inactive form in the pollen grain, and are activated on germination. Here is a fertile field for good biochemical work.

In summary, the activation of the pollen grain during germination is accompanied by the initiation of protein synthesis. A sizable fraction of ribosomes in ungerminated pollen grains is in polysomes. During germination there is a very rapid formation of additional polysomes. A fraction of the ribosomes in ungerminated pollen grains appears to be prepackaged with messenger RNA. Experiments with protein synthesis inhibitors show that it is on the translation of this message that germination and early tube growth are dependent.

RNA SYNTHESIS DURING POLLEN GRAIN DEVELOPMENT

We have already seen that no ribosomes are made in the germinating pollen grain and the growing pollen tube. Let us now examine the periods in the development of the pollen grain when ribosomes are synthesised.

The first biochemical study of ribosomal RNA synthesis in developing pollen was that of Steffensen [63], who placed lily buds of different lengths for three days in a ^{32}P solution. The buds were then removed and grown in non-radioactive media until the pollen matured. The resulting pollen was analysed for its total RNA content, and the results were expressed as specific activity of RNA. The RNA with the highest specific activity was synthesised after DNA synthesis, but prior to actual microspore mitosis (bud length approximately 55–62 mm). A second peak of synthesis much lower in specific activity than the first followed microspore division (bud length approximately 65–75 mm).

Mature pollen was obtained from lily plants grown in ^{32}P-containing nutrient solutions to label microspores at early stages (bud lengths from 25 to 35 mm). The pollen RNA was extracted and analysed on a MAK column, and almost 75% of the total labelled RNA was found to be ribosomal. Steffensen [63] concluded that the analyses of total RNA synthesis reflected primarily the patterns of ribosome production rather than production of transfer RNA or messenger RNA.

In *Tradescantia paludosa* a similar pattern of ribosome synthesis is seen [44]. *Tradescantia* flowers open in the morning, the anthers dehisce and within a few hours the flowers have wilted. If conditions of light and temperature are kept constant, the period of day when the anthers dehisce is also constant, thus making it possible to label the pollen at a certain stage in development with ^{32}P and chase for various periods of time thereafter. Pollen development in the anthers of *Tradescantia* is also synchronous [77].

Inflorescence cuttings were placed in ^{32}P solutions for 24 h and then transferred to cold solutions for a chase of different periods of time. Every 24 h pollen from open flowers was collected, purified and analysed on sucrose density gradients. The results are shown in Fig. 8. During the last 48 h of pollen maturation no ribosomal RNA is synthesised. During the period from 48 to 72 h after anther dehiscence, labelled ribosomal RNA, both 16 and 25S, is seen, although in very small amounts. Most of the ribosomal RNA is synthesised earlier than 96 h from the time the pollen is shed. This labelled 16 and 25S rRNA is present in ribosomes at all periods when it is observed. The smaller species of RNA, 4S transfer, and 5S ribosomal, also seem to follow the same pattern of synthesis as the 16S and 25S rRNA [44].

Bryan [9], Taylor [67] and Woodard [77] have described the time sequence of development of *Tradescantia paludosa* pollen and also the stages of total RNA and DNA synthesis as measured by cytochemical methods. Incorporation ^{32}P into RNA, as measured by autoradiography, was found to occur in the microspores after their separation from tetrads. No ^{32}P incorporation took place during DNA synthesis, but incorporation began again before the mitotic prophase and became very active following microspore mitosis [67].

Woodard [77] has shown that there is a post-mitotic increase of RNA in the chromosomes, nucleolus and cytoplasm of the vegetative cell of *Tradescantia paludosa*. In contrast, there is a rapid diminution in RNA of both the nucleolus and chromosomes in the generative nucleus. At approximately 43 h after mitosis the RNA of the vegetative nucleolus and chromosomes begins to decrease, although the cytoplasmic RNA continues to increase. The decrease in RNA content of the nucleolus persists and the latter eventually disappears at 87 h. At this time the generative nucleolus is absent or minute and the chromosomes of the generative nucleus are

216

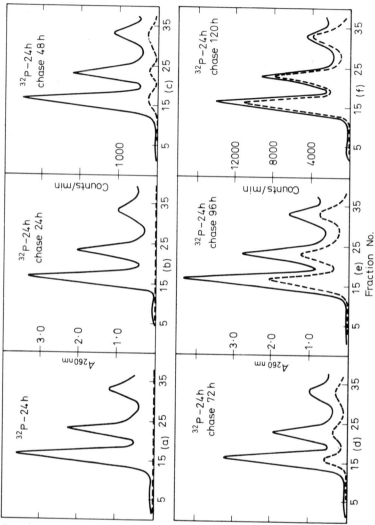

Figure 8. Synthesis of ribosomal RNA in developing pollen. Inflorescence cuttings were placed for 24 h in a solution containing ^{32}P. The cuttings were then removed, washed and placed in cold nutrient solutions for various periods of time (chase). Pollen was collected every 24 h for the next 120 h and diluted with carrier pollen, and cytoplasmic RNA was extracted. The RNA in SDS buffer was layered on a 15–30% sucrose gradient in SDS buffer and centrifuged in the Spinco SW 41 rotor at 24 000 rev/min at 25°C. (a) RNA from pollen from cuttings placed in ^{32}P for 24 h. (b) RNA from pollen from cuttings placed in ^{32}P for 24 h; followed by chase for 24 h. (c) RNA from pollen from cuttings placed in ^{32}P for 24 h; followed by chase for 48 h. (d) RNA from pollen from cuttings placed in ^{32}P for 24 h; followed by chase for 72 h. (e) RNA from pollen from cuttings placed in ^{32}P for 24 h; followed by chase for 96 h. (f) RNA from pollen from cuttings placed in ^{32}P for 24 h; followed by chase for 120 h. Solid line, $A_{260 \, nm}$; dashed line, ^{32}P counts/min. (From Ref. 44)

azure B negative. In the absence of the nucleolus and during the final decline of chromosomal RNA, the cytoplasmic RNA increased 50%. Woodard concluded that the nucleus is not the sole source of cytoplasmic RNA, the other source being the cytoplasm itself.

Our own results [44] discussed earlier agree very well with the presence of the nucleoli and the periods of RNA synthesis described by Woodard [77]. There is one exception. We do not find a large increase in cytoplasmic RNA during the last 24 h of pollen development, or at least we find no new cytoplasmic RNA synthesised as measured by incorporation of $[^{32}P]$.

Most of the rRNA is found to be made between 96 and 144 h before anthesis. This corresponds approximately to the period from 24 h before microspore mitosis to 24 h after microspore mitosis. As Woodard [77] has pointed out, since the generative nucleus loses its nucleolus very soon, it seems likely that it is primarily the vegetative nucleus that is involved in the synthesis of the RNA.

PROTEIN SYNTHESIS DURING POLLEN GRAIN DEVELOPMENT

The protein changes during *Tradescantia* pollen grain maturation approximated, in general, that of RNA [77]. In corn protein synthesis as measured by naphthol yellow S staining was found to increase rapidly after the tetrad stage until pollen grain maturity. A gel electrophoretic study of the proteins extractable at pH 6 from developing microspores of *Lilium henryi* by Linskens [36] showed a continuous change in the pattern, according to the cytological stage.

In *Lilium longiflorum*, during the life of the microspore, which extends for several weeks, thymidine kinase activity appears at a precisely defined time and lasts for no more than 24 h [23]. The transient appearance of thymidine kinase probably represents a *de novo* synthesis of at least a part of the enzyme protein.

More such studies on specific protein synthesis during microspore maturation are needed for an understanding of the development of the pollen grain and the pollen tube.

Nuclear proteins both basic and acidic are important in the regulation of gene activity [1, 3, 6, 24, 25, 31, 53, 62, 64]. There are a few studies on the types of proteins associated with pollen nuclei. Sauter and Marquardt [57, 58] have found that meiotic cells of *Paeonia tenuifolia* lose most of their cytoplasmic RNA and protein in the process of development to form microspores. After mitosis in the microspores, cytoplasmic RNA and protein increase markedly. The stages of high RNA and protein synthesis coincide with low or no nucleohistone staining, while those with reduced RNA and protein synthesis coincide with high nucleohistone staining. In the two-celled pollen grain the generative nucleus showed a high nucleohistone staining, whereas the vegetative nucleus gave no nucleohistone stain. The histone pattern correlated with high RNA and protein synthesis in the vegetative cell and no detectable synthesis in the generative cell.

Bryan [9] found that the vegetative nucleus of *Tradescantia* pollen possessed twice as much acid-insoluble protein as the generative nucleus. On the basis of fast green staining it was concluded that much of the protein in vegetative nuclei differed in its chemical constitution from that contained in somatic and generative nuclei, the indication being that it is more acid in character. Rasch and Woodard [54] observed that histone content as measured by fast green staining of the vegetative nucleus of *Tradescantia* was about half of that of the generative nucleus.

It is quite apparent from the above studies and also from the differential fluorescence of vegetative and generative nuclei when stained with acridine orange [5] that these two nuclei are different. This is an exciting area of work, as it might shed some light on the reasons for the completely different behaviour and functions of the two pollen nuclei.

REFERENCES

1 ALLFREY, V. G. and MIRSKY, A. E., 'Mechanisms of synthesis and control of protein and ribonucleic acid synthesis in the cell nucleus', *Cold Spring Harb. Symp. quant. Biol.*, **28**, 247–62 (1963)

2 ARONSON, A. I. and SPIEGELMAN, S., 'On the nature of the ribonucleic acid synthesized in the presence of chloramphenicol', *Biochim. biophys. Acta*, **53**, 84–95 (1961)

3 BENJAMIN, W. and GELLHORN, A., 'Acidic proteins of mammalian nuclei: isolation and characterization', *Proc. natn. Acad. Sci. U.S.A.*, **59**, 262–8 (1968)

4 BENNETT, L. L., SMITHERS, D. and WARD, C. T., 'Inhibition of DNA synthesis in mammalian cells by actidione', *Biochim. biophys. Acta*, **87**, 60–9 (1964)

5 BHADURI, P. N. and BHANJA, P. K., 'Fluorescence microscopy in the study of pollen grains and pollen tubes', *Stain Technol.*, **6**, 351–5 (1962)

6 BLOCH, D. P., 'Genetic implications of histone behavior', *J. cell. comp. Physiol.*, **62**, 87–94 (1963)

7 BREWBAKER, J. P. and EMERY, G. C., 'Pollen radiobotany', *Radiation Botany*, **1**, 101–54 (1962)

8 BROWN, D. D. and GURDON, J. B., 'Absence of ribosomal RNA synthesis in the anucleolate mutant of *Xenopus laevis*', *Proc. natn. Acad. Sci., U.S.A.*, **51**, 139–46 (1964)

9 BRYAN, J. H. D., 'DNA–protein relations during microsporogenesis of *Trade-scantia*', *Chromosoma*, **4**, 369–92 (1951)

10 CASHEL, M. and GALLANT, J., 'Control of RNA synthesis in *Escherichia coli*. I. Amino acid dependence of the synthesis of the substrates of RNA polymerase', *J. molec. Biol.*, **34**, 317–30 (1968)

11 DEXHEIMER, J., 'Sur les modification du reticulum endoplasmique des grains de pollen de *Lobelia erinus* L. traites par le chloramphenicol', *C.r. hebd. Séanc. Acad. Sci., Paris*, **262**, 853–5 (1966)

12 DEXHEIMER, J., 'Sur les modifications, sous l'action de diverses substances, du reticulum endoplasmique des tubes polliniques de *Lobelia erinus* L.', *C.r. hebd. Séanc. Acad. Sci., Paris*, **263**, 1703–5 (1966)

13 DEXHEIMER, J., 'Sur la synthèse d'acide ribonucleique par les tubes polliniques en croissance', *C.r. hebd. Séanc. Acad. Sci., Paris*, **267**, 2126–8 (1968)

14 DICKINSON, D. B., 'Permeability and respiratory properties of germinating pollen', *Physiologia Pl.*, **20**, 118–27 (1967)

15 FERGUSON, M. C., 'Contributions to the knowledge of the life history of *Pinus* with special reference to sporogenesis, the development of the gametophytes and fertilization', *Proc. Wash. Acad. Sci.*, **6**, 1–202 (1904)

16 FURTH, J. J., KAHAN, F. M. and HURWITZ, J., 'Stimulation by ribonucleic acid (RNA) polymerase of amino acid incorporation into proteins by extracts of *Escherichia coli*', *Biochem. biophys. Res. Commun.*, **9**, 337–43 (1962)

17 GIRARD, M., LATHAM, H., PENMAN, S. and DARNELL, J. E., 'Entrance of newly formed messenger RNA and ribosomes into He La cell cytoplasm', *J. molec. Biol.*, **11**, 187–201 (1965)

18 GOLDBERG, I. H., RABINOWITZ, M. and REICH, E., 'Basis of actinomycin action II. Effect of actinomycin on the nucleoside triphosphate-inorganic pyrophosphate exchange', *Proc. natn. Acad. Sci. U.S.A.*, **49**, 226–9 (1963)

19 GORSKA-BRYLASS, A., 'Hydrolases in pollen grains and pollen tubes', *Acta Soc. Bot. Pol.*, **34**, 589–604 (1965)

20 GUHA, S. and MAHESHWARI, S. C., '*In vitro* production of embryos from anthers of *Datura*', *Nature, Lond.*, **204**, 497 (1964)

21 GUHA, S. and MAHESHWARI, S. C., 'Cell division and differentiation of embryos in the pollen grains of *Datura in vitro*', *Nature, Lond.*, **212**, 97–8 (1966)

22 HAIDLE, C. W. and STORCK, R., 'Inhibition by cycloheximide of protein and RNA synthesis in *Mucor rouxii*', *Biochem. biophys. Res. Commun.*, **22**, 175–80 (1966)

23 HOTTA, Y. and STERN, H., 'Transient phosphorylation of deoxyribosides and regulation of deoxyribonucleic acid synthesis', *J. biophys. biochem. Cytol.*, **11**, 311–19 (1961)

24 HUANG, R. C. and BONNER, J., 'Histone, a suppressor of chromosomal RNA synthesis', *Proc. natn. Acad. Sci. U.S.A.*, **48**, 1216–22 (1963)

25 HUANG, R. C. and BONNER, J., 'Histone bound RNA, a component of native nucleohistone', *Proc. natn. Acad. Sci. U.S.A.*, **4**, 960–7 (1965)

26 JENSEN, W. A. and FISHER, D. B., 'Cotton embryogenesis: the sperm', *Protoplasma*, **65**, 277–86 (1968)

27 JENSEN, W. A., FISHER, D. B. and ASHTON, M. E., 'Cotton embryogenesis: the pollen cytoplasm', *Planta*, **81**, 206–28 (1968)

28 KATSUMATA, T. and TOGASAWA, Y., 'Biochemical studies on pollen Part IX. The enzymes related to the metabolism of carbohydrates in pollen. (1) On amylase of pollen', *J. agric. Chem. Soc. Japan*, **42**, 1–7 (1968)

29 KATSUMATA, T. and TOGASAWA, Y., 'Biochemical studies on pollen Part X. The enzymes related to the metabolism of carbohydrates in pollen. (2) On invertase and maltase of pollen', *J. agric. Chem. Soc. Japan*, **42**, 8–12 (1968)

30 KATSUMATA, T. and TOGASAWA, Y., 'Biochemical studies on pollen Part XI. The enzymes related to the metabolism of carbohydrates in pollen. (3) On phosphorylase and hexokinase of pollen', *J. agric. Chem. Soc. Japan*, **42**, 13–17 (1968)

31 KLEINSMITH, L. J., ALLFREY, V. G. and MIRSKY, A. E., 'Phosphorylation of nuclear protein early in the course of gene activation in lymphocytes', *Science, N.Y.*, **154**, 780–1 (1966)

32 KNOWLTON, H. E., 'Studies in pollen, with special reference to longevity', *Mem. Cornell Univ. agric. Exp. Stn*, **52**, 746–93 (1922)

33 KONAR, R. N., 'A haploid tissue from the pollen of *Ephedra foliata* Boiss', *Phytomorphology*, **13**, 170–4 (1963)

34 KORNER, A., 'Effect of cycloheximide on protein biosynthesis in rat liver', *Biochem. J.*, **101**, 627–31 (1966)

35 LARSON, D. A., 'Fine structural changes in the cytoplasm of germinating pollen', *Am. J. Bot.*, **52**, 139–54 (1965)

36 LINSKENS, H. F., 'Die Änderung des Protein- und Enzymmusters während der Pollenmeiose und Pollenentwicklung', *Planta*, **69**, 79–91 (1966)

37 LINSKENS, H. F., 'Isolation of ribosomes from pollen', *Planta*, **73**, 194–200 (1967)

38 LINSKENS, H. F., KOCHUYT, A. S. L. and SO, A., 'Regulation der nucleinsauren— Synthese durch Polyamine in Keimendem Pollen von *Petunia*', *Planta*, **82**, 111–22 (1968)

39 MÄKINEN, Y. and BREWBAKER, J. L., 'Isoenzyme polymorphism in flowering plants I. Diffusion of enzymes out of intact pollen grains', *Physiologia Pl.*, **20**, 477–82 (1967)

40 MARUYAMA, K., GAY, H. and KAUFMANN, B. P., 'The nature of the wall between generative and vegetative nuclei in the pollen grain of *Tradescantia paludosa*', *Am. J. Bot.*, **52**, 605–10 (1965)

41 MASCARENHAS, J. P., 'Pollen tube growth and ribonucleic acid synthesis by vegetative and generative nuclei of *Tradescantia*', *Am. J. Bot.*, **53**, 563–9 (1966)

42 MASCARENHAS, J. P. and BELL, E., 'A polysaccharide containing a uridine phosphate residue—a possible intermediate in pollen tube wall synthesis', *Pl. Physiol.*, *Lancaster*, **43**, S-24 (1968)

43 MASCARENHAS, J. P. and BELL, E., 'Protein synthesis during germination of pollen: studies on polyribosome synthesis', *Biochim. biophys. Acta*, **179**, 199–203 (1969)

44 MASCARENHAS, J. P. and BELL, E., 'RNA synthesis during development of the male gametophyte of *Tradescantia*', *Devl. Biol.*, **21**, 475–90 (1970)

45 MASCARENHAS, J. P., this volume, 230

46 MOSS, G. I. and HESLOP-HARRISON, J., 'A cytochemical study of DNA, RNA, and protein in the developing maize anther II. Observations', *Ann. Bot.*, **31**, 555–72 (1967)

47 NAKATA, K. and TANAKA, M., 'Differentiation of embryoids from developing germ cells in anther culture of tobacco', *Jap. J. Genet.*, **43**, 65–71 (1968)

48 NEIDHART, F. C., 'The regulation of RNA synthesis in bacteria', *Prog. nucleic Acid Res.*, **3**, 145–81 (1964)

49 NITSCH, J. P. and NITSCH, C., 'Haploid plants from pollen grains', *Science, N.Y.*, **163**, 85–7 (1969)

50 NOMURA, M. and WATSON, J. D., 'Ribonucleoprotein particles with chloromycetin-inhibited *Escherichia coli*', *J. molec. Biol.*, **1**, 204–17 (1959)

51 PENMAN, S., SMITH, I., HOLTZMAN, E. and GREENBERG, H., 'RNA metabolism in the He La cell nucleus and nucleolus'. In: *International Symposium on the Nucleolus, its Structure and Function*, U.S. Govt. Printing Office, Wash., D.C., 489–512 (1966)

52 PERRY, R. P., 'The nucleolus and the synthesis of ribosomes'. In: *International Symposium on Genes and Chromosomes—Structure and Function*, U.S. Govt. Printing Office, Wash., D.C., 325–40 (1965)

53 POGO, B. G. T., POGO, A. P., ALLFREY, V. G. and MIRSKY, A. E., 'Changing patterns of histone acetylation and RNA synthesis in regeneration of the liver', *Proc. natn. Acad. Sci. U.S.A.*, **59**, 1337–44 (1968)

54 RASCH, E. and WOODARD, J. W., 'Basic proteins of plant nuclei during normal and pathological cell growth', *J. biophys. biochem. Cytol.*, **6**, 263–76 (1959)

55 RHOADES, M. M., 'Meiosis'. In: J. BRACHET and A. E. MIRSKY (Eds.), *The Cell*, Vol. 3, 1–75, Academic Press, New York (1961)

56 ROSEN, W. G. and GAWLIK, S. R., 'Fine structure of lily pollen tubes following various fixation and staining procedures', *Protoplasma*, **61**, 181–91 (1966)

57 SAUTER, J. J. and MARQUARDT, H., 'Nucleohistone und Ribbonucleinsäure—Synthese während der Pollenentwicklung', *Naturwissenschaften*, **55**, 546 (1967)

58 SAUTER, J. J. and MARQUARDT, H., 'Die Rolle des Nucleohistons bei der RNS- und Proteinsynthese während der Mikrosporogenese von *Paeonia tenuifolia*', *Z. Pfl. Physiol.*, **58**, 126–37 (1967)

59 SPENCER, D., 'Protein synthesis by isolated spinach chloroplasts', *Archs Biochem. Biophys.*, **111**, 381–90 (1965)

60 STANLEY, R. G., YOUNG, L. C. T. and GRAHAM, J. S. D., 'Carbon dioxide fixation in germinating pine pollen *(Pinus ponderosa)*', *Nature, Lond.*, **182**, 1462–3 (1958)

61 STANLEY, R. G. and LINSKENS, H. F., 'Enzyme activation in germinating *Petunia* pollen', *Nature, Lond.*, **203**, 542–4 (1964)

62 STEDMAN, E. and STEDMAN, E., 'Cell specificity of histones', *Nature, Lond.*, **166**, 780 (1950)

63 STEFFENSEN, D. M., 'Synthesis of ribosomal RNA during growth and division in *Lilium*', *Expl Cell Res.*, **44**, 1–12 (1966)

64 SWIFT, H., 'Nucleic acids and cell morphology in Dipteran salivary glands'. In: J. M. ALLEN (Ed.), *The Molecular Control of Cellular Activity*, 73–125, McGraw-Hill, New York (1962)

65 TANAKA, M. and NAKATA, K., 'Tobacco plants obtained by anther culture and the experiment to get diploid seeds from haploids', *Jap. J. Genet.*, **44**, 47–54 (1969)

66 TANO, S. and TAKAHASHI, H., 'Nucleic acid synthesis in growing pollen tubes', *J. Biochem., Tokyo*, **56**, 578–80 (1964)

67 TAYLOR, J. H., 'Autoradiographic detection of incorporation of P^{32} into chromosomes during meiosis and mitosis', *Expl Cell Res.*, **4**, 164–73 (1953)

68 TSINGER, N. Y. and PETROVSKAYA-BARANOVA, T. P., 'The pollen-grain wall—a living physiologically active structure', *Dokl. Akad. Nauk SSSR*, **138**, 466–9 (1961)

69 TULECKE, W., 'A tissue derived from the pollen of *Ginkgo biloba*', *Science, N.Y.*, **117**, 599–600 (1953)

70 TULECKE, W., 'The pollen cultures of C.D. Larue: a tissue from the pollen of *Taxus*', *Bull. Torrey bot. Club*, **82**, 283–9 (1959)

71 TULECKE, W. and SEGHAL, N., 'Cell proliferation from the pollen of *Torreya nucifera*', *Contr. Boyce Thompson Inst. Pl. Res.*, **22**, 153–63 (1963)

72 TUPÝ, J., 'Changes in pollen proteins during tube growth from incompatibility point of view', *Genetics Today*, **1**, 212–13 (1963)

73 TUPÝ, J., 'Metabolism of proline in styles and pollen tubes of *Nicotiana alata*'. In: H. F. LINSKENS (Ed.), *Pollen Physiology and Fertilization*, 86–94, North-Holland, Amsterdam (1964)

74 TUPÝ, J., 'Synthesis of protein and RNA in pollen tubes stimulated with 2-thiouracil', *Biologia Pl.*, **8**, 398–410 (1966)

75 VASIL, I. K., 'Physiology and cytology of anther development', *Biol. Rev.*, **42**, 327–73 (1967)

76 WATSON, J. D. and RALPH, R. K., 'Changes in RNA synthesis following transfer of Sarcoma 180 ascites cells from the mouse to synthetic medium', *Cancer Res.*, **26**, 2362–7 (1966)

77 WOODARD, J. W., 'Intracellular amounts of nucleic acids and protein during pollen grain growth in *Tradescantia*', *J. biophys. biochem. Cytol.*, **4**, 383–90 (1958)

78 YOUNG, L. C. T. and STANLEY, R. G., 'Incorporation of tritiated nucleosides, thymidine, uridine, and cytidine in nuclei of germinating pine pollen', *Nucleus, Calcutta*, **6**, 83–90 (1963)

Ribosome Synthesis Compared During Pollen and Pollen Tube Development*

Dale M. Steffensen, *Department of Botany, University of Illinois, Urbana, Illinois, U.S.A.*

The unique properties of developing pollen and pollen tubes are their dramatic changes in morphology and metabolic activity, which occur within the same cell wall. Our interests have focused on their varying capacity to synthesise ribosomes at different stages in development. The first effort [5] noted the presence and absence of nucleoli in lily pollen and pollen tubes, respectively. Ribosomal RNA (rRNA) was synthesised in developing pollen, while none was detected in pollen tubes when nucleoli were inactive.

The present report extends the comparative study of ribosome synthesis, dealing with the other two components, the ribosomal proteins and 5S RNA. A similar approach is taken with transfer RNA (tRNA) to ascertain where it is synthesised and modified during development, examining specifically new chain formation, end-group addition and methylation. A by-product of the methylation study has uncovered a simple method to isolate ^{3}H-methyl-pectin from pollen tubes.

Lilium longiflorum (varieties Ace and Nellie White) is used. The procedures for culturing buds and pollen tubes have been presented elsewhere [4, 5]. The procedure for the electrophoresis of proteins, slicing and determining radioactivity in acrylamide gels has been given in detail [2]. The methods for the separation of RNA on gels are similar to those of Loening [3]. Procedures for the isolating and chromatography of RNA as well as the purification of ribosomes have been described [5].

Ribosomal proteins. An earlier experiment [5] had indicated that most of the rRNA was synthesised just before and after the first microspore mitosis in lily, while rRNA (18S and 28S) was

* Abstract.

223

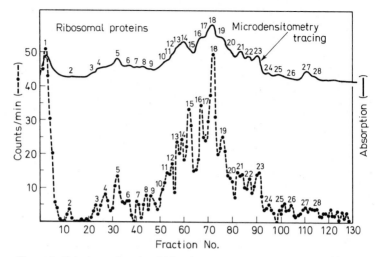

Figure 1. Gel electrophoresis of lily ribosomal proteins labelled with $^{14}CO_2$ at microspore through early pollen stages. Fraction numbers start at the origin of the 15% acrylamide gel. Each fraction consists of a 250 μm slice

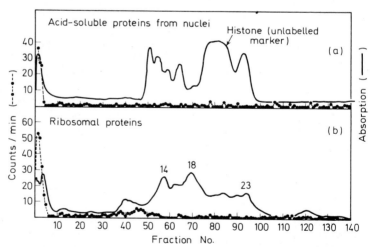

Figure 2. Gel electrophoresis of proteins labelled in pollen tubes with ^{35}S. (a) Split gel comparing acid-soluble proteins from nuclei against unlabelled lily histone as reference proteins. (b) Ribosomal proteins from pollen tube labelling. The peaks, numbers 14, 18 and 23, correspond to another ribosome preparation (Fig. 1), as does the over-all microdensitometer tracing. The traces of radioactivity between fractions 35 and 55 are attributed to protein contamination from the supernatant

probably not being made in pollen tubes. In order to confirm this observation, the ribosomal proteins were studied to see whether they conformed to the previous observation. Microspores in detached buds were labelled *in vitro* with $^{14}CO_2$ at the peak of ribosome synthesis. The ribosomal proteins were prepared from the ribosomes of pollen tubes after the ^{14}C pollen had been grown in unlabelled media. The ribosomal proteins were separated by gel electrophoresis and, as indicated in Fig. 1, there are approximately 29 proteins. The radioactive peaks coincide perfectly with the microdensitometer tracings of the protein bands. Other experiments using ^{35}S

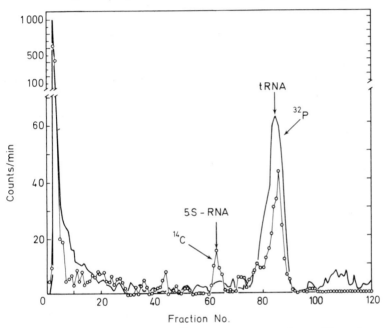

Figure 3. Electrophoretic separation of ^{32}P-RNA from pollen tubes on 9% acrylamide gel. Uniformly labelled ^{14}C-RNA from E. coli was used as reference markers. (Note: the ^{32}P-5S RNA from lily is not radioactive)

to label ribosomal proteins at microspore stages gave similar results.

By contrast, if pollen tubes were labelled with ^{35}S and the ribosomes were extracted, the resulting ribosomal proteins are not radioactive at all (Fig. 2b) nor are the histones (Fig. 2a). The only proteins which are synthesised and become labelled in pollen tubes are those in the supernatant of the post-ribosomal fraction. Apparently neither the proteins nor the rRNA of ribosomes are being made *de novo* in pollen tubes. A detailed account will be presented elsewhere regarding the comparative aspects of protein synthesis between pollen and pollen tubes.

5S RNA. A final test of the last conclusion about ribosomes was done to see if the third component, 5S RNA, is labelled or not in pollen tubes when high levels of ^{32}P (38 μCi/ml) are used. Limitations of space necessitate our presenting only the data where RNA was separated on acrylamide gels and omitting a number of experiments where low molecular weight RNA was separated by different column chromatography methods. From Fig. 3 it can be seen that the 5S RNA is unlabelled. (The 5S RNA molecule is always associated with the 60S ribosomal particle.) As will be discussed later, the tRNA is radioactive (Fig. 3). The experiment with 5S RNA confirms the contention that none of the components of the ribosome are synthesised in pollen tubes.

Transfer RNA. The pollen tube offers a unique system to study

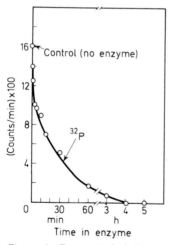

Figure 4. Enzymatic hydrolysis of ^{32}P-tRNA with snake venom phosphodiesterase (1 mg/ml)). The tRNA had been labelled in pollen tubes. Nearly half of the radioactivity was removed in 5 min, indicating addition to the 3' end

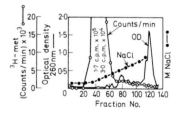

Figure 5. Separation of ^{3}H-methylated RNA and pectin on a MAK column. Each had been labelled in pollen tubes with ^{3}H-methyl methionine (^{3}H-met). The broad radioactive peak eluting at low salt concentrations is pectin. In this experiment the radioactivity profile of tRNA is obscured by the tailing counts from pectin

different RNA molecules because the 'noise' from ribosome synthesis is shut off. For these and other reasons, tRNA has occupied much of our attention. Our unpublished results of separating tRNA on a methylated albumin on kieselguhr (MAK) column and on benzoylated DEAE cellulose [1] have shown that tRNA is always

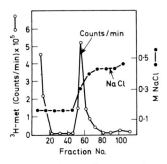

Figure 6. Elution of ³H-pectin from a MAK column with 0·35 M NaCl before starting a gradient to fractionate the RNA

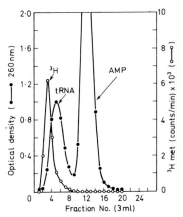

Figure 7. Further chromatography of ³H-pectin from a MAK column on Sephadex G 75. Unlabelled tRNA and adenylic acid (AMP) were used as markers. The radioactivity was in the void volume and there were no counts in the AMP peak, where small labelled molecules might appear.

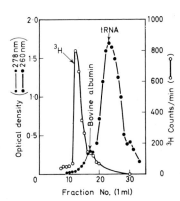

Figure 8. Separation of ³H-pectin on Sephadex G 100. As in Fig. 7, the majority of the radioactivity is in the void volume and thus larger than a molecular weight of 100 000. Bovine serum albumin and tRNA were used as markers

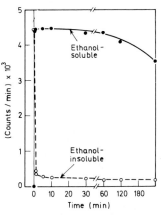

Figure 9. Removal of the ³H-methyl label from pectin with pectin methyl esterase. This reaction is: methylated pectin→methyl alcohol+pectin. The decline of radioactivity in the ethanol-soluble phase was due to evaporation of methanol from the reaction vessel

labelled with ^{32}P, ^{3}H-cytidine and ^{3}H-uridine just as in Fig. 3. We still are not certain whether the label represents only addition to the 3'-end or whether some small proportion of tRNA has its complete chain synthesised. To test each possibility, tRNA was subjected to snake venom phosphodiesterase. In Fig. 4 about half of the radioactivity is removed in 5 min of cleavage at the 3'-end, terminating in CpCpA. At this writing we still do not know whether the remaining radioactivity represents contamination from low molecular weight messenger RNA, as nucleotide base ratios indicate, or whether a small amount of tRNA is really being made. The critical experiment has still to be done, where a brief cleavage is carried out with phosphodiesterase and the remaining chain is separated by gel electrophoresis as in Fig. 3.

The most interesting finding about tRNA concerns its methylation in the pollen tube when ^{3}H-methyl methionine is used. A preliminary nucleotide analysis of ^{3}H-tRNA on a Dowex 1-formate shows at least 10 radioactive peaks, distributed over the four major nucleotids: 2 with C, 3 with A, 2 with U and 3 with G. The tRNA is the only RNA methylated in the pollen tube, as examined by column chromatography or acrylamide gel electrophoresis. Our present excitement relates to the possibility that the methylation of tRNA might be a primary regulatory mechanism to control protein synthesis in pollen tubes. This hypothesis is being tested to see whether the start of methylation coincides with the beginning of protein synthesis.

Methylation of pectin. A finding which was unexpected was that pectin synthesised in pollen tubes can be isolated on a MAK column just before the tRNA begins to elute with salt. ^{3}H-methyl methionine was used to label both pectin and tRNA in growing pollen tubes. The standard SDS–phenol method for removing protein and other substances still leaves both RNA and pectin in the aqueous phase. When the resulting ethanol precipitate is loaded with 0·1 M NaCl on a MAK column and the salt gradient is started, the ^{3}H-pectin begins to come off with 0·1 M NaCl (Fig. 5). If one loads as before and then elutes only with 0·35 M NaCl, then all of the ^{3}H-pectin comes off, leaving the RNA on the column (Fig. 6). The molecular weight of the pectin must be greater than 100000, since it remains in the void volume during column chromatography with either G75 or G100 Sephadex (Figs. 7 and 8). The proof that the ^{3}H-labelled compound under discussion is really ^{3}H-methylated pectin comes from enzymatic hydrolysis with pectin methyl esterase shown in Fig. 9. All but 2% of the ^{3}H-methyl groups are removed and converted to ^{3}H-methanol.

This research was carried out with the aid of Mrs. L. Cooper and Mr. G. Holmquist, as well as by the support of the National Science Foundation (GB 3981).

REFERENCES

1 GILLIAM, I., MILLWARD, S., BLEW, D., VON TIGERSTROM, M., WIMMER, E. and TENNER, G. M., 'The separation of soluble ribonucleic acids on benzoylated diethylamino-ethylcellulose', *Biochemistry*, **6**, 3043–56 (1967)

2 GRAY, R. H. and STEFFENSEN, D. M., 'A high-resolution method for slicing and counting radioactivity in polyacrylamide gels', *Analyt. Biochem.*, **24**, 44–53 (1968)

3 LOENING, U. E., 'The fractionation of high-molecular weight ribonucleic acid by polyacrylamide-gel electrophoresis', *Biochem. J.*, **102**, 251–7 (1967)

4 STEFFENSEN, D. M., 'Proteins of sperm nuclei examined by autoradiography at fertilization and subsequent nuclear divisions in Lilium', *Genetics*, **52**, 631–41 (1965)

5 STEFFENSEN, D. M., 'Synthesis of ribosomal RNA during growth and division in Lilium', *Expl Cell Res.*, **44**, 1–12 (1966)

Lack of Transfer RNA Synthesis in the Pollen Tube of *Tradescantia paludosa**

Joseph P. Mascarenhas, *Department of Biological Sciences, State University of New York at Albany, New York, U.S.A.*

When pollen of *T. paludosa* is grown either in [³H]-uridine or [³²P]-labelled growth medium and the cytoplasmic RNA is purified by phenol extraction and analysed on sucrose density gradients, one obtains labelled RNA which peaks between 4 and 8S. Transfer RNA (tRNA) sediments at 4S. The base composition of pollen tube 4–8S RNA is very high in adenylic acid (A) and is quite different from that expected for tRNA. The following experiments were performed to demonstrate whether or not some of the material sedimenting between 4 and 8S was tRNA which was being masked by other species of RNA of similar size, with a high A content.

When 2–8S ³H-uridine-labelled RNA was analysed on a methylated albumin kieselguhr column (MAK) with *E. coli* tRNA. marker, radioactivity was found in the region of tRNA elution. The absorbance and radioactivity profiles, however, did not coincide, which indicated a probable absence of tRNA synthesis. Most of the labelled material was found to elute at higher concentrations of NaCl than required for tRNA elution.

When similarly sedimenting ³²P-labelled cytoplasmic RNA was analysed on a MAK column, the radioactivity seemed partly to follow the tRNA absorbance profile, although most of the radioactivity eluted later than the tRNA. Base analyses after alkaline hydrolysis and high-voltage electrophoresis of fractions from the MAK column separation are shown in Table 1A. The base analyses of fractions eluting between 0·3 and 0·5M MaCl are high in A and C (cytidylic acid), and low in guanylic acid (G). To compare these analyses with those of tRNA, inflorescence cuttings of *Tradescantia* were labelled in [³²P] nutrient solution for 48 h. The cytoplasmic

* Abstract.

230

RNA from whole anthers of young flower buds was purified and analysed on sucrose gradients. RNA sedimenting between 1 and 9S was pooled and analysed on a MAK column, and fractions from the MAK separation were analysed for their base composition (see Table 1B). Transfer RNA has a low A content and comparatively high and equal content of G and C.

The high C content indicated that some of the ^{32}P label was probably being incorporated into the terminal –CCA ends of pre-existing tRNA molecules. It has been computed that after alkaline hydrolysis of tRNA so labelled, the label in the hydrolysate should be in the ratio 10:3:1:1 for C:A:G:U (Glisin, V. R. and Glisin, M. V., *Proc. natn. Acad. Sci. U.S.A.*, **52**, 1548–53 (1964)). The base analyses of pollen tube RNA do not agree with this ratio. As the C content is high, however, some of the label is probably due to end-labelling of already existing tRNA. The high A content, however, cannot be explained on the basis of either end-labelling or complete labelling of tRNA.

Table 1

RNA eluting between	A. Pollen tube RNA % ^{32}P counts in				B. Anther RNA % ^{32}P counts in			
	C	A	G	U	C	A	G	U
0·3M NaCl→0·4M	33·8	26·0	17·3	22·9	26·2	23·9	26·1	23·8
0·4M NaCl→0·5M	30·2	40·7	12·5	16·6	25·1	24·6	24·6	25·7
0·5M NaCl→0·6M	23·5	29·5	20·8	26·2	21·6	27·3	23·5	27·4

From these data it is concluded that in the growing pollen tube of *Tradescantia paludosa* no tRNA is synthesised.

This work was supported by a grant from Research Foundation of the State University of New York.

In vitro Synthesis of Proteins by Polysomes from Germinating Petunia Pollen*

H. F. Linskens, *Department of Botany, University, Nijmegen, The Netherlands*

In order to analyse specific proteins which originate during the interaction of pollen tubes with the style tissue in the case of incompatible pollination, efforts were made to establish a cell-free protein synthesis system by making use of the ribosomal fraction from pollen grains. Following synchronised mass germination for different periods of time, polysomes were extracted in a 50 mM TRIS–HCl buffer (pH 7·6) with 10 mM NH$_4$Cl, 5 mM Mg acetate and 300 mM sucrose. After discontinuous sucrose gradient centrifugation and cleaning, the fraction was layered on to a linear sucrose gradient (10–30%) in TRIS buffer with ammonia and acetate as above.

Attempts to use extracts of germinating pollen as an enzyme system for *in vitro* protein synthesis were unsuccessful because of the high RNase concentration of this plant material. The 10 500 g supernatant from rat liver was found to perform satisfactorily. The incubation medium contained, besides the polysomes from pollen and the enzymes from the animal system, a mixture of amino acids, pyruvate kinase, ATP, GTP and phosphoenolpyruvate tricyclohexyl ammonium salt.

Ungerminated pollen contains only monosomes which show no incorporation activity of labelled amino acids into protein. Formation of polysomes starts immediately with the germination and is accompanied by a decrease of the monosomal pool. Amino acid incorporation reaches its maximum about 60 min after the start of germination. At that moment about 90% of the germinable pollen grains have germinated. The instantaneous activation of protein synthesis enables pollen germination to proceed at a stage at which pollen genome is not yet activated.

* Abstract.

The rapid synthesis of polysomes suggests the presence of a preformed pool of a kind of masked messenger RNA in the resting pollen grain.

In vitro protein synthesis of the mixed plant–animal system shows the protein-synthesising machinery on 80S ribosomes of plant cells to be identical with that of animal cells.

The next step is to be an investigation of the specific proteins formed based on the various genomes in pollen and after interaction of different S-alleles during and after the incompatible interaction.

The Production of Haploid Embryos from Pollen Grains*

J. P. Nitsch, *Laboratoire de Physiologie Pluricellulaire, C.N.R.S., Gif-sur-Yvette, France*

INTRODUCTION

While culturing *Datura* stamens *in vitro*, Guha and Maheshwari [2] observed the formation of embryo-like structures, the cells of which were found to be haploid and which, presumably, were derived from microspores. These authors, however, did not grow such 'embryos' into mature plants. In 1967 we observed the same phenomenon with stamens of *Nicotiana*. Plantlets were obtained from the cultures and grown to the flowering stage in the greenhouse. Chromosome counts made from the root tips of these plants gave in all cases the haploid number. Since then we have made extensive studies on the origin, the cultural conditions and the possible applications to experimental embryology and to genetics of the process of causing microspores to develop into embryos instead of pollen grains [3–6]. A summary of the results obtained is presented here.

ORIGIN OF THE EMBRYOS

Stage of development. Extensive trials with both *N. tabacum* and *N. sylvestris* have shown that the most important point in the technique is to excise the stamens at the proper stage of development. Stamens planted at the tetrad stage or when starch was being produced in the pollen grains failed to produce embryos. The crucial stage occurs at the moment microspores are nearing the first mitosis. This stage can be recognised from the development of the petals in the floral bud.

Origin of the embryos. The earliest event we have seen until now

* Abstract.

234

is the formation of pollen grains with three nuclei. With the subsequent increase in cell number and volume, the exine bursts and liberates round masses of cells which represent the globular stage of young embryos. Soon a radicle end becomes apparent, and cotyledons begin to differentiate at the other extremity. The 'heart', 'torpedo' and 'cotyledonary' stages follow as in seed embryos. Germination takes place when the embryos become responsive to geotropism, when the cotyledons synthesise chlorophyll and the radicle elongates.

Haploidy. Chromosome counts made on root tips of either plantlets or adult plants have always given the haploid number, that is, $n = 24$ for *N. tabacum* and $n = 12$ for *N. sylvestris*.

CULTURAL CONDITIONS

The minimal medium. The minimal medium is quite simple: 2% sucrose solidified by $0.8-1\%$ agar. On this medium a relatively low percentage of stamens form embryos, and these embryos do not develop beyond the globular stage.

The optimal medium. The optimal medium consists, in addition to 2% sucrose, of the mineral salts and the vitamins indicated by Nitsch and Nitsch [3]. The absolute necessity of all these ingredients has not yet been checked individually. Ammonium ions (which are present in the medium) are not indispensable. Indole-3-acetic acid at $50-100$ $\mu g/l$ is beneficial, but not necessary.

Effect of growth substances and nucleic acid derivatives. Neither cytokinins nor gibberellins had any beneficial effect. The first tended to cause the formation of undifferentiated calli instead of plantlets. Purines and pyrimidines, alone or in combination, tended to be inhibitory rather than beneficial. Various amino acids or amides gave no improvement over the medium without them.

Role of iron. One key element for the complete development of the embryos was iron. Without iron they did not grow beyond the globular stage. Zinc, manganese or other minor elements could not substitute for iron.

Iron-deficiency in higher plants is known to bring about an abnormally high level of arginine. In fact, when arginine was added to the optimal medium (around 10^{-3} M), production of embryos was nearly suppressed. This effect occurred at a particular stage, namely the second week of culture. If 3×10^{-3} M arginine was given to the cultures during the first or the third or fourth weeks only, embryo development occurred normally. Thus it is not arginine as such but the presence of arginine at a particular stage of development which blocks embryo formation. It should be noted that high levels of arginine-rich histones occur in the cytoplasm of the pollen grain soon after mitosis [7].

Effect of abscisic acid. Abscisic acid (10^{-5} M) does not suppress development, at least until the cotyledonary stage, but it slows it down and prevents germination. Thus embryos remained small and chlorophyll-less for months when abscisic acid was present. They turned green and germinated when transferred to a medium without abscisic acid. As with dormant buds such as those of the potato, glutathione (10^{-3} M) is able to reverse the effect of abscisic acid.

APPLICATIONS

Experimental embryology. The production of free embryos in large quantities will enable embryologists to experiment more conveniently than hitherto. For example, we have started to study the morphogenesis of cotyledons and obtained embryos with no cotyledons (with 10^{-3} M lysine), with 3–4 cotyledons, with misshapen cotyledons or with cotyledons fused in a ring which developed into a cup (with morphactin, for example).

Mutants. By irradiating haploid plantlets with a cobalt source, various mutants have been obtained, e.g. plants with narrow leaves, with malformed leaves, with petals that were more separated than in normal flowers or which had a different colour (stronger or lighter pink, or white colour in the case of *N. tabacum*).

Diploidisation. Taking advantage of the endomitosis which goes on in tissue cultures, we have doubled the chromosome number of many of the haploid plants obtained, by culturing *in vitro* stem or petiole fragments. Homozygous strains of the mutants have thus been obtained, a method which opens new horizons to the genetics of higher plants.

REFERENCES

1 BOURGIN, J. P. and NITSCH, J. P., *Annls Physiol. vég., Paris*, **9**, 377–82 (1967)
2 GUHA, S. and MAHESHWARI, S. C., *Nature, Lond.*, **212**, 97–8 (1966)
3 NITSCH, J. P. and NITSCH, C., *Science, N.Y.*, **163**, 85–7 (1969)
4 NITSCH, J. P., NITSCH, C. and HAMON, S., *C.r. Séanc. Soc. Biol.*, **162**, 369–72 (1968)
5 NITSCH, J. P., NITSCH, C. and HAMON, S., *C.r. hebd. Séanc. Acad. Sci., Paris*, **269** (D), 1275–8 (1969)
6 NITSCH, J. P., NITSCH, C. and PÉREAU-LEROY, P., *C.r. hebd. Séanc. Acad. Sci., Paris*, **269** (D), 1650–2 (1969)
7 SAUTER, J. J., *Z. Pfl. Physiol.*, **60**, 434–49 (1969)

Pistil–Pollen Interactions

Pistil–Pollen Interactions in *Lilium*

W. G. Rosen, *State University of New York, Buffalo, New York, U.S.A.*

'The gradual progress of the tube down the conducting tissue of the style appears to be attended by the absorption of nourishment as it passes, for in the case of such flowers as the lily, the length of the tube is far too great for its cellulose to be supplied from the comparatively small store of carbohydrate in the pollen grain itself. We must look, therefore, to the tissue of the style as the seat of some metabolism, having for its purpose the feeding of the pollen tube during, at any rate, the latter part of its growth.' So wrote Green in 1894 [9].

THE LILY PISTIL

Although variations and combinations are known, flower pistils can be roughly divided into those possessing a canal through which the pollen tubes grow to the ovules and those lacking such a canal. In pistils possessing a canal the cells lining the canal have been referred to most frequently as *epidermal* or *stigmatoid* cells. In pistils lacking a canal the tissue through which the pollen tubes grow is distinct from the surrounding tissue and has been termed *transmitting tissue*, because of its function in transmitting the pollen tubes on their way to the ovary, or *stigmatoid tissue*, on the basis of its presumed functional and morphological relationship and resemblance to the cells of the surface of the stigma itself.

To the best of my knowledge, studies of hollow pistils, at least in recent years and with the aid of electron microscopy, have been limited to lilies.

The stigma in lilies is three-lobed, each lobe being divided by a suture (Fig. 1). In the mature pistil the stigma surface is covered with viscous, opalescent exudate. The exudate has a dramatic promotive effect on pollen germination and tube growth, but in our early and crude assays of chemotropism it appeared to be chemo-tropically inactive [25]. The surface of the stigma itself is covered by numerous papillae, each consisting of from one to several cells (Fig. 2). Beneath the papillae is a several-layered zone of par-enchymatous cells which form the domed roof of the bulbous chamber at the top of the triradiate canal of the style. At the ovary the canal separates into three individual canals, one in each lobe.

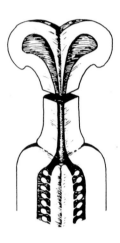

Figure 1. Semi-diagrammatic representation of the pistil of Lilium longiflorum. *The long style has been fore-shortened.* ×7

From the bulbous upper end in the stigma, through the style and into the separate canals of the ovary, this passage is lined by unique cells, termed *epidermal* cells by Yamada [37], and *stigmatoid* cells by Welk, Millington and Rosen [34], adhering to the terminology of Esau [8]. These cells, which line the entire route traversed by the pollen tubes after they leave the stigma surface, are distinct from other cells of the pistil in that they are smaller, more densely cyto-plasmic, and singularly rich in polysaccharide, protein and RNA. Their inner wall, that facing the canal, is strikingly thicker than the walls which abut the neighbouring cells (Fig. 4).

The cells of the stigma surface and the cells which line the canal are secretory. The secretion products of these cells provide the medium through which the pollen tubes grow from the stigma to the embryo sacs. Thus, when we consider pistil–pollen interactions, we are in fact considering interaction between the secretory products of these cells and the pollen tubes. The present discussion will therefore first consider cells which are the source of these secretions, and will then turn to an examination of the various roles of the secretion products of these cells in pollen tube growth.

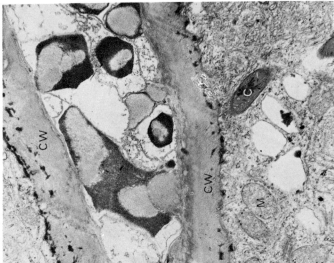

Figure 2. Section (ca. 1·0 μm) through a stigma of L. longiflorum showing part of the stigma surface, with papillae, and a portion of the top of the canal. Fixed (as in Fig. 3) for electron microscopy but stained with methylene blue and azure II for viewing in the light microscope. Original magnification: ×265. Reproduced at ⁴⁄₅

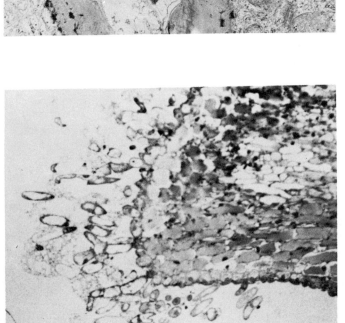

Figure 3. Section through cells of two adjacent papillae on the stigma surface, 0–1 day after anthesis. The space between the walls of the two cells is occupied by what is presumed to be fixed stigmatic exudate. Cell wall, CW; chloroplast, C; mitochondrion, M; elements of endoplasmic reticulum can also be seen. Cells fixed in phosphate-buffered 6% glutaraldehyde, post-fixed in osmium tetroxide, both pH 7·3; stained with uranyl acetate and lead hydroxide, 2 min each. Original magnification: ×12 100. Reproduced at ⁴⁄₅

FINE STRUCTURE OF THE SECRETORY CELLS

The cells of the stigma surface (Fig. 3) have lateral walls which are relatively thick and densely textured. External to the walls is a complexly structured, foamy-appearing material which is presumed to be fixed exudate. Dense zones are seen in the walls which may represent material in the process of transfer to the outside. The interior surface of the walls is often irregularly thickened. The cytoplasm is rich in mitochondria, smooth and rough endoplasmic reticulum, and free ribosomes. Poorly developed but densely staining chloroplasts are occasionally seen. The abundant dictyosomes appear to be actively producing vesicles from the ends of their cisternae.

The cells which line the canal display much greater structural complexity than the cells of the stigma surface. As mentioned, their most striking feature is the highly elaborated and thickened wall which faces the canal (Figs. 4 and 5). The differences in morphology in the cells of these two regions are sufficient to recommend against the use of the term *stigmatoid* for describing the cells which line the canal. For this reason, we propose to refer to the latter as *canal cells*. Jensen and Fisher [13] also recognised the need for this distinction in cotton, and recommended *transmitting tissue* in preference to *stigmatoid tissue* or other terms. Consistent with the cytochemical observations [34], the canal cells are densely cytoplasmic and give the appearance of intense metabolic activity. They are rich in smooth and rough endoplasmic reticulum, free ribosomes, polysomes, mitochondria, dictyosomes, microtubules and lipid droplets. In addition, they have large nuclei and often contain starch grains and chloroplasts. Although they are vacuolated, they possess a relatively much greater volume of cytoplasm than the underlying parenchymal cells.

The most striking features of these cells are (a) the very elaborate wall on the side of the cell facing the canal, which, by inference from its appearance and presumed function, we have termed the *secretion zone*, and (b) the numerous inclusions in wall, vacuole and cytoplasm which are not readily identified as familiar types of organelles or other structures.

Just beneath the loosely textured wall of the secretion zone is an elaborate spongy zone of material of various textures and densities. Beneath this zone one finds the plasmalemma, but this membrane is difficult to trace and the cytoplasm appears to interpenetrate the spongy zone (Fig. 6). Aggregates of vesicles and tubules are encountered in the region between the plasmalemma and the spongy zone. These aggregates, which may be similar to the

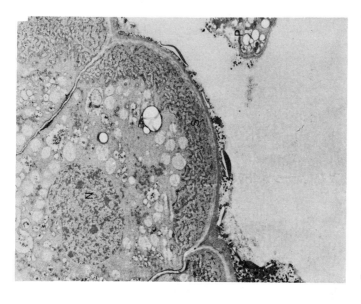

Figure 5. Canal cells, L. longiflorum style. Fixation and staining as in Fig. 3. Nucleus, N; secretion zone, S; portion of a pollen tube, P. Original magnification: ×400. Reproduced at ⅘

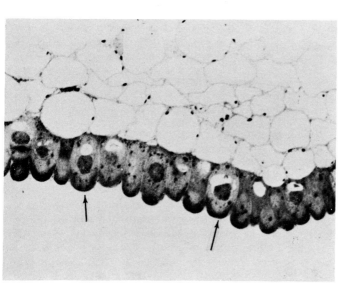

Figure 4. The canal cells from the style of L. longiflorum. Preparation as in Fig. 3. Note dense staining of the canal cells, their large nuclei, and the thickened wall of the secretion zone (arrows). Original magnification: ×425. Reproduced at ⅘

plasmalemmasomes or lomasomes reported in fungi and higher plants, are also seen in between the plasmalemma and lateral walls of these cells, as well as in the cytoplasm and in vacuoles. Until more is known about their origin and function it seems well to refer

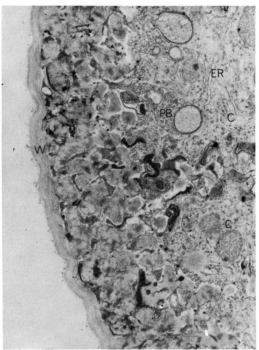

Figure 6. Higher magnification of a portion of a canal cell, showing some detail of the secretion zone. Note irregular and indistinct boundary between cytoplasm (C) and wall (W); paramural bodies (PB); endoplasmic reticulum (ER). Original magnification: × 18 960. *Reproduced at* $\frac{4}{5}$

to them as *paramural bodies*, a term suggested by Marchant and Robards [21] in their recent summary of the distribution of such puzzling structures.

ORIGIN, FUNCTION AND COMPOSITION OF THE SECRETORY PRODUCTS

Origin

Does the exudate which we observe on the stigma originate there, or is it produced in the canal and transported upwards? Conversely, what is the origin of the exudate which we encounter in the stylar canal after pollination? Although it would probably take rather sophisticated experiments to get direct answers to these questions, we can draw tentative conclusions from indirect evidence. The stigma surface becomes covered with exudate before pollination, while the canal appears to be empty until pollination has triggered the release of secretion product. Thus we can conclude that the exudate on the stigma surface is produced by the cells of the surface [29, p. 185]. Pollen tubes appear to be strongly attracted by papillae, which suggests that these cells are the source of chemotropic activity (Fig. 7). Since chemotropic activity appears first in the stigma and upper style, and is encountered in successively lower portions of the pistil as the pistil matures [34], it would seem that the cells lining the canal are the source of the secretion product which is encountered in the canal and of the chemotropic activity of stylar tissues.

What is the relation between the secretory function and the fine structure of the cells which appear to be the source of the secretion product? We have searched for fine-structural evidence of the functional differences between secreting and non-secreting canal cells, between cells of pollinated and non-pollinated pistils, and between self- and cross-pollinated pistils. Regrettably, we can see no clear difference between these various conditions. We do, however, notice some suggestive trends. There appears to be an increase in the thickness and complexity of the secreting surface with increasing age of the pistil, from one or two days before anthesis to 3–4 days after the flower opens. Further, there is some indication that when it is first elaborated the secretion product is retained beneath a thin cuticle which later either dissolves or ruptures, much as in the situation described by Konar and Linskens [14] for the stigma surface in *Petunia*.

Yamada [37] has reported the occurrence of 'colloidal bodies' in the parenchyma cells of mature *Lilium longiflorum* pistils, especially in the region of the stigma. He observed these bodies by light microscopy in pistils fixed in formalin–acetic acid–alcohol or neutral formalin. The colloidal bodies stained with aniline blue and haematoxalin, as did the secretory material which filled the stylar canal after pollination, and he suggested that the material somehow moved from the parenchyma cells into the canal, where it accumulated on the secreting surface of the canal cells. Yamada suggested that the canal cells are somehow active in the absorption and transfer of the colloidal material from the parenchyma cells to the canal, but it is not clear whether he considers that the canal cells actively secrete this material or merely accumulate it on their surfaces.

The colloidal bodies are readily identified in thick sections of plastic embedded material (Fig. 2). In electron micrographs they appear as fibrous inclusions in apparently disintegrating protoplasts (Fig. 8).

Roles of pistil secretion products

Elsewhere in this symposium Martin and Brewbaker (p. 262) discuss the various roles which have been attributed to the stigmatic exudate. These range from exclusion of fungi, bacteria and foreign pollen from growth on the pistil, to promotion, in various ways, of the growth of the native pollen. I will limit myself in the present discussion to those aspects of pistil–pollen interactions which relate directly to pollen tube growth. They are as follows.

(1) *Chemotropism.* Pollen tubes show a positive directional growth response to tissues of the pistil. This positive tube chemotropism is generally assigned a role in the guidance of the pollen tube to the embryo sac.

(2) *Incompatibility.* By definition this is a phenomenon requiring a juxtaposition of male and female tissue for its expression. There can be no incompatibility reaction unless there are interactions or blocked interactions between pollen and pistils.

(3) *Growth regulation.* The pollen tubes of many species are incapable of growth *in vitro* sufficient to permit them to cover the distance between stigma surface and ovule *in situ*—at least with culture methods currently employed. It is therefore inferred that when growing *in situ* the pollen tube must derive at least a portion of its sustenance from the female tissue. The pistil may simply provide substrates for energy metabolism, but other possibilities must also be considered. These include essential minerals, specific structural components or their precursors, enzymes and hormones.

Chemotropic activity

Initially, Rosen [25] found that chemotropic activity was present in stigma and style, weak or absent in the ovary, and absent from the stigmatic exudate of *L. longiflorum.* Using a more sensitive assay, Welk *et al.* [34] found that chemotropic activity could be demonstrated to all tissue fragments containing secretory cells, including those of the ovary, as well as to the stigmatic exudate of *L. leucanthum* and *L. regale.* Chemotropic activity was absent in tissue fragments lacking secretory cells. Using a more sensitive bioassay,

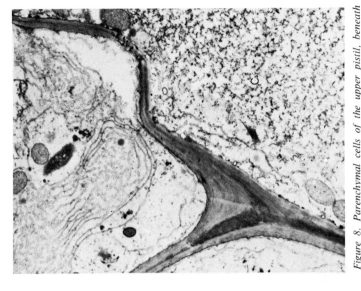

Figure 8. Parenchymal cells of the upper pistil, beneath stigma surface, showing presumed colloidal bodies (C). Original magnification: ×12100. Reproduced at $\frac{4}{5}$

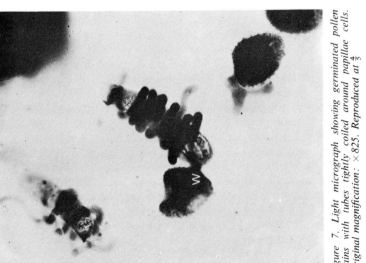

Figure 7. Light micrograph showing germinated pollen grains with tubes tightly coiled around papillae cells. Original magnification: ×825. Reproduced at $\frac{4}{5}$

we have subsequently observed chemotropic activity to stigmatic exudate of *L. longiflorum*. It thus appears safe to conclude that the canal cells and the cells of the stigma surface are the sources of chemotropic activity, and that this activity is found in the secretion products of these cells. Whether the colloidal bodies discussed in the preceding section provide precursors for the synthesis of chemotropically active substances remains to be determined.

At the present time the nature of the chemotropic agent remains unknown. Rosen [25] described a procedure for the preparation of a chemotropically active cell-free extract from lily pistils. The extract was partially purified by Rosen and Kress [29], but identification of the active components has been hampered by the lack of a bioassay which is quantitative and which clearly distinguishes between chemotropism and non-directional growth promotion [26, 27].

Calcium (Ca^{++}) is chemotropically active for *Antirrhinum* pollen [22] but is chemotropically inactive for *Lilium* [26] and for *Oenothera* (p. 255). The speculation of Mascarenhas and Machlis [22] that Ca^{++} might prove to be a universal chemotropic agent in flowering plants thus seems not to have been fulfilled.

Incompatibility

The mechanism of the self-incompatibility reaction is not yet known, but there is ample evidence that it involves at least one, perhaps many, pistil–pollen interactions. Geneticists describe two distinct incompatibility systems. Sporophytic incompatibility is associated with species which shed trinucleate pollen grains and involves inhibition of germination on the stigma or interference with the early stages of tube growth while the tube is still in the stigma. Gametophytic incompatibility, the type found in the lilies, occurs primarily in species with binucleate pollen and is manifested during growth of the tubes through the style. In the incompatible pistils the growth rate of the pollen tubes is slower, or growth stops before the tubes reach the ovary, or both. The inhibited tubes may show morphological aberrations such as branching or swelling, particularly of the tube tips, thickening of the walls, increased deposition of callose, and changes in behaviour of the nuclei.

Interspecific crosses within the genus *Lilium* result in a number of gross abnormalities in pollen tube morphology [3]. In contrast, incompatible (self-) pollination in *L. longiflorum* results in changes in tube morphology which have only been detected with the aid of the electron microscope.

A few years ago Rosen and Gawlik [28] reported that tubes growing *in vitro* and in incompatible styles are characterised by a compartmented cap over their growing tips; compatible tubes appear to undergo a transition from compartmented cap to deep embayments at the tips. The embayments resemble, in texture and stainability, the material in the spongy zone of the secreting surface of the canal cells. Rosen and Gawlik suggested that compatible tubes take up secretion product for continued growth, whereas incompatible tubes and tubes growing in culture are unable to switch from autotrophic to heterotrophic nutrition.

We have attempted to explore this possibility through the use of electron-opaque and, more recently, radioactive markers. First we attempted to demonstrate the uptake of material from the stylar canal into the embayments by injecting opaque markers such as saccharated iron and ferritin into the stylar canal. These efforts failed because the tubes stopped growing as soon as they reached the injected material, even when it consisted merely of the sucrose and boron on which they seem to thrive *in vitro*.

In these proceedings Ascher and Drewlow (p. 267) report that under appropriate conditions the injection of stigmatic exudate before pollination stimulates the growth of both compatible and incompatible tubes, and they suggest that exudate can be used as a carrier for other substances. Our experiments will therefore be repeated with exudate as diluent for electron-dense markers.

The incompatibility barrier in the Easter Lily disappears about five days after floral anthesis [1]. It can be abolished by subjecting the pistils to heat [2] or to X-radiation [12]. These properties give added support to the suggestion that the barrier is a substance produced by the pistil which might be expected to be secreted in the exudate.

It is therefore frustrating to have to report that we have been unable to detect differences in the growth of pollen *in vitro* in exudate from compatible as opposed to incompatible pistils, even when we took pains to collect the exudate only during the first 5 days in order to ensure that we were working with exudate from pistils which were fully self-incompatible. There are at least three rather likely ways of explaining these results if one recalls that the incompatibility reaction normally manifests itself not on the stigma but in the style, and not at the outset of tube growth but after growth has proceeded for some time: (1) possibly the pollen tube is indifferent to the substances causing incompatibility for as long as it has sufficient reserves to draw upon for its continued growth; (2) the substances causing the incompatibility reaction may be present only in the secretion product of the canal cells and absent in the stigma exudate;

(3) the incompatibility factors may be produced only in response to self-pollination. If any of these assumptions is correct, one should not expect to detect an incompatibility reaction with freshly germinating pollen growing in stigmatic exudate.

Growth regulation

From *in vitro* experiments with labelled myo-inositol (MI) we knew that tubes are able to incorporate this pectin precursor into

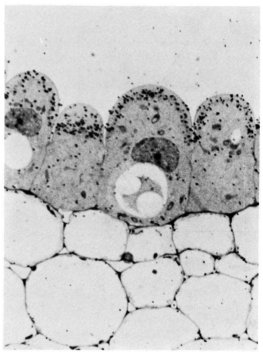

Figure 9. Radioautograph of canal cells from pistil fed 48 h with 50 μCi myo-inositol-²H; fixation and staining as in Fig. 2; Kodak NTB 2 emulsion. Note accumulation of grains over secretion zone. Original magnification: × 1060. Reproduced at ⅘

their walls at the tip [6]. We therefore proceeded to label the pistils and, hopefully, the exudate, by feeding labelled MI to cut pistils. The label accumulated dramatically in the secreting surface of the canal cells (Fig. 9), as well as in the papillae of the stigma surface and the surrounding exudate.

In collaboration with Dr. F. Loewus and his group we have undertaken a biochemical analysis of the role of the exudate in tube nutrition [15]. The lily stigmatic exudate consists primarily of carbohydrate and protein, which comprise approximately 70% and 7% of its dry weight, respectively [17]. Label from MI appears in the exudate, where it is found in both a high and a low molecular weight fraction [18]. From the pistil the label enters the tubes and appears to accumulate most abundantly in their walls, near the tips [16]. Label is accumulated by tubes in compatible as well as incompatible pistils. By itself, however, this does not obviate the suggestion that compatible pistils permit the pollen to switch to more heterotrophic nutrition. Tubes in incompatible crosses may be taking up label from the low molecular weight component of the exudate by diffusion only, while tubes in compatible pistils may be able to incorporate large molecular precursors of their wall substance via the embayments. Loewus and his associates are attempting to identify the labelled components of the exudate.

Our thinking regarding the role of the exudate in tube growth must not be restricted to a consideration of wall carbohydrates (polysaccharides). Boron has been known for many years to be a powerful stimulant of the growth of pollen tubes, and Thomas [31] has shown that this element is accumulated in the pistils of *Lilium*.

Calcium is often cited as playing a major role in pollen tube growth. In *Lilium* Dickinson [7] reported that this ion is necessary to prevent the leakage of enzymes from pollen tubes growing *in vitro*. But whereas Ca^{++} mimics the population effect in stimulating the germination of numerous species [4] and acts as a chemotropic factor in *Antirrhinum* [22], we find it has neither of these actions in the Easter Lily [26].

Yamada [35] has shown that lily pollen growth *in vitro* is stimulated by cobalt, and that this ion accumulated in pistils, from which the growing tubes can apparently accumulate it [36]. Umebayashi [33] has shown that Co^{++} activates an aminoacylase which is found in pollen. The stigma exudate of Easter Lily contains a considerable amount of γ-aminobutyric acid in a bound form which seems to be an *N*-acylated or peptide-like derivative, and which may be the substrate for this enzyme (M. Umebayashi, Personal communication).

Differences in amino acid levels before and after pollination [32] and between compatible and incompatible pollinations [20] indicate that amino acid metabolism and protein synthesis comprise an important aspect of pistil–pollen interactions.

γ-Aminobutyric acid [32] and proline [5] are among the amino acids whose levels change significantly in response to pollination.

The recent interest in cell wall proteins as possible regulators of cell elongation, sparked largely by Lamport's studies of the wall protein [19] which he terms 'extensin', reminds us that amino acids may function not merely in energy metabolism but also as specific factors in wall growth. Hydroxyproline is an important component of extensin, where it arises from hydroxylation following incorporation of proline into a precursor or component of the wall which appears to be a glycopeptide. In another paper in this volume Dashek, Harwood and Rosen (p. 194 report the apparent regulation of lily pollen tube growth by a hydroxyproline-containing glycopeptide [11]. If this scheme can be confirmed, there is every reason to suspect that it is operative *in situ* as well as *in vitro*.

To the extent that pistil–pollen interactions are related to enzymatic activity it would appear that either (*a*) the pollen secretes enzymes which act on substrates in the pistil or (*b*) the pistil may provide an activator for a pollen enzyme (which in turn may act upon a pistil substrate). Evidence for pistil enzymes acting on pollen substrates seems to be lacking.

Roggen and Stanley [24] have concluded that cellulase and pectinase are important in wall extension in pear pollen. In *Hemerocallis*, however, the same authors found that the changes in cell wall hydrolytic enzymes of the pistil were independent of pollen growth.

Roggen [23] described increases in the levels of activity of a number of enzymes in the styles of *Petunia hybrida* after pollination. His data must be treated with caution, however, since while he removed the stigmas and their adhering pollen before assaying the styles for enzyme activity, the styles were extracted along with the pollen tubes which they contained. This otherwise excellent procedure might be applied to pistils such as that of the lily, from which the pollen tubes can be removed without much difficulty.

The role of growth hormones in pistil–pollen interactions related to fruit ripening is well known. There may also be a role in relation to tube growth. Search and Stanley [30] claim that ethylene stimulates the growth of pollen tubes (species not mentioned) albeit at higher optimal concentrations than have been reported for other tissues. Ethylene is reported to be released from flowers, particularly from stigmas and styles, following their treatment with growth regulators [10]. From this, and from what is previously known about ethylene production, it is not unreasonable to postulate that ethylene may be yet another factor contributed by the pistil to the regulation of growth of the pollen tube.

Work from the author's laboratory was supported by a grant (GB 5402) from the National Science Foundation.

REFERENCES

1 ASCHER, P. D. and PELOQUIN, S. J., 'Influence of temperature on incompatible and compatible pollen tube growth in *Lilium longiflorum*', *Can. J. Genet. Cytol.*, **8**, 661–4 (1966)

2 ASCHER, P. D. and PELOQUIN, S. J., 'Effect of floral aging on the growth of compatible and incompatible pollen tubes in *Lilium longiflorum*', *Am. J. Bot.*, **53**, 99–102 (1966)

3 ASCHER, P. D. and PELOQUIN, S. J., 'Pollen tube growth and incompatibility following intra and inter-specific pollinations in *Lilium longiflorum*', *Am. J. Bot.*, **55**, 1230–4 (1969)

4 BREWBAKER, J. and KWACK, B. H., 'The essential role of calcium ion in pollen germination and pollen tube growth', *Am. J. Bot.*, **50**, 859–65 (1963)

5 BRITIKOV, E. A., VLADIMIRSETAVA, S. V. and MUSATAVA, N. A., 'Transformation of proline in germinating pollen', *Fiziologiya Rast.*, **12**, 839–50 (1965)

6 DASHEK, W. V. and ROSEN, W. G., 'Electron microscopical localization of chemical components in the growth zone of lily pollen tubes', *Protoplasma*, **56**, 192–204 (1966)

7 DICKINSON, D. B., 'Permeability and respiratory properties of germinating pollen', *Physiologia Pl.*, **20**, 118–27 (1967)

8 ESAU, K., *Plant Anatomy* (2nd edn.), Wiley, New York (1965)

9 GREEN, J. R., 'Researches on the germination of the pollen grain and the nutrition of the pollen tube', *Phil. Trans. R. Soc. B*, **185**, 385–409 (1894)

10 HALL, I. U. and FORSYTH, F. R., 'Production of ethylene by flowers following pollination and treatments with water and auxin', *Can. J. Bot.*, **45**, 1163–6 (1967)

11 HARWOOD, H. I., DASHEK, W. V. and ROSEN, W. G., 'Does "Extensin" regulate pollen tube growth?: Studies with hydroxylation inhibitors', *Pl. Physiol., Lancaster*, **44**, (Suppl.) (1969)

12 HOPPER, J. E. and PELOQUIN, S. J., 'X-ray inactivation of the stylar component of the self-incompatibility reaction in *Lilium longiflorum*', *Can. J. Genet. Cytol.*, **10**, 941–4 (1968)

13 JENSEN, W. A. and FISHER, D. B., 'Cotton embryogenesis: The tissues of the stigma and style and their relation to the pollen tube', *Planta*, **84**, 97–121 (1969)

14 KONAR, R. N. and LINSKENS, H. F., 'The morphology and anatomy of the stigma of *Petunia hybrida*', *Planta*, **71**, 356–71 (1966)

15 KROH, M., and MIKI-HIROSIGE, H., 'Conversion of myo-inositol to pectic substance of pollen tubes', *Pl. Physiol., Lancaster*, **43** (Suppl.) 1, S-52 (1968)

16 KROH, M., MIKI-HIROSIGE, H., ROSEN, W. and LOEWUS, F., 'Incorporation label into pollen tube walls from myo-inositol labeled *Lilium longiflorum* pistils', *Pl. Physiol., Lancaster*, **45**, 92–4 (1970)

17 LABARCA, C., KROH, M., CHEN, M. and LOEWUS, F., 'Studies on stigmatic exudate from *Lilium longiflorum*', *Pl. Physiol., Lancaster*, **44** (Suppl.), 2 (1969)

18 LABARCA, C., KROH, M. and LOEWUS, F., 'Pectic nature of pistil exudate in *Lilium longiflorum*', *Fedn Proc. Fedn Am. Socs exp. Biol.*, **28**, 869 (1969)

19 LAMPORT, D., 'The protein component of primary cell walls', *Adv. Bot. Res.*, **2**, 151–218 (1965)

20 LINSKENS, H. and TUPÝ, J., 'The amino acids pool in the style of self-incompatible strains of *Petunia* after self- and cross-pollination', *Züchter*, **36**, 151–8 (1966)

21 MARCHANT, R. and ROBARDS, A. W., 'Membrane systems associated with the plasmalemma of plant cells', *Ann. Bot.*, **32**, 457–71 (1968)

22 MASCARENHAS, J. P. and MACHLIS, L., 'Chemotropic response of the pollen of *Antirrhinum majus* to calcium', *Pl. Physiol., Lancaster*, **39**, 70–7 (1964)

23 ROGGEN, P., 'Changes in enzyme activities during the programme phase in *Petunia hybrida*', *Acta. bot. neerl.*, **16**, 1–31 (1967)

24 ROGGEN, P. and STANLEY, R. G., 'Cell-wall hydrolysing enzymes in wall formation as measured by pollen-tube extension', *Planta*, **84**, 295–303 (1969)

25 ROSEN, W. G., 'Studies on pollen-tube chemotropism', *Am. J. Bot.*, **48**, 889–95 (1961)

26 ROSEN, W. G., 'Chemotropism and fine structure of pollen tubes'. In: H. F. LINSKENS (Ed.), *Pollen Physiology and Fertilization*, North-Holland, Amsterdam (1964)

27 ROSEN, W. G., 'Ultrastructure and physiology of pollen', *A. Rev. Pl. Physiol.*, **19**, 435–62 (1968)

28 ROSEN, W. G. and GAWLIK, S. R., 'Relation of lily pollen tube fine structure to pistil compatibility and mode of nutrition', *Electron Microscopy*, **2**, 313 (1966) (Proc. Int. Conf. Electron Microsc., 6th, Kyoto (Maruzen, Tokyo))

29 ROSEN, W. G. and KRESS, L. F., 'Studies on pistil extracts causing pollen tube chemotropism in *Lilium longiflorum*', *Pl. Physiol., Lancaster*, **36** (Suppl.), XXVIII (1961)

30 SEARCH, R. W. and STANLEY, R. G., 'The effects of ethylene on pollen tube elongation', *Pl. Physiol., Lancaster*, **43** (Suppl.), S-52 (1968)

31 THOMAS, W. H., 'Boron contents of floral parts and the effects of boron on pollen germination and tube growth of lilium species', Master's Thesis, University of Maryland (1952)

32 TUPÝ, J., 'Investigation of free amino-acids in cross-, self- and non-pollinated pistils of *Nicotiana Alata*', *Biologia Pl.*, **3**, 47–64 (1961)

33 UMEBAYASHI, M., 'Aminoacylase in pollen and its activation by cobaltous ion', *Pl. Cell Physiol., Tokyo*, **9**, 583–6 (1968)

34 WELK, M. SR., MILLINGTON, W. F. and ROSEN, W. G., 'Chemotropic activity and the pathway of the pollen tube in lily', *Am. J. Bot.*, **52**, 774–81 (1965)

35 YAMADA, Y., 'The effect of cobalt on the growth of pollen I', *Bot. Mag., Tokyo*, **71**, 319–25 (1958)

36 YAMADA, Y., 'The effect of cobalt on the growth of pollen. II Differential acquisition of cobalt-60 in the style of *Lilium longiflorum*', *Bot. Mag., Tokyo*, **73**, 417–21 (1960)

37 YAMADA, Y., 'Studies on the histological and cytological changes in the tissue of pistil after pollination', *Jap. J. Bot.*, **19**, 69–82 (1965)

Can Ca^{++} Ions Act as a Chemotropic Factor in *Oenothera* Fertilisation?*

H. O. Glenk, W. Wagner and O. Schimmer, *Botanisches Institut der Universität, Erlangen-Nürnberg, West Germany*

From many studies we know that calcium ions play an important part in normal physiological processes of pollen such as germination, tube elongation and chemotropism of pollen tubes [1-4]. Furthermore the well-known pollen 'population effect' *in vitro* is overcome by Ca^{++} in many species [4] and Ca^{++} ions have a 'protective action' against some pollen growth inhibitors [1, 5].

During our investigations of pollen physiology in the genus *Oenothera* in the years 1957–58 we found that an addition of calcium chloride to the germination medium could enhance the germination rate of pollen from underdeveloped buds [6]. In later flowering periods an addition of calcium was needed in some cases for successful germination even of mature pollen grains [7].

As was proved in earlier papers [8], combinations of amino acids, sugars and K^+ ions are chemotropically active in *Oenothera* fertilisation. However, we do not know yet whether the chemotropism of *Oenothera* pollen tubes is influenced in any way by Ca^{++} ions, as is the case in *Antirrhinum* pollen tubes [4, 9]. The distribution of calcium in the tissues, especially in the gynoecium of *Oenothera*, is also still unknown.

For these reasons we carried out some investigations on chemotropism by Ca^{++} ions *in vitro*, and an analysis of calcium content in different parts of certain *Oenothera* species during the years 1963–69.

* Abstract.

255

IN VITRO EXPERIMENTS ON CHEMOTROPISM

In 14 series of experiments we tested eight different calcium compounds for their chemotropic activity on *Oenothera* pollen tubes on our gelatin medium 334 [6]. We worked with the test method of Schneider [10], which Rosen called the 'angle test' [11]. The calcium compounds tested and the results are shown in Table 1.

Table 1. CHEMOTROPISM TESTS WITH CALCIUM COMPOUNDS *in vitro* (ANGLE TEST)
(*Oenothera longiflora, O. argentinae, O. hookeri*; medium 334, no Ca^{++} added for germination; germination, 40–60%)

Ca compound	Maximum positive chemotropism (%)	Number of tests positive	total
$CaCl_2$	33+4	(1)	14
$Ca(NO_3)_2$	33+20	2	14
$Ca(H_2PO_4)_2$	33+2	(1)	12
Ca HPO_4		0	12
$Ca_3(PO_4)_2$	33+0	0	12
Ca SO_4	(at random)	0	14
Ca-lactate	33+25	2	14
Ca-citrate	33+0	0	10
	(at random)		
	Total	4 (6)	102

In only very few cases was a straight growth direction of the tubes towards the calcium source observed. These few positive results could neither be reproduced nor statistically validated. On the average, the direction of the tube growth was random with all calcium compounds tested.

It is therefore unlikely that Ca^{++} ions alone act on *Oenothera* pollen tubes *in vitro* as chemotropic factor. In accordance with our results, Rosen [12] did not find a positive chemotropic response towards calcium in *Lilium longiflorum* pollen tubes.

However, an absolute chemotropic ineffectiveness of calcium on *Oenothera* pollen tubes cannot be inferred from the negative results of our *in vitro* experiments. *In situ*, mixtures of different substances may be involved in the chemotropic reaction [8]. Thus it may be possible that *in situ* perhaps existing calcium gradients in co-operation with other ions or dissolved substances still have a chemotropic effect. In order to decide upon the role of calcium as a chemotropic factor or co-factor in *Oenothera*, it was necessary to determine the calcium distribution in the gynoecium.

CONTENT OF CALCIUM IN DIFFERENT PARTS OF *OENOTHERA* FLOWERS

Distribution of total calcium

The content of total calcium was determined after wet incineration by flame photometry [13]; the results are shown in Table 2.

Of all parts of the gynoecium the placenta shows the highest content of total calcium in relation to dry weight (3·65%) and to

Table 2. DISTRIBUTION OF TOTAL CALCIUM IN DIFFERENT FLOWER PARTS OF *Oenothera* (*hook.*, *Oenothera hookeri*; *long.* (old), *O. longiflora*, plants several years old; *long.* (young), *O. longiflora*, plants 6 months old)

Flower part	Ca (% dry weight) hook.	long. (old)	Ca (% d.w.)	long. (young) % H$_2$O content	Ca (% f.w.)
Hypanthium	0·73	0·67	0·63	88·5	0·072
Sepals	0·99		1·88	82·6	0·327
Petals	0·29		0·29	92·3	0·022
Filaments	0·23		0·16	89·0	0·018
Anthers + pollen	0·41		0·22	64·7	0·078
Stigma	0·49	0·58	0·60	91·1	0·052
Style, total	0·51	0·51	0·48	90·6	0·046
Style, upper third			0·61	91·3	0·053
Style, middle third			0·56	91·7	0·046
Style, lower third			0·39	90·7	0·036
Ovary, total	1·75	2·54	2·57	82·3	0·455
Ovary wall, upper third			2·43	80·0	0·486
Ovary wall, middle third			2·55	81·0	0·484
Ovary wall, lower third			2·20	80·6	0·427
Placenta (+ sterile column—see Ref. 17)			3·65	81·3	0·683
Ovules			1·25	74·7	0·316
Fruits (ripe capsules with seeds)		2·00	2·33	17·6	1·920

fresh weight (0·683%). In general, the percentages of calcium found by us correspond well with the calcium contents given by Mascarenhas and Machlis [9] for *Antirrhinum majus*.

For the path taken by the pollen tubes on their way from the stigma to the ovules, Mascarenhas actually found an increasing calcium gradient down to the ovary in the case of *Antirrhinum* flowers—contrary to our observations on *Oenothera*. This gradient could have a positive effect on the tube direction. However, in the analysis of *Antirrhinum*, no distinction was made between placenta and ovules. In *Oenothera longiflora* the placenta contains more than twice as much calcium as do the ovules (Table 2). In *Oenothera*, if a positive chemotropic reaction of the pollen tubes were caused by calcium, it would not be possible for the tubes to grow from the placenta (with high calcium content) to the ovules (with lower content).

In later investigations Mascarenhas [14] also found a large calcium decrease from the placenta to the ovules in *Antirrhinum*. He stained the ionic calcium in the tissues with glyoxal-bis(2-hydroxy-anil). From that fact he concluded that calcium ions could only be effective as a chemotropicum down from the stigma to the placenta. Then, in directing the tubes to the ovules other factors must be involved.

Amounts of soluble calcium compounds in the gynoecium

Probably only the soluble part of the total calcium amount is effective in pollen physiological processes. As the ratios between soluble and insoluble calcium in the different parts of the plants are certainly quite unequal, we finally had to test whether a calculation of the soluble part of calcium alone could reveal an effective gradient from the placenta to the micropyles.

In *Oenothera* we find insoluble calcium as precipitates of calcium oxalate in the form of rhaphides, larger single crystals or smaller particles of microcrystals. We counted these precipitates in freeze sections of the gynoecium using polarised light, and then calculated their volumes and masses.

The relationships between the amounts of total calcium, calcium oxalates and soluble calcium are shown in Table 3.

Table 3. TOTAL CALCIUM, CALCIUM OXALATE AND SOLUBLE CALCIUM (CA^{++} IONS) IN FLOWER PARTS OF *Oenothera longiflora*

Flower part	Fresh weight (mg)	Total Ca (% f.w.)	Weight of calcium oxalate (μg)	Calcium oxalate (% f.w.)	Soluble calcium (Ca ions; % f.w.)
Style (lower third)	7·0	0·036	6·11	0·087	0·012
Placenta (one ovary)	8·1	0·683	71·15	0·878	0·443
Ovule	0·04	0·316	0·25	0·625	0·145

In the soluble calcium immediately available to the pollen tubes we see about three times more in the placenta (0·443%) than in the ovules (0·145%). This means that even soluble calcium (= Ca^{++} ions) does not form gradients able to lead the pollen tubes to the ovules.

CONCLUSIONS

In our experiments the pollen tubes of *Oenothera* did not grow towards a source of calcium *in vitro*. *In situ*, as is to be seen from Fig. 1, there are no calcium gradients from the stigma to the lower

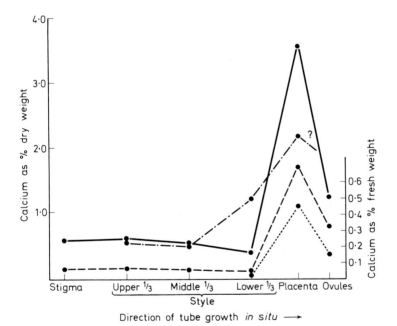

Figure 1. Calcium content of stigma, style and ovary of Oenothera longiflora. —————— *total calcium,* % *dry weight;* — — — — — *total calcium,* % *fresh weight;* .... *soluble calcium,* % *fresh weight;* —.—.—.— *calcium,* % *dry weight in* Antirrhinum majus. *(From Mascarenhas and Machlis, Ref. 9)*

third of the style. In the placenta, as compared to the style, a large increase of calcium is to be noted, but on the way from the placenta to the ovules a very pronounced decrease of calcium content takes place. The curves of both total calcium and soluble calcium follow the same pattern.

Accordingly, we cannot imagine that calcium plays a role in the chemotropism of *Oenothera* pollen tubes. Since calcium lacked chemotropic action in the genus *Oenothera* as well as in the genus *Lilium* [12], it is not likely to be the general principle in pollen tube chemotropism of higher plants, as was suggested in earlier papers [4, 9].

Chemotropically active substances, which direct the pollen tubes in the gynoecium, and gamones, which lead to successful fertilisation [15–17], are probably mixtures of different inorganic and organic substances (according to Schildknecht and Benoni [8], K^+ ions, amino acids, peptides, amines, sugars). They may differ in their concentration relationship, as suggested by the results of the extensive hybridisation experiments of Schwemmle [16], or they may be species-specific.

REFERENCES

1 BRINK, R. A., 'Preliminary study of role of salts in pollen tube growth', *Bot. Gaz.*, **78**, 361–77 (1924)

2 BERG, H. VOM, 'Beiträge zur Kenntnis der Pollenphysiologie', *Planta*, **9**, 105–43 (1930)

3 BREWBAKER, J. L. and KWACK, B. H., 'The essential role of calcium ion in pollen germination and pollen tube growth', *Am. J. Bot.*, **50**, 859–65 (1963)

4 MASCARENHAS, J. P. and MACHLIS, L., 'Chemotropic response of the pollen of *Antirrhinum majus* to calcium', *Pl. Physiol., Lancaster*, **39**, 70–7 (1964)

5 KWACK, B. H. and KIM, J. H., 'Effects of calcium ion and the protective action on survival and growth inhibition of pollen', *Physiologia Pl.*, **20**, 73–82 (1967)

6 GLENK, H. O., 'Keimversuche mit *Oenothera*-Pollen in vitro', *Flora, Jena*, **148**, 378–433 (1960)

7 GLENK, H. O., SCHIMMER, O. and WAGNER, W., 'Die Calcium-Verteilung in *Oenothera*-Pflanzen und ihr möglicher Einfluss auf den Chemotropismus der Pollenschläuche und auf die Befruchtung', *Phyton (Austria)*, **14**, 97–111 (1970)

8 SCHILDKNECHT, H. and BENONI, H., 'Über die Chemie der Anziehung von Pollen-schläuchen durch die Samenanlagen von *Oenotheren*', *Z. Naturforsch.*, **18b**, 45–54 (1963)

9 MASCARENHAS, J. P. and MACHLIS, L., 'Chemotropic response of *Antirrhinum majus* pollen to calcium', *Nature, Lond.*, **196**, 292–3 (1962); 'The pollen tube chemotropic factor from *Antirrhinum majus*: Bioassay, extraction and partial purification', *Am. J. Bot.*, **49**, 482–9 (1962)

10 SCHNEIDER, G., 'Wachstum und Chemotropismus von Pollenschläuchen', *Z. Bot.*, **44**, 175–205 (1956)

11 ROSEN, W. G., 'Chemotropism and fine structure of pollen tubes'. In: H. F. LINSKENS (Ed.), *Pollen Physiology and Fertilization*, 159–66, North-Holland, Amsterdam (1964)

12 ROSEN, W. G., 'Studies on pollen tube chemotropism', *Am. J. Bot.*, **48**, 889–95 (1961)

13 LINSKENS, H. F. and TRACEY, M. V., *Moderne Methoden der Pflanzenanalyse*, Vol. V, Springer-Verlag, Berlin (1962)

14 MASCARENHAS, J. P., 'The distribution of ionic calcium in the tissues of the gynoe-cium of *Antirrhinum majus*', *Protoplasma*, **62**, 53–8 (1966)

15 SCHWEMMLE, J., ARNOLD, C. G. and GLENK, H. O., 'Preuve et bases chimiques de la fertilisation sélective chèz les *Oenotheres*', *Proc. X Int. Congr. Genet., Montreal*, Vol. II (1958)

16 SCHWEMMLE, J., 'Selective fertilization in *Oenothera*', *Adv. Genetics*, **14**, 225–324 (1968)

17 GLENK, H. O., 'Untersuchungen über die sexuelle Affinität bei *Oenotheren*'. In: H. F. LINSKENS (Ed.), *Pollen Physiology and Fertilization*, 170–81, North-Holland, Amsterdam (1964)

The Nature of the Stigmatic Exudate and its Role in Pollen Germination*

F. W. Martin, *Federal Experiment Station, Mayaguez, Puerto Rico* and
J. L. Brewbaker, *Department of Horticulture, University of Hawaii, Honolulu, Hawaii*

ROLE OF THE STIGMATIC EXUDATE

The stigma provides a physically and chemically suitable medium for pollen germination. On its surface substances produced within, the stigmatic exudate, react with substances from the pollen to convey, or not, the stimulation to germinate. Studies of pollen germination *in vitro* show that in some species the uptake of water is sufficient stimulation to initiate the germination process. In others, sugars and a few minerals are required [1]. Still other pollens (e.g. *Ipomoea* species) do not germinate at all *in vitro*, which suggests that their germination requirements are more complex.

In general, pollens do not germinate well on stigmas of unrelated species. In one study only 5% of interfamily cross-pollinations resulted in germination [9]. Less than half of the cross-pollinations within *Ipomoea* and *Solanum* germinated. In bizarre cross-pollination combinations excellent germination rarely occurs. The stigmas from some species are better than others as sites for germination of foreign pollens, and similarly, some pollens germinate better on foreign stigmas than do other pollens. Pollen germination failures after distant cross-pollinations are probably related to differences in germination requirements and in stigmatic environments.

Stigmatic surfaces differ in two respects. Firstly, their size, degree of branching, and type of papillae may vary. Secondly, the composition of the exudate is frequently different. The various compounds of the stigmatic exudate probably control the specificity of pollen–stigma interactions. However, in interpreting the functions of such compounds one must remember that stigmatic substances may play other roles also. For example, the sticky nature of the exudate helps

* Abstract.

262

capture pollen. The stigma resists attacks of fungi, bacteria and insects, maintaining such properties by its resistance to wetting. The stigma also resists desiccation.

LIPID COMPOUNDS

Lipids of the stigmatic surface were first noticed by Vasil and Johri [14], and first analysed by Konar and Linskens [7]. Martin analysed lipids of the stigmatic exudate of *Strelitzia reqinae* Banks [12] and *Zea mays* [13]. The lipids of *Ipomoea batatas* (L.) Poir are reported

Table 1. COMPARISONS OF THE LIPIDS OF THE STIGMATIC EXUDATE OF FOUR SPECIES

Species	Amount of exudate produced	Amount of free fatty acids	Number of fatty acids after esterification	Range of chain lengths	Chief fatty acid
Petunia hybrida	Much	None	8	11–20	Linoleic
Strelitzia reqinae	Much	Much	14	8–18	Oleic
Zea mays	Little	Little	12	6–18	Oleic
Ipomoea batatas	Little	Little	2	10–12	Capric

in Table 1. Stigmatic lipids occur as free fatty acids or esters of intermediate chain lengths. They are neither volatile liquids nor hard waxes.

Of the species in Table 1, two have stigmatic exudates with large quantities of lipids, while two produce very little. In *Petunia* no free fatty acids occur, while in *Strelitzia* fatty acids are abundant. Lipids which occur on leaf surfaces associated with cutin and other compounds [2] are usually of longer chain length (20–34) than those of stigmatic exudate. Apparently the stigma, as well as the leaf, is protected from desiccation by its lipid coating.

PHENOLIC COMPOUNDS

Phenolic substances occur in exudates of stigmas of most if not all species [10]. When dissolved in alcohol, exudates show characteristic absorption peaks (Table 2). Sometimes these peaks are similar to those of simple hydroxycinnamic acids and their derivatives. On addition of NaOH, the peaks are usually shifted to a higher wavelength. The magnitude of the shift helps to determine the nature of the phenolic compound.

Using UV spectroscopy, Martin [11] showed that some exudates include from one to several derivatives of a single cinnamic acid. In some cases, however, such phenolics can only be revealed after separation of the principal components of the exudate by thin-layer chromatography.

The principal phenolic compounds of the exudate of *Strelitzia reqinae* Banks are UV-absorbing, and non-fluorescing. Each shows an absorption peak at 310 nm and a shift of 55. The peak and its shift are characteristic of the esters of *p*-coumaric acid [12] but not of the glycosides and glucose esters [3]. They may be esters with aliphatic side chains which increase their miscibility with lipids.

Table 2. ABSORPTION PEAKS (nm) OF SIMPLE PHENOLIC COMPOUNDS AND OF ALCOHOLIC EXTRACTS OF SOME STIGMATIC EXUDATES

Solution tested	Absorption peak	
	No NaOH	*With* NaOH
Compound		
p-coumaric acid	310	336
esters of *p*-coumaric acid	311	366
caffeic acid	325	Breaks down
chlorogenic acid	324	354
Extract, *Strelitzia*		
crude extract	310	365
principal compound	310	365
Extract, *Ipomoea*		
crude extract	328	386
principal compound	326	380
Extract, *Zea*		
crude extract	270–285	Same
compound 1	275	290
compound 2	346	405
compound 3	324	378

The exudate of *Ipomoea batatas* (L.) Poir contains two principal compounds with bluish fluorescence, absorbing at a peak of 326 nm. The peak is shifted on ionisation to 386 nm. Hydrolysis yields caffeic acid, which breaks down on addition of NaOH. The properties and chromatographic mobilities of these compounds are similar to those of glycosides of caffeic acid [3] and of chlorogenic acid (Table 2).

The exudate of *Zea mays* L. is more complex [13]. Alcoholic extract absorbs at a peak of 270–5 nm. This peak is not shifted by NaOH. After thin-layer chromatography three principal compounds are found (Table 2). That absorbing at 346 nm has a light blue fluorescence. At least 12 other compounds are present. Hydrolysis yields *p*-coumaric acid, probably ferulic acid, and other compounds.

The phenolic compounds of the stigmatic exudates tested are thus chiefly derivatives of hydroxy-cinnamic acids.

OTHER COMPONENTS

Although very little exudate can be washed off the stigma, stigmatic exudates do contain appreciable amounts of water [10]. In *Petunia* this water is below the superficial layers of the stigmatic exudate [7]. Hydroxyl and carboxyl groups of fatty acids and hydroxy-cinnamic acids assure the presence of water. In turn, water makes possible the presence of polar compounds, such as sugars and amino acids.

Konar and Linskens [7] found traces of five sugars in stigmatic exudate of *Petunia*. Aqueous extractions from stigmas of 10 species [10] responded negatively or weakly to tests for reducing sugars. Thus the stigmatic exudate cannot really be considered a sugary solution.

Stigmatic exudates of *Ipomoea* contain little or no enzymes or proteins [8]. Only traces of free amino acids were found in exudates of *Petunia* [7], and no protein was detected. However, stigmatic exudates of *Epipactis* entrap and digest insects [6], and presumably contain proteolytic enzymes.

No systematic searches have been made for growth-regulating compounds in stigmatic exudates.

ROLES OF THE INDIVIDUAL COMPONENTS OF THE STIGMATIC EXUDATE

The lipids of the stigma may serve chiefly as protection from desiccation. Pollen grains placed on stigmatic exudate of *Petunia* germinated only when moisture was made available [7]. The possible influence of lipids of the stigma on permeability of pollen and tube membranes should be investigated.

Phenolic compounds play a variety of roles in other plant tissues, and may have several functions in the stigma. They may, for example, repel insects and impede germination of spores [5]. Phenolic glycosides could possibly serve as sources of sugars for germinating pollen. Simple phenolic compounds may stimulate and inhibit indole acetic acid oxidase and thus influence growth processes [4]. Such functions have not yet been clearly demonstrated on the stigma, and indeed the possible presence of growth substances in stigmatic exudates now requires investigation.

Stigmatic exudate apparently varies in amount with age and condition of the stigma. The amount of caffeic acid derivatives of sweet potato stigma increases up to 24 h before anthesis and then decreases to about 25% of the peak by anthesis. After anthesis the amount decreases slowly. Neither compatible nor incompatible pollination significantly affects this pattern. The absorption peak remains constant. Thus the phenolic compounds are not changed chemically with time, or by action of the pollen. The apparent decrease in phenolic content could reflect the presence of increased lipids impeding alcoholic extraction.

Alcoholic extracts of stigmatic exudates impede germination and growth *in vitro* of pollen of *Crinum asiaticum* and *Kiqelia pinnata*, and phenolic compounds eluted from thin layers of silica gel after chromatography also inhibit germination. The inhibiting substances and their effective concentrations need to be better defined.

REFERENCES

1 BREWBAKER, J. L. and KWACK, B. H., 'The essential role of calcium ion in pollen germination and pollen tube growth', *Am. J. Bot.*, **50**, 859–65 (1963)

2 EGLINTON, G., GONZALEZ, A. G., HAMILTON, R. J. and RAPHAEL, R. A., 'Hydrocarbon constituents of the wax coatings of plant leaves: A taxonomic survey', *Phytochemistry*, **1**, 89–102 (1962)

3 HARBORNE, J. B. and CORNER, J. J., 'Plant Polyphenols 4: Hydroxy-cinnamic acid-sugar derivatives', *Biochem. J.*, **81**, 242–50 (1961)

4 GALSTON, A. W. and DAVIES, P. V., 'Hormonal regulation in higher plants', *Science, N.Y.*, **163**, 1288–97 (1969)

5 HARE, R. C., 'Physiology of resistance to fungal diseases in plants', *Bot. Rev.*, **32**, 95–137 (1966)

6 JOST, L., 'Zur Physiologie des Pollen', *Ber. dt. bot. Ges.*, **23**, 504–15 (1905)

7 KONAR, R. N. and LINSKENS, H. F., 'Physiology and biochemistry of the stigmatic fluid of *Petunia hybrida*', *Planta*, **71**, 372–87 (1966)

8 MARTIN, F. W., 'Some enzymes of the pollen and stigma of the sweet potato', *Phyton, B. Aires*, **25**, 97–102 (1968)

9 MARTIN, F. W., 'Pollen germination on foreign stigmas', *Bull. Torrey bot. Club*, **97**, 1–6 (1970)

10 MARTIN, F. W., 'Compounds from the stigma of 10 species', *Am. J. Bot.*, **56**, 1023–7 (1969)

11 MARTIN, F. W., 'The ultraviolet absorption profile of stigmatic extracts', *New Phytologist*, **69**, 425–30 (1970)

12 MARTIN, F. W., 'The stigmatic exudate of *Strelitzia*', *Phyton, B. Aires*, **27**, 47–53 (1970)

13 MARTIN, F. W., 'Compounds of the stigmatic surface of *Zea mays* L.' (In press)

14 VASIL, I. V. and JOHRI, M. M., 'The style, stigma, and pollen tube. I', *Phytomorphology*, **14**, 352–69 (1964)

Effect of Stigmatic Exudate Injected into the Stylar Canal on Compatible and Incompatible Pollen Tube Growth in *Lilium longiflorum* Thunb.*

P. D. Ascher and L. W. Drewlow, *Department of Horticultural Science, University of Minnesota, St Paul, Minnesota, U.S.A.*

Yamada [2] observed colloidal bodies in the parenchymatous cells of both stigma and style of *Lilium longiflorum*. With maturation of the pistil the colloidal material appeared to pass from the parenchyma into the epithelium cells and then to the surface of stigmatic papillae or stylar canal cells as exudate. Staining properties of the intracellular colloidal material were not different from those of exudate on the stigmatic surface or in the stylar canal.

Presumably stylar exudation places metabolic products at the disposal of growing pollen tubes. In self-incompatible, hollow-styled *L. longiflorum*, stylar exudate might also be the vehicle of transfer of the stylar self-incompatibility substances to the pollen tubes. Incompatible pollen tubes in the Easter Lily are inhibited in the style, about half-way between the stigma and the ovary. If, as Yamada's observations suggest, stigmatic exudate and stylar canal exudate are similar, and if the stylar self-incompatibility identity is expressed in canal exudate, stigmatic exudate might also contain the incompatibility specificities of the style. This hypothesis can be tested by filling the stylar canal of compatibly pollinated styles with stigmatic exudate from flowers of the same genotype as the pollen and observing pollen tube growth. The presence of stylar incompatibility substances in the exudate would be manifest as an inhibition of pollen tube growth compared with tube growth in styles filled with exudate from the same cultivar as the style.

Greenhouse grown plants of *L. longiflorum* Thunb. cultivars Ace and Nellie White provided the plant materials used in this

* Abstract.

267

experiment. These cultivars are self-incompatible but cross-compatible. Flowers were cut 1 day after anthesis and placed in water in the laboratory for collection of stigmatic exudate. As exudate accumulated in drops on the stigma, it was gathered with a hypodermic needle and syringe, and stored at $-20°F$. Styles were prepared for injection by excising the pistil from the flower 1 day after anthesis, snapping the ovary from the style, and injecting fresh or thawed exudate through the stigma with a hypodermic syringe and a 22 gauge needle. When a drop of exudate appeared at the ovarian end of the style, we assumed that the stylar canal was filled. Non-injected (control) styles were pierced through the stigma with a clean hypodermic needle and the ovary was removed. The styles were pollinated and placed on moistened filter paper in a 150 mm Petri dish and incubated at 25°C for 48 h. After incubation, styles were cut longitudinally with a razor-blade and stained with aceto-carmine and aniline blue, and the longest pollen tube in each half-style was measured to the nearest millimetre.

Table 1. POLLEN TUBE GROWTH AS AFFECTED BY STIGMATIC EXUDATE INJECTED INTO THE STYLAR CANAL OF *Lilium longiflorum* BEFORE COMPATIBLE OR INCOMPATIBLE POLLINATION. TREATMENTS WERE REPLICATED SIXTEEN TIMES

Cultivar used as source of style	Source of exudate injected	Cultivar used as source of pollen	Type of pollina- tion	Length of pollen tubes in injected styles (mm)	Length of pollen tubes (mm) in non-injected styles (control)
Nellie White	Nellie White	Nellie White	Self	67·4	51·3*
Nellie White	Nellie White	Ace	Cross	99·4	89·6
Ace	Nellie White	Nellie White	Cross	96·6	93·2
Ace	Ace	Ace	Self	63·6	54·2
Ace	Ace	Nellie White	Cross	91·1	80·3
Nellie White	Ace	Ace	Cross	97·4	88·5

* Differences between means in horizontal rows are significant at the 1% level (F test).

The effect of stigmatic exudate in the stylar canal on pollen tube growth *per se* was tested by injecting exudate from the same cultivar as the style, and self- or cross-pollinating. To test for the presence of incompatibility specificity in the exudate, styles were injected with exudate from flowers of the cultivar used as pollen source for cross-pollination. Exudate alone or exudate diluted with an equal volume of distilled water was injected into styles before, 6, 12 or 24 h after compatible or incompatible pollination.

Stigmatic exudate injected into the stylar canal of L. *longiflorum* before pollination significantly increased the length that compatible and incompatible pollen tubes grew in 48 h (Table 1). Presence of stigmatic exudate in the stylar canal during pollen tube growth did not affect the incompatibility reaction, whether the exudate originated from stigmas of the cultivar used as pollen source or from the same cultivar as the style.

Dilution of stigmatic exudate with an equal volume of distilled water before injection into the style removed the stimulatory effect

Table 2. LENGTH OF POLLEN TUBES AFTER 48 h GROWTH IN L. *longiflorum* THUNB. STYLES INJECTED BEFORE POLLINATION WITH STIGMATIC EXUDATE, EXUDATE DILUTED WITH AN EQUAL VOLUME OF WATER, OR NON-INJECTED. EACH TREATMENT WAS REPLICATED EIGHT TIMES

| | Treatment | | |
Pollination	Stylar canal filled with exudate	Stylar canal filled with exudate and water 1:1 by volume	Non-injected (control)
Compatible	99·4 a*	99·8 a	93·9 a
Incompatible	77·8 b	55·2 c	57·4 c

* Means followed by different letters are significantly different at 1% (Duncan's new multiple range test).

of exudate on pollen tube growth (Table 2), as did injection of un-diluted exudate 6 or 12 h after pollination (Fig. 1). Exudate placed in the style 24 h after pollination or diluted exudate injected into the style 12 or 24 h after pollination retarded both compatible and incompatible pollen tube growth (Fig. 1). The greatest inhibition of pollen tube growth occurred on introduction of the extraneous materials into the style 24 h after pollination.

The often-reported observation that compatible and incompatible pollen tubes grow side by side in a style, the former proceeding to the ovary while the latter are inhibited, suggests that the reaction resulting in self-incompatibility occurs within an individual pollen tube, not in the style. Therefore metabolic products reflecting the stylar self-incompatibility genotype must pass from the style into the pollen tube. The pollen–stylar interaction leading to the self-incompatibility reaction in L. *longiflorum* appears to be mediated by a potent entity from the style, one that is not diluted or diminished by the presence of extraneous liquid material in the stylar canal. Although stigmatic exudate added to the stylar canal before or after pollination stimulated or retarded pollen tube growth, the self-incompatibility reaction was not obscured.

That stylar incompatibility identity was not expressed in the stigmatic exudate of Easter Lily suggests a fundamental difference

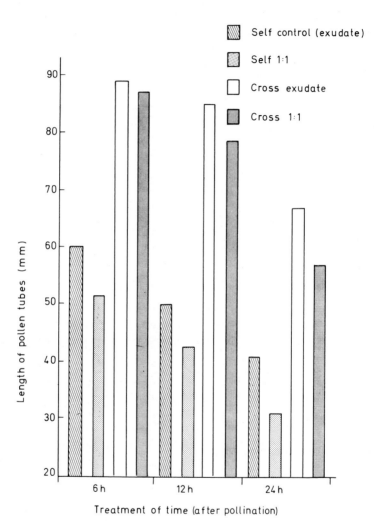

Figure 1. Compatible and incompatible pollen tube growth in L. longiflorum, *Thunb. styles injected with stigmatic exudate or exudate and water (1:1 by volume) 6, 12 or 24 h after pollination. Pollen tube length was determined 48 h after pollination*

between stigmatic and stylar canal exudates, if, indeed, the self-incompatibility specificity of the style is passed to the pollen tubes via canal exudate. However, since pollen tubes in *Lilium* grow closely appressed to the epithelium cells of the stylar canal, stylar incompatibility substances may pass directly into the pollen tube without being released into the exudate.

An effective carrier of exogenous substances into the style would be most useful in the study of pollen tube growth and the self-incompatibility reaction. To be effective, a carrier must not adversely affect pollen tube growth. Sucrose or sucrose–mineral solutions [1], successfully used for *in vitro* germination of pollen, retard pollen tube growth when introduced into the style of *L. longiflorum* (Drewlow and Ascher, Unpublished). The normal growth of pollen tubes in styles filled 6 or 12 h after pollination with exudate, or before or 6 h after pollination with exudate diluted with an equal volume of distilled water, and the expression of self-incompatibility as differential pollen tube growth in all styles containing exudate or exudate and water, indicate that pollen tube metabolism and the self-incompatibility reaction can be studied in *L. longiflorum* with stigmatic exudate as a carrier of metabolic inhibitors.

The retarding effect of diluted exudate placed in the style after pollination exposed a limitation in the utility of stigmatic exudate as a carrier. To minimise dilution of exudate, metabolic inhibitors used at low concentrations must be added directly to exudate rather than diluted in water. However, the acidic nature of the exudate of *L. longiflorum* (pH about 4·5) may prohibit dissolution of acid-insoluble substances in exudate alone.

Exudate-injected styles may be manipulated with ease for pollination and incubation, as the viscosity of exudate prevents drainage of the injected material. The Easter Lily produces stigmatic exudate abundantly; several drops may be gathered from each flower daily for up to 9 days after anthesis. Repeated freezings and storage in the frozen state did not damage the function of exudate in the style as compared to fresh exudate. Stigmatic exudate appears to fulfil satisfactorily the criteria of an effective carrier of exogenous substances into the environment of growing pollen tubes in *L. longiflorum*.

Paper number 7034, Scientific Journal Series, Minn. Agric. Exptl. Stn. Research was supported in part by funds provided by the Graduate School, University of Minnesota. The authors wish to thank the United Bulb Co., Mount Clemen, Michigan, for lily bulbs. Mr Drewlow is a National Science Foundation predoctoral trainee.

REFERENCES

1 BREWBAKER, J. L. and KWACK, B. H., 'The essential role of calcium ion in pollen germination and pollen tube growth', *Am. J. Bot.*, **50**, 859–65 (1963)
2 YAMADA, YOSHIO, 'Studies on the histological and cytological changes in the tissues of pistil after pollination', *Jap. J. Bot.*, **19**, 69–82 (1965)

Use of Pistil Exudate for Pollen Tube Wall Biosynthesis in *Lilium longiflorum**

M. Kroh, C. Labarca and F. Loewus, *Department of Biology, State University of New York, Buffalo, New York, U.S.A.*

Considerable cell wall formation accompanies pollen tube development, and the major components of the walls produced are polysaccharides. Two sources of carbohydrate are available for wall formation: reserves stored in the pollen grain, and substances provided by the pistil. Pollen tubes grown *in vitro* seldom attain the full length needed to effect fertilisation and, hence, it is assumed that the pistil provides a portion of the carbohydrate needed for wall biosynthesis by developing tubes. Labelling experiments with *Lilium longiflorum* have demonstrated that carbohydrate material derived from germinating medium or pistil is used by pollen during tube wall formation [3, 4]. In this plant, germinating pollen tubes are surrounded by a secretion product from the stigma and style— specifically, from those cells which cover the distal surface of the stigma and line the stylar canal [6]. We have shown that stigmatic exudate (that portion of the secretion product which accumulates and is shed from the surface of the stigma) is predominantly carbohydrate, of which 95% is in the form of high molecular weight compounds [5]. Presumably it is this pistil-derived secretion product that is used, in part, by the developing pollen tube for wall biosynthesis.

The present report concerns the role of exudate as a source of carbohydrate precursors for pollen tube wall formation. In this study advantage was taken of the fact that the carbohydrate components of the exudate became labelled when detached pistils are fed myo-inositol-U-^{14}C or D-glucose-1-^{14}C [4, 5]. Only data from glucose-labelled experiments are included here. The uptake and

* Abstract.

273

incorporation of label into pollen tube walls of pollen grown *in vitro* and *in situ* in the presence of labelled exudate are described.

Flowers of Easter Lily, *Lilium longiflorum* (cvs. Ace and Nelly White) from plants grown under greenhouse conditions were used. Labelled exudate was obtained from stigmas of excised pistils which had been fed D-glucose-1-^{14}C. A portion of the exudate was fractionated on Sephadex G-100 into a high and a low molecular weight fraction. Methods of labelling and fractionation will be described elsewhere.

Pollen was germinated and grown *in vitro* by suspending 100 mg of Ace pollen in 1·8 ml of Dickinson's pentaerythritol medium [2] and incubating the suspension on a rotatory shaker (100 rev/min) at 30°C. After incubating 1·5 h, 0·2 ml of unfractionated exudate containing $1·6 \times 10^6$ counts/min was added to the medium. After an additional 13 h of incubation, the suspension was centrifuged, washed and analysed as described below.

Pollen was germinated and grown *in situ* by pollinating detached Ace pistils (100 mg of Nelly White pollen/pistil). Prior to pollination, unfractionated labelled exudate (3 pistils), labelled high molecular weight fraction (2 pistils) or labelled low molecular weight fraction (3 pistils) was introduced through the central opening of the stigma into the empty style canal of each pistil with the aid of a hypodermic syringe and needle. Pistils were held in a moist chamber with their pedicles submerged in water for 65 h at 28°C. Pollen tubes were recovered by sectioning the style into 3 or 4 cm pieces and excising the tubes from the central canal. Pollen tissues from each set of pistils were combined, rinsed seven times with Dickinson's medium (2 ml portions) to remove adhering exudate and ground in water with the aid of a conical glass homogeniser. Tube fragments were again rinsed three times with water (2 ml portions) and then hydrolysed with 2 N trifluoroacetic acid for 1 h at 121°C in a sealed tube [1]. Labelled components which were solubilised by this treatment were separated by descending paper chromatography (ethyl acetate–pyridine–water, 8:2:1, v/v) and paper electrophoresis (0·1 M ammonium formate, pH 3·8, 3 h, 20 V/cm). Radioactive regions on each chromatogram or pherogram were located by scanning and each region was then eluted and counted separately in liquid scintillation fluid on a spectrometer with an efficiency of about 80% for ^{14}C.

Pollen tubes germinated *in vitro* retained about 1% of the ^{14}C present as labelled exudate. Of this, 38% was incorporated into water-washed tube wall residue, all of which could be solubilised by treatment with trifluoroacetic acid.

Pollen tubes germinated *in situ* retained about 4% of the ^{14}C from pistils injected with unfractionated labelled exudate and about 3% of the ^{14}C from pistils injected with high molecular weight fraction. Percentage incorporation was calculated from the total ^{14}C recovered from stigma, style and pollen at the end of the experiment, about 50% of the radioactivity injected. Most of the ^{14}C not recovered in these pistil parts was found in the ovary. Tubes germinated *in situ* from pistils injected with low molecular weight fraction of labelled exudate contained no detectable ^{14}C. In the case of *in situ* labelled pollen tubes, about 70% of the ^{14}C was incorporated into tube wall residue but in contrast with what was found with tubes germinated *in vitro*, only three-fourths of the label could be released by acid hydrolysis.

Table 1 records the distribution of counts recovered by eluting radioactive regions of chromatograms on which the labelled hydrolysates were separated. These regions were the origin, galactose

Table 1. DISTRIBUTION OF ^{14}C AMONG CARBOHYDRATE RESIDUES IN TRIFLUOROACETIC ACID HYDROLYSATES OF TUBE WALL RESIDUES AFTER LABELLING WITH EXUDATE DERIVED FROM PISTILS FED WITH D-GLUCOSE-1-^{14}C

	Whole exudate in vitro	*Whole exudate in situ*	*High molecular weight fraction in situ*
^{14}C in tube residues (counts/min)	3500	11 000	5000
^{14}C solubilised by acid (counts/min)	3500	6900	3750
^{14}C recovered after paper chromatography (counts/min)	3490	5800	3610
Distribution on paper (%)			
Origin (galacturonic acid)	9	15	22
Galactose–glucose	63	44	38
Arabinose	24	33	30
Xylose–fucose	0	3	3
Rhamnose	4	5	7

and glucose combined, arabinose, xylose and fucose combined, and rhamnose. The labelled constituent at the origin was identified as galacturonic acid by electrophoresis.

Although the chromatographic separation used in this study did not fully resolve labelled galactose and labelled glucose, it was possible to show that fractionation of the exudate removed almost all labelled glucose from the high molecular weight fraction (compare scans E and B in Fig. 1). Pollen tubes recovered from pistils containing the labelled high molecular weight fraction yielded a hydrolysate that contained labelled glucose as well as labelled

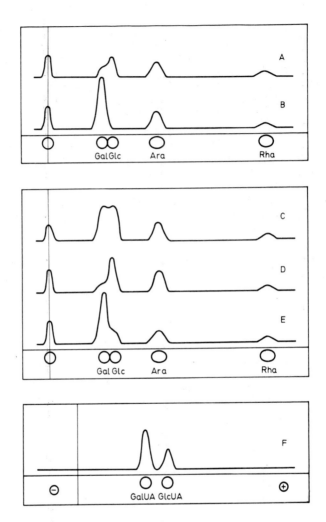

Figure 1. Radiochromatogram scans showing the distribution of labelled components after acid hydrolysis of stigmatic exudate, high molecular weight component of stigmatic exudate or pollen tube residues. In all cases label was ultimately derived from lily pistils (detached) fed with D-glucose-1-^{14}C. Relative peak areas within a single scan have been faithfully reproduced, but scales of individual scans have been adjusted to allow intercomparisons. A, tube residues from pistils injected with high molecular weight fraction of labelled exudate; B, high molecular weight fraction of labelled exudate; C, tube residues grown in vitro in the presence of whole exudate; D, tube residues from pistils injected with whole exudate; E, whole labelled exudate; F, electrophoretic separation of radioactivity remaining at the origin in scans B or E. Abbreviations: Gal = galactose, Glc = glucose, Ara = arabinose, Rha = rhamnose, GalUA = galacturonic acid and GlcUA = glucuronic acid. Origins are identified by the vertical lines

galactose (scan A). These results suggest that at least a portion of the polysaccharide substance in labelled exudate is broken down and resynthesised into tube wall by growing pollen.

Note in Fig. 1 that labelled material retained at the origin after paper chromatography of hydrolysed exudate or its high molecular weight component (scans E and B) could be resolved by paper electrophoresis into galacturonic acid and glucuronic acid with a ^{14}C ratio, 3:1 (scan F). As indicated above, only labelled galacturonic acid was found in radioactive material retained at the origin after chromatography of hydrolysed tube wall residues (scans A, C and D).

Results obtained from the *in vitro* and *in situ* experiments indicate that growing pollen tubes use pistil exudate—specifically, the high molecular weight fraction—for tube wall biosynthesis. Results also show that the low molecular weight component is not used in significant amounts by growing pollen. Use of labelled exudate by pollen tubes grown *in situ* greatly exceeded that observed in the *in vitro* experiment, but this may be due, in part, to the fact that some tubes grown *in situ* reached a length of 10 cm, whereas those grown *in vitro* seldom exceeded 1 cm.

Pollen tubes rinsed in Dickinson's medium to remove all traces of labelled exudate, then ground in water and thoroughly washed with more water yielded water-insoluble residues which contained 40–70% of the ^{14}C present in the tubes. Acid hydrolysis of these residues revealed interesting differences in the distribution of ^{14}C as compared to that found in labelled exudate. The latter contained both labelled galacturonic acid and labelled glucuronic acid, whereas the former contained only labelled galacturonic acid. Moreover, labelled glucose, which was not detected in hydrolysates of the high molecular weight fraction of exudate, was present in tube residues. Galactose, which was so prominently labelled relative to arabinose and rhamnose in hydrolysates of exudate, gave a much reduced radioactive peak relative to the latter sugars in tube residues.

The process or processes involved in transfer of labelled polysaccharides from exudate to growing pollen tube wall must still be explored. It might involve direct use of high molecular weight material for tube wall formation or it might involve breakdown, partial or complete, of high molecular weight substance provided by the exudate followed by resynthesis of fragments into new polysaccharides. Detailed knowledge of the molecular structure of exudate and pollen tube wall substances is necessary in order to solve the problem of carbohydrate transfer from exudate to tube wall.

This work was supported by a grant (GM-12422) from the U.S. Public Health Service and by a postdoctoral fellowship (to C.L.) from the Graduate School, State University of New York at Buffalo.

REFERENCES

1 ALBERSHEIM, P., NEVINS, D. J., ENGLISH, P. D. and KARR, A., 'A method for the analysis of sugars in plant cell-wall polysaccharides by gas-liquid chromatography', *Carbohydrate Res.*, **5**, 340–5 (1967)
2 DICKINSON, D. B., 'Rapid starch synthesis associated with increased respiration in germinating Lily pollen', *Pl. Physiol., Lancaster*, **43**, 1–8 (1968)
3 KROH, M. and LOEWUS, F., 'Biosynthesis of pectic substance in germinating pollen: Labelling with *myo*-inositol-2-^{14}C', *Science, N.Y.*, **160**, 1352–4 (1968)
4 KROH, M. and MIKI-HIROSIGE, H., 'Conversion of *myo*-inositol to pectic substance of pollen tube walls', *Pl. Physiol., Lancaster*, **43**, S-52 (1968)
5 LABARCA, C., KROH, M. and LOEWUS, F., 'The pectic nature of pistil exudate in *Lilium longiflorum*', *Fedn Proc. Fedn Am. Socs exp. Biol.*, **28**, 869 (1969)
6 WELK, M., SR., MILLINGTON, W. F. and ROSEN, W. G., 'Chemotropic activity and the pathway of the pollen tube in lily', *Am. J. Bot.*, **52**, 774–81 (1965)

Incompatibility

Advances in the Study of Incompatibility

C. E. Townsend, *Crops Research Division, Agricultural Research Service, U.S.D.A., Department of Agronomy, Colorado State University, Fort Collins, Colorado, U.S.A.*

INTRODUCTION

There are several general reviews on self-incompatibility, the most recent being that of Arasu [3]. The role of self-incompatibility in the physiology and genetics of embryology was discussed by Maheshwari and Rangaswamy [64]. Therefore the purpose of this article is to review some of the data which have appeared primarily during the past decade. The gametophytic homomorphic type of incompatibility will be stressed more than the sporophytic homomorphic type. The heteromorphic type of incompatibility, which is associated with different flower forms on plants of the same species, will not be discussed.

Self-incompatibility was defined by Brewbaker [11] as 'the inability of a plant producing functional male and female gametes to set seed when self-pollinated'. According to the oppositional S allele hypothesis proposed by East and Mangelsdorf [23], pollen grains that possess S alleles identical to one of those in the pistil will not be functional on that particular pistil. This condition holds only for the gametophytic homomorphic type. In contrast, the genotype of the sporophyte, rather than that of the gametophyte, determines the incompatibility reaction of the pollen in the sporophytic homomorphic type of incompatibility.

PHYSIOLOGY

Grafting and related effects

Evans [26] reported that self-compatibility in *Trifolium pratense* was influenced by grafting, with homografts and heterografts giving

281

4·8% and 9·2% seed set, respectively. The percentage seed set for the stocks was 1·0, and for the ungrafted controls it was 0·8. In similar studies Denward [18] found that grafting alone increased the degree of pseudo-self-compatibility in *T. pratense*; thus fertility following self-pollination without grafting was 0·22%, while fertility following self-pollination with grafting was 8·4%. He considered pseudo-self-compatibility to be controlled genetically.

In stylar grafting experiments with *Oenothera organensis*, Hecht [33] obtained pollen tube growth in incompatible styles. The lactose–gelatin 'splint' technique was used. Stylar and 'tandem' grafts in *O. rhombipetala* Nutt [8] and *O. organensis* [34] showed that the incompatibility reaction was complete in the stigma but was incomplete at all levels of the style.

Linskens [58] observed the effect of castration on pollen tube growth following self-pollination in *Petunia hybrida*. Emasculation of immature flowers reduced pistil elongation and weakened the incompatibility barrier. Pollen tube growth of mature grains was inhibited less on very immature styles of castrated flowers than on mature styles of similarly treated flowers. Immature pollen grains were capable of developing the incompatibility reaction. In similar studies with *O. organensis* Kumar and Hecht [43] could not find a relation between stamen development and elongation of the style and its associated incompatibility substance.

Nutritional

Studies with excised pistils of *Trifolium pratense in vitro* revealed that pollen tubes would grow through the styles of compatible matings, and that certain treatments stimulated pollen tube growth in styles of incompatible matings [39]. Various carbohydrates, boric acid, calcium nitrate and several plant hormones were used in the cultural media. Media containing di- and trisaccharides usually gave better pollen tube growth in comparison to the monosaccharides, but none overcame the incompatibility reaction. Boric acid stimulated pollen tube growth in both compatible and incompatible matings. However, calcium enhanced pollen tube growth in the compatible matings only. Gibberellic acid, α-naphthalene acetamide, traumatic acid and 3-indolebutyric acid did not stimulate pollen tube development at the concentrations used in either compatible or incompatible matings. There was some evidence that the incompatibility reaction was affected by the application of relatively large amounts of pollen to the stigma. Kendall and Taylor [40] noted that zinc inhibited pollen tube elongation.

Kendall and Taylor [41] studied the effects of certain chemicals in the medium on the pseudo-self-compatibility reaction of excised flowering stems of *Trifolium pratense*. The chemicals which influenced pollen tube growth in excised pistils of a previous experiment [39] did not affect selfed seed set of the excised flowering stems. Extracts of compatible styles and stigmas, yeast extract, boric acid or sucrose did not affect the incompatibility reaction of *Oenothera rhombipetala* [8]. In contrast, studies by Pandey [82] revealed that an application of a stigmatic secretion from mature stigmas of *Nicotiana alata* to immature stigmas, followed by bud pollination, markedly increased selfed seed set.

In physiological studies of the self-incompatibility reaction of *Oenothera organensis* Kwack [44] found that calcium enhanced incompatible pollen tube development. If the styles had been pre-soaked in a solution containing calcium or were embedded in a medium containing calcium, incompatible pollen tubes grew to some extent. In comparison, no growth occurred in treatments without calcium. The site of the incompatibility reaction was considered to be the pollen grain wall.

Temperature

Temperature has been one of the variables in incompatibility studies for many years. In the earlier studies temperature effects were generally measured in terms of compatible and incompatible pollen tube growth in styles of whole flowers which had been removed from the plant. In relatively recent years the experiments have been designed to determine what effect temperature has prior to, and following, incompatible and compatible pollinations of intact and detached styles, and on whole plants. Such studies differentiate the site (pollen or style) of the temperature effect. In general, only relatively high temperatures have been found to inactivate the incompatibility mechanism, and the site of inactivation has been the style and not the pollen.

The effect of temperature on pollen tube growth of compatible and incompatible matings varies with the species. Lewis [46] showed that incompatible pollen tube growth in *Oenothera organensis* and *Prunus avium* was best at 15°C, but was retarded at 20°C. Compatible pollen tube growth was best at 25–30°C. Since incompatible tube growth was inhibited at higher temperatures, Lewis suggested that the incompatibility reaction and the immunity reaction were similar, as both required heat. Compatible pollen tube growth in detached stigmas and styles of *O. rhombipetala* increased with an increase in temperature from 10°C to 30°C [9]. However, temperature did not influence the growth of incompatible tubes, and many incompatible pollen grains did not germinate. Temperatures of 10–25°C, applied at the time of pollination, usually increased pollen tube growth, while temperatures of 30–35°C usually decreased pollen tube growth in both compatible and incompatible matings with excised pistils of *Trifolium pratense* [39].

Ascher and Peloquin [6] reported a linear relation between compatible pollen tube growth and temperature in detached styles of four cultivars of *Lilium longiflorum*. The temperature treatments were applied following self-pollination. Incompatible pollen tube growth also increased with temperature, but not to the same extent as that of compatible tubes until a temperature of 39°C was reached. At the latter temperature incompatible and compatible pollen tube growths were indistinguishable, as pollen tubes had grown completely through the style in 48 h.

Ascher and Peloquin [7] observed two types of interspecific incompatibility in styles of *Lilium longiflorum* which had been pollinated with pollen from various *Lilium* spp. and incubated at 24°C. Pollen tubes of *L. formosanum* and *L. regale*, self-compatible species closely related to *L. longiflorum*, ceased growth at the stylar canal and were of abnormal morphology. Such tubes did not continue to grow with additional incubation. When *L. longiflorum* was crossed with more distantly related *Lilium* spp., pollen tube growth and morphology were similar to that following intraspecific incompatible matings. However, incubation at 39°C did not overcome the incompatibility reaction of the interspecific matings.

Hecht [34] reported that intact styles and stigmas of *Oenothera organensis* which had been placed in water at 50°C for 5 min were receptive to both compatible and incompatible pollinations. Similar results were obtained with *O. rhombipetala* by Bali [8], who also reported that the inactivation of the incompatibility reaction was not permanent. If pollination of treated styles was delayed, pollen tube growth was retarded in comparison to that following immediate pollination. Contrasting studies by Hecht [35] showed that complete inactivation of the incompatibility reaction by temperature required several hours to reach completion. Kwack [44] found that a temperature of 5°C, prior to and after pollination, reduced self-incompatibility in *O. organensis* slightly. Preheating the styles at 50°C for 3 min also reduced the self-incompatibility reaction but not as much as the 5°C treatment. She concluded that temperature had inactivated a protein which was associated with the incompatibility reaction. The protein was considered to have enzymatic properties.

The self-incompatibility reaction in *Lilium longiflorum* was inactivated by heating detached and intact styles in water at 50°C for 6 min prior to pollination [37]. Twenty fruits were set following 48 pollinations of pretreated intact styles. Seed set of these fruits ranged from 6 to 114 seeds. In comparison, seed yield of normally cross-pollinated flowers ranged from 75 to 300 seeds. Pretreatment at temperatures of 25°C and 45°C did not inactivate the incompatibility reaction. A 55°C treatment apparently injured the stylar canal cells, a factor which prohibited both compatible and incompatible pollen tube growth. Greenhouse temperatures at which the plants were grown prior to treatment were believed to influence the stylar response to temperature.

Incompatibility was reduced in the styles of *Trifolium pratense* by a temperature of 40°C [39]. Incompatible pollen tube growth was significantly greater in excised styles of flowers which had developed at 40°C than in styles developed at 25°C. The temperature effect was confined to the styles. When flowering stems of *T. pratense*, excised at the bud stage, were held at 40°C and self-pollinated, varying degrees of selfed seed set were observed on 10 clones, and a maximum of 20% was obtained [41].

The physiological age of the style had a pronounced effect on the growth of incompatible pollen tubes in detached styles of *Lilium longiflorum* [5]. Growth was restricted during the first 4 days following anthesis but increased considerably thereafter until floral senescence.

Self-incompatibility in *Lilium longiflorum* was overcome by an interaction of temperature and a 1% solution of naphthalene acetamide in lanolin [25]. The growth regulator was applied at the wound where a petal had been removed. Greenhouse temperatures were 11, 17, 22, 28 and 33°C. The highest yield of selfed seed was obtained at 17°C when the growth regulator was applied 3 days before self-pollination.

Seed set following cross-pollination of radish was higher at 17°C than at 26°C; however, seed set following selfing was increased at the higher temperature [24].

Many investigators have shown that seasonal variation under field conditions influenced the self-compatibility reaction. Temperature was suspected of contributing to the variability. Generally, self-compatibility studies with whole plants were conducted in a greenhouse environment until recently, when controlled environmental chambers became available for such purposes. Under greenhouse conditions pseudo-self-compatibility in four clones of *Trifolium pratense* was highest during the month of July (4·5 seeds/head) followed by September (1·3 seeds), June (1·0 seeds) and May

(0·4 seeds) [45]. Cohen and Leffel [17] reported that pseudo-self-compatibility in *T. repens* was influenced by season of year. The self-compatibility reaction of 10 clones of tetraploid *T. hybridum* whose percentage of florets setting seed ranged from 0 to 85 also was affected under greenhouse conditions by the season of year [98]. The highest average percentage of florets setting seed was in the spring (28%), with no difference between summer (22%) and fall (23%). In general, the effect was not of sufficient magnitude to cause misclassification of individual plants; i.e. plants classified as self-compatible (SC) in one season would not be classified as self-incompatible (SI) in another season, nor vice versa.

Controlled environmental chambers have assisted materially in studying the effect of temperature on the compatibility reaction. Pseudo-self-compatibility in *Trifolium pratense* was increased at a 32°C temperature treatment (5·9 seeds/head) in growth chambers, as compared with 15°C (0·6 seeds) and 24°C (2·7 seeds) treatments [45]. The 10 clones of tetraploid *T. hybridum* used in the above greenhouse studies were evaluated for their response to three temperatures in controlled environmental chambers [98]. The three temperatures were constant 15°C, 24°C day–17°C night, and constant 32°C. Two highly SC clones responded the same at all three temperatures. The percentage of florets setting seed for three moderately SC clones increased with each increase in temperature. One highly SC clone showed a marked decrease in seed set at 32°C. Three SI clones responded slightly to the 32°C treatment. One clone, 3-7B, that was highly SI at the two lower temperatures, responded markedly to the 32°C temperature.

The physiology of the SC reaction of clone 3-7B was investigated further in controlled environmental chambers [100]. From one to two days of constant 32°C or 32°C day–27°C night temperature treatments changed the incompatibility reaction (7% of the florets setting seed) to one of compatibility (50% of the florets setting seed). Constant 27°C and 32°C day–21°C night temperature treatments also affected the incompatibility reaction, but no more than 20% of the florets set seed. After the incompatibility reaction had been changed to one of compatibility at 32°C, it took 24 h at 21°C to restore the compatibility reaction to its original incompatibility status. The site for the change in the compatibility reaction was the style. Temperature had no effect on the compatibility reaction of the pollen. Inheritance studies showed that the SC response to temperature (or temperature-sensitivity) was due to a dominant gene designated as T. One dose of the T gene was sufficient to give the response.

In similar physiological studies with diploid *Trifolium hybridum*,

Townsend [102] found that the compatibility reactions of two I_1 clones were temperature-sensitive in controlled environmental chambers. From one to three days at 32°C were required to change the incompatibility reaction to one of compatibility. The maximum response to temperature was obtained on the fourth day of treatment; 35% and 62% of the florets set seed for clones 6-5 and 7-1, respectively. At 32°C day–21°C night, 11% of the florets set seed for clone 6-5, but for clone 7-1 only 3% of the florets set seed. The site of the change in the compatibility reaction was the style for 7-1, but it was not possible to locate the site for clone 6-5.

Studies by Townsend and Danielson [104] showed that the SC reaction of tetraploid *T. hybridum* clone 3-7B was affected only by air temperature and that soil temperature had no effect on the reaction. They also studied the possible translocation of the temperature-induced self-compatibility substance(s) in tetraploid clone 3-7B and diploid clones 6-5 and 7-1. Flowering stems from each propagule of the three clones were arranged so that they were exposed to two air temperatures, 32°C and 23°C, simultaneously. The SC response of individual florets depended on the air temperature at which they were held; i.e. at 23°C only a few selfed seeds were formed but at 32°C many florets contained seeds. The authors concluded that the substance(s) responsible for the change in the compatibility reaction was not translocated from one flowering stem to another on the propagule. The conclusion that the incompatibility substance(s) is not translocated is supported by immunochemical studies with SI and SC inbreds of *Brassica oleracea* var. *capitata* [76]. Stigma extracts possessed specific antigenic properties which were not found in other plant parts.

GENETICS

Gene models

The structure and general nature of the S gene has been the subject of investigation by various researchers. Lewis [49] proposed two gene models as possible explanations for the results obtained from mutation experiments with *Oenothera organensis*. One hypothesis held that the S gene consisted of two parts or cistrons, one controlling specificity of pollen antigen and the other controlling specificity of stylar antigen. The second hypothesis held that one cistron controlled specificity and the other controlled activity of the antigen in both pollen and style. By using the S_4 and S_6 alleles and their mutated forms, $S_{4'}$ and $S_{6'}$, in diploid and autotetraploid forms of *O. organensis*, Lewis found that the evidence supported the second hypothesis. Diploid plants possessing either the $S_{4'}$ or $S_{6'}$ allele produced the usual S_4 or S_6 incompatibility substance in the style but not in the pollen. The S_2 allele restored the activity of the $S_{4'}$ in the diploid pollen, which indicated that the mutation affected only the activity but not the specificity of the S_4 allele.

In a discussion on theory of S gene structure Pandey [79] proposed that 'the physiology of the S gene is controlled by 4 components: growth substance, protective substance, primary specificity and secondary specificity, each with corresponding pollen and stylar units attached in that order, the component of secondary specificity being added last'. According to such a scheme the usual SI alleles (S_1) would possess all of the above features in both style and pollen; an S_F allele would have lost the property of secondary specificity in both style and pollen; and an S_f allele would have lost secondary specificity for the pollen only. He also indicated how other types of self-incompatibility might fit into the scheme.

Lewis [52] and Ascher [4] have proposed gene action models to explain gametophytic incompatibility. The dimer hypothesis of Lewis suggests that an identical polypeptide molecule is produced in the pollen and style by the S gene complex, and that each allele has its characteristic molecule. The two molecules or monomers polymerise to form a dimer protein in both pollen and style. Following an incompatible mating, a tetramer is formed by the polymerisation of identical dimers in pollen and style with the aid of an allosteric molecule; and 'the tetramer acts as a genetic regulator either to induce the synthesis of an inhibitor or to repress the synthesis of an auxin of pollen tube growth'. Lewis discussed this hypothesis in relation to the experimental evidence, particularly that from the S and Z genes found in the Gramineae and that from the competition interaction of alleles in diploid pollen grains. He also concluded that the data support the induction of an inhibitor rather than the repression of an auxin. Ascher's model was very similar to the above model except that he considered the active components to be monomers and dimers rather than dimers and tetramers.

Sampson [92] proposed a hypothesis to explain various gene actions found in the gametophytic and sporophytic incompatibility systems. Basically, his suggestion held that production of an incompatibility substance, specific for one S allele, must reach a certain threshold level before that particular S allele could be expressed.

Mutations

In a theoretical discussion on mutation of the S gene and pseudo-compatibility, Pandey [78] cites the occurrence of four types of S gene mutations: revertible, pollen-part, stylar-part and those affecting both pollen and style. Pseudo-compatibility was described as being due to environmental, artificial and genetical causes. He also discussed the possible effects that various types of crossing over and mutations might have on pseudo-compatibility.

In mutation studies with selected homozygous and heterozygous S genotypes of *Petunia inflata* L., Brewbaker and Shapiro [15] obtained considerable seed set following incompatible matings with irradiated pollen from the heterozygotes, but not with similarly treated pollen from the homozygotes. They suggested that such seed set may have been due to a genetic interchange within the S locus. Brewbaker and Natarajan [14] found that self-compatible plants of *P. inflata* obtained by irradiation carried a centric chromosome fragment. The fragment was presumed to carry the S locus; thus compatibility was attributed to the competition interaction between the normal S allele of the chromosome complement and the one carried on the centric fragment. They indicated that results of other mutation studies might be explained on a similar basis.

Lewis [50] discounted the centric S fragments and their associated competition interaction as an explanation of his mutation results. The two basic reasons were that (*a*) the competition interaction in diploid pollen of *Oenothera organensis* was not sufficient to permit self-fertilisation; and (*b*) completely SC homozygous S mutants were obtained. Pandey [84] also claimed that his mutation data with *Trifolium pratense* and *T. repens* could not be explained by centric S fragments.

Cytogenetic studies with irradiation-induced mutants of *Nicotiana alata* revealed that 42 of the 44 plants examined contained a centric chromosome fragment [84]. One of the two plants without the fragment was considered to be a pollen-part mutant heterozygote. Most of the selfed progeny from this plant were heterozygous at the S locus but there were a few homozygotes. The homozygous plants examined, as well as their progeny, were SC. Pandey [84] suggested that the centric fragment observed in *P. inflata* and *N. alata* may complement a deficiency due to mutant alleles as well as producing a competitive interaction.

Pandey [87] reported that true S gene mutations occurred in *Nicotiana alata,* but that they differed from those in *Oenothera, Prunus* and *Trifolium* in that pollen-part mutations in *N. alata* were lethal unless the S-bearing centric fragment was present. Consequently he indicated that the S-bearing fragment had two roles: one was to provide a competitive interaction with the normal S allele in heterogenic pollen grains, and the second was complementation of the mutated S allele in the normal chromosome complement which otherwise would be lethal. He stated that complementation was the true role of the centric fragment and that the competition interaction phenomenon was a nuisance. Pandey also observed two stylar-part mutations which did not need the centric fragment for viability. This was interpreted as meaning that complementation occurred in the style between the normal S allele and the mutated S allele.

Colchicine treatment produced genetic and non-genetic changes in the SI reaction of *Nicotiana* plants [88]. The genetic effects were due to the competitive interaction of S alleles in the diploid pollen following chromosome doubling. All except one heterozygous genotype was SC at the tetraploid level ($S_{F10}S_{F10}S_fS_f$). Non-genetic changes in the SI reaction were believed to be due to cytoplasmic constituents which influenced the regulation of S gene activity.

The self-incompatibility-inducing substance in the style is sensitive to irradiation. Viable selfed seed was produced in *Petunia hybrida* by X-irradiation of the styles just prior to pollination [59]. Ultra-violet irradiation of styles of *Oenothera organensis* prior to pollination enhanced self pollen tube growth in comparison with pretreatment with visible light [42].

General

The different types of self-incompatibility systems which had been discovered by 1957 were summarised by Pandey [77]. The most recently discovered systems were the two-loci gametophytic types found in the Gramineae and in one species of Solanaceae, *Physalis ixocarpa* Brot. The genetics of incompatibility was reviewed by Lundqvist [61], who also designated the two loci occurring in the Gramineae as S and Z. For incompatibility to occur in the Gramineae, alleles at both loci must be matched in both pollen and style. Lundqvist hypothesised that 'the loci act cooperatively in the elaboration of pollen specificities'. Thus each pair of alleles produces a unique specificity, the number of which would be limited only by the number of S and Z combinations. No evidence was found in polyploids for epistasis or dominance interaction between alleles at the same locus. Lundqvist concluded, therefore, that the Z locus originated from the S locus by duplication and that complementary interaction evolved thereafter. In other studies Lundqvist [62] observed an association between S and Z alleles and pseudo-compatibility in *Festuca pratensis*. In I_1 and I_2 progenies of an $S_1S_2Z_3Z_4$ plant, 'pseudo-compatibility was constantly increased by the presence of S_1 relative to S_2, and Z_4 relative to Z_3'. Lundqvist [63] considered that the S gene specificities of the pollen and style were determined by separate gene segments. Loss of self-incompatibility might be due to the 'unbalanced recombination' between the two segments.

The two-loci gametophytic system found in *Physalis ixocarpa* [77] was similar to that found in the Gramineae, except that (a) the alleles of the two loci may have individual action or epistatic interaction; and (b) if alleles at either one or both loci are matched in the pollen and style, incompatibility results.

In studies with *Solanum pinnatisectum*, Pandey [80] accepted the proposal that incompatibility was controlled by two independent loci, S and R. In addition to the normal S alleles, he found two mutant alleles at the R locus. The latter were designated R_F and R_{IC}. When they were homozygous, they were epistatic over the S alleles. They were not expressed if heterozygous. Elements of sporophytic control were also detected with dominance between S alleles.

Rowlands [91] discussed the relation between the self-incompatibility system and the genetic system itself in sexually propagated species in which self-incompatibility cannot be explained on the basis of S alleles. From studies with *Vicia faba* he concluded 'that the basic genetic system of a species can take over the control of the level of hybridity following the breakdown of a previously efficient S locus self-incompatibility system'.

Identical S alleles were found in different botanical varieties of *Brassica oleracea* [95]. One S allele was detected in all varieties examined. They suggested that the wide distribution of this S allele was probably due to its occurrence in the ancestral form of *B. oleracea* rather than to mutation or other means. In related studies Thompson and Taylor [96] reported that several pollen recessive S alleles occurred in a wide range of botanical varieties of *Brassica oleracea*. In general, such alleles were associated with populations that contained higher frequencies of partially and completely self-compatible plants.

Self-incompatibility in *Tradescantia paludosa* was found to be controlled by the *Nicotiana* type incompatibility system by Anner-stedt and Lundqvist [2]. Induced autotetraploids were self-incompatible. They suggested that there was a basic difference between monocotyledons and dicotyledons in regard to S allele dominance and competition interaction.

Genetic control of SI in pineapple, *Ananas comosus*, and in *Gasteria* spp. conformed to the *Nicotiana* scheme [13]. Pollen tube growth was inhibited in the upper portion of the style in *Ananas* and in the ovaries of *Gasteria*. Two sites of inhibition, stylar and ovarian, were found in the genus *Hermerocallis*.

In a review of incompatibility in the sweet potato, *Ipomoea batatas* (L.) Lam., a hexaploid species, Martin [70] reported the genetics to be very complex. Later Martin [73], using *Ipomoea setifera* Poir, a diploid species, concluded that incompatibility was controlled by the multi-allelic, sporophytic type. By using the information obtained from the diploid species, he suggested that incompatibility in the hexaploid species was due to two loci which were 'epistatically inter-related'.

Work on Trifolium

The frequency of self-compatible plants in populations of diploid and tetraploid *Trifolium hybridum* L. was determined by Townsend [97, 99]. Percentage of florets setting seed for individual plants ranged from 0 to 95%. There were approximately twice as many plants in the diploid population (68%) as in the tetraploid population (31%) that did not set a single seed when selfed. Breeding studies have indicated that plants with a 10% or less seed set always produced highly SI plants, or plants that represented very low levels of SC; therefore plants with a 10% or less seed set were classified as SI. Twenty-five per cent of the tetraploid population was SC, whereas only two diploid plants or 1% of the population were actually SC, because breeding tests indicated that the seed set of other plants was probably due to pseudo-self-compatibility. The inheritance of SC in tetraploid *T. hybridum* was explained on the basis of the competition interaction hypothesis. There was evidence that the strength of the competition interaction between the two S alleles in the diploid pollen was influenced by modifiers or polygenes. After three generations of inbreeding the level of SC remained high in selected progenies.

In similar studies with tetraploid *Trifolium hybridum*, Brewbaker [12] also found continuous variation in expression of SC. He reported that 56% of the population examined was SC. However, he used three seeds per selfed head as the arbitrary division of SC from SI. Brewbaker also suggested that the competition class of alleles might have a selective advantage. If this class of alleles did monopolise the population, the level of SC could increase with subsequent loss of plant vigour due to inbreeding depression. The results of Townsend [97] did not indicate that the competition class of alleles would predominate. New combinations of S alleles should constantly be formed, owing to segregation and recombination. Consequently, old combinations of alleles giving competition interaction should be broken up while new combinations are being formed.

Williams [105], working with diploid *Trifolium hybridum*, observed one pseudo-compatible plant that was homozygous at the S locus. The behaviour of this plant could not be explained with the available data. Williams proposed several possible explanations, one of which was that 'the high pseudo-compatibility shown may have been brought about by a spontaneous mutation either at the S locus itself, or else in a locus directly modifying its expression of pseudo-compatibility'. Pandey [78] reinterpreted ·Williams' data on the basis of male and female part mutations in the S gene.

One of the two SC diploid *Trifolium hybridum* plants found by Townsend [99] carried the self-compatibility (S_f) factor. This factor, inherited as a simple Mendelian character, was allelic to the multiple S series of alleles. The self-compatibility reaction of the second SC plant had gametophytic as well as sporophytic characteristics [101]. For ease of discussion and reference these two kinds of SC were designated as Type I and Type II, respectively.

A genetic locus non-allelic to, but modifying, the S locus was postulated to explain the inheritance of the Type II self-compatibility [101]. This locus, designated A, suppressed S gene action when A was heterozygous. The alleles of A acted independently. The self-compatibility genotype of the SC plant used to develop this hypothesis was considered to be $A_1A_2S_1S_2$. The A_1 allele suppressed the action of the S_1 allele in the pollen but not in the style. The A_1 allele had no apparent effect on the action of the S_2 allele. Likewise, the A_2 allele of the A locus had no apparent effect on either of the two alleles at the S locus. Therefore only A_1S_1-bearing pollen grains were assumed compatible under selfing; thus the expected I_1 genotypes upon selfing a $A_1A_2S_1S_2$ plant would be $A_1A_1S_1S_1$ (SI), $A_1A_1S_1S_2$ (SI), $A_1A_2S_1S_1$ (SC) and $A_1A_2S_1S_2$ (SC) in equal frequencies. The sporophytic effect was shown by the SI × SC cross, $A_1A_1S_1S_1 \times A_1A_2S_1S_1$. All F_1 plants were SI and S_1S_1. The reciprocal cross, $A_1A_2S_1S_1 \times A_1A_1S_1S_1$, was incompatible. Townsend concluded that this type of self-compatibility was similar to that reported by Williams [105].

Inheritance of Type II self-compatibility in diploid *Trifolium hybridum* was studied in different genetic backgrounds [103].

Results from a series of crosses between SC and unrelated SI plants showed that the original hypothesis needed modification. In certain genetic backgrounds the A_1 allele suppressed the action of S alleles other than the S_1. When these SI F_1 plants were backcrossed to a SC plant, only two of 46 backcross plants were SC. In other crosses the numbers of SI and SC F_1 plants that were obtained agreed with those expected according to the hypothesis that the A_1 allele suppressed the action of only the S_1 allele. In general, the deviation of the results obtained from those expected, and the complete absence of SC plants in some progenies, can be explained by the presence of modifying genes or polygenes. Genetic background had tremendous influence on the expression of Type II self-compatibility.

According to Pandey [78], compatibility due to a mutation in the S gene can be distinguished from that due to environmental or other genetic effects by observation of the frequency of S genotypes in the resulting progeny. If compatibility was due to environmental or other genetic effects, the selfed progeny would consist of two homozygous and one heterozygous genotypes (1 S_1S_1: 2 S_1S_2: 1 S_2S_2). If a mutation in the S allele was involved, two genotypes (S_1S_1 and S_1S_2 or S_1S_2 and S_2S_2) would most likely be observed. If all three genotypes were observed, they would probably not be in a 1:2:1 ratio.

In pseudo-self-compatibility (PSC) studies with *Trifolium pratense*, Leffel [45] reported that 32 of 79 I_1 plants classified for S genotype were homozygous. All homozygotes were of the same S genotype. Inheritance of PSC was studied in *T. pratense* by making diallel crosses between clones which were rated high and low for PSC [10]. PSC was found to be a heritable character. They also reported that one homozygous S genotype was missing from I_1 progenies of two PSC clones. The ratio of heterozygous to homozygous plants in these progenies fits a 1:1 hypothesis better than a 2:1 hypothesis. Therefore they concluded that the absence of the second homozygous class was due to selective fertilisation rather than to zygotic lethals.

The segregation of S alleles in I_1 progenies of PSC I_0 plants of *Trifolium pratense* was studied by Johnston *et al.* [38]. Five hundred I_1 plants from 15 I_0 plants were classified for S genotype. One hundred and fifty-nine plants were homozygous and 341 were heterozygous at the S locus. Intercrosses within three of the I_1 progenies showed that both homozygous genotypes were present in two progenies, but only one homozygous genotype was present in the third progeny. The over-all deficiency of homozygous genotypes was believed to be due to a heterozygotic advantage or to zygotic lethality.

Denward [19, 20] conducted a series of compatibility studies with *Trifolium pratense*. New S specificities were observed, but he discounted crossing over in the S locus as a possibility for their origin. He concluded that the S-specificity was due to a locus that was closely linked to, if not identical with, the flower colour locus. Independently segregating modifier genes were believed to influence S specificity and pollen tube growth promoting substances. In addition to being very sensitive to environmental conditions, the modifiers were thought to act on the sporophytic level. There was no difference between the self-incompatibility system and the basic genetic system of the plant itself. Denward indicated that his data supported the hypothesis proposed by Mather [74] in that the strength of the incompatibility reaction was controlled by polygenes and could range from complete self-incompatibility to complete self-compatibility.

In studies with *Trifolium repens*, Cohen and Leffel [17] concluded that PSC was independent of the S locus. In six I_1 progenies the expected $1:2:1$ ratio of S allele genotypes was obtained.

Cohen and Leffel [16] made a cytological investigation of PSC in white clover, *Trifolium repens* L., with two plants of identical S allele genotype. One relatively 'high' PSC plant set 3·45 seeds per head when selfed, while the second plant, which had a 'low' level of PSC, set only 0·17 seed per head. Self pollen germinated rapidly on stigmas of the 'high' PSC plant, but self pollen grains germinated slowly on the stigmas of the 'low' PSC plant. Cohen and Leffel hypothesised that the early pollen germination on the 'high' PSC stigma was accompanied by a release of large quantities of an antigen-like substance. In turn, this substance absorbed the preformed antibody in the style, thus permitting sufficient pollen tube growth to effect limited fertilisation.

The segregation of S alleles was studied in the I_2 progenies derived from two temperature-sensitive I_1 clones of diploid *Trifolium hybridum*, as well as in the F_1 and F_2 progenies of crosses involving these two clones with a temperature-insensitive tester (Townsend, Unpublished). The two temperature-sensitive clones were heterozygous (S_3S_4 and S_5S_6) and the tester was homozygous (S_1S_1) at the S locus. The S alleles of one clone segregated in a $1\ S_3S_3:2\ S_3S_4$: $1\ S_4S_4$ ratio in the I_2 progeny even though there were relatively fewer plants in the S_4S_4 class. One homozygous class (S_6S_6) was completely absent in the I_2 progeny from the second clone, but the other two classes (S_5S_5 and S_5S_6) segregated in a $1:1$ ratio.

The expected S genotypes (S_1S_3, S_1S_4, S_1S_5, and S_1S_6) were obtained in equal frequencies in the two F_1 progenies ($S_1S_1 \times S_3S_4$, $S_1S_1 \times S_5S_6$, and reciprocals). In two F_2 progenies of F_1 plants with an S_1S_3 genotype only two of 54 plants were homozygous S_1S_1. The other two classes (S_1S_3 and S_3S_3) were obtained in equal frequency. None of the 40 plants from three F_2 progenies whose F_1 parents were either S_1S_5 or S_1S_6, was homozygous S_1S_1. There were equal frequencies of S_1S_5 and S_5S_5 or S_1S_6 and S_6S_6 classes. The data indicated that the deficiency or complete absence of one homozygous class in each of the progenies studied was due to selective fertilisation rather than to zygotic lethals.

The inheritance of temperature-sensitivity also was studied in the above I_2, F_1 and F_2 progenies. The data were subjected to several genetic hypotheses, and they were interpreted best by a two-gene model with an interaction between S allele genotype and temperature-sensitivity.

Modifier genes influenced the intensity and stability of self-incompatibility in *Brassica oleracea* var. *capitata* [76c]. Four inbred lines, two homozygous for the S_1 allele and two homozygous for the S_2 allele, were investigated. One inbred for each S allele was highly self-incompatible, while the other inbred of a pair was less self-incompatible. Variation in seed set due to different flowers and plants and sensitivity of the incompatibility reaction to temperature were less for the highly self-incompatible inbred than for the less self-incompatible inbred of the S_1 pair.

Number of S alleles

The frequency of S alleles is high in the populations examined. This finding, taken with the inability to generate new S alleles by mutation, has been of some concern. Several investigators, including Wright [106, 107], Fisher [29], Lewis [51], Ewens [27], Ewens and Ewens [28] and Mayo [75], have considered the problem involving the number of S alleles. In general, the large number of alleles are believed to be due to mutation and a drastic decrease in population size. A statement made by East [22] concerning the number of S alleles probably is still appropriate: 'As to the number of factors existing in material collected from the wild, one can make no valid estimate.'

PLANT BREEDING

The subject on application of incompatibility in plant breeding has been reviewed by Lewis [48], Reimann-Philipp [90], Duvick [21] and Wallace and Nasrallah [104a]. As was pointed out by Reimann-

Philipp, incompatibility studies generally have been aimed at breaking down the incompatibility system rather than at attempting to use the system in a practical breeding programme. Thus far, the application of incompatibility in a breeding programme has been most successful with the Cruciferae. However, there have been problems of detecting and producing homozygous genotypes because members of the Cruciferae possess the sporophytic system. Detecting homozygous genotypes is also a problem with the gametophytic system. In addition, there are relatively few characters known to be closely linked to the S locus. To be used successfully in such a scheme, the species must be easily propagated vegetatively.

One of the primary objectives of the incompatibility work being done with the *Trifolium* spp. is to adapt the incompatibility system for effective and efficient use in an improvement programme. Whether or not these objectives will be realised is problematical. There is the possibility, as mentioned by Reimann-Philipp [90], that other sterility systems, such as cytoplasmic male sterility, will be discovered. If so, cytoplasmic male sterility will probably be used in place of the incompatibility system because of ease of manipulation, provided that insect pollination is not a problem.

UNILATERAL INCOMPATIBILITY

The subject of unilateral incompatibility has received considerable attention in recent years. Such studies have provided information concerning the nature of the self-incompatibility S locus in regard to its structure and evolutionary role. The term 'unilateral hybridisation' was used by Harrison and Darby [32] to describe the condition where species hybrids could be made in one direction only. For example, pollen of the SI species functioned in the styles of the SC species, but pollen of the SC species was inhibited in the styles of the SI species of the reciprocal cross. Unilateral incompatibility and unilateral hybridisation are considered to be one and the same phenomenon.

Restricted pollen tube growth in the SI × SC crosses in unilateral incompatibility (UI) studies was similar to that of the self-pollen of the SI species. This observation suggested that unilateral incompatibility and self-incompatibility were controlled by the S locus; thus the S locus had a dual role. Lewis and Crowe [54] studied the unilateral interspecific incompatibility relationships in species of the families Cruciferae, Onagraceae and Solanaceae, and reviewed the pertinent literature on the subject. They reported that unilateral incompatibility occurred in the gametophytic homomorphic one-locus system, in the sporophytic homomorphic system and in the heteromorphic system. However, it has not been definitely established as occurring in the gametophytic homomorphic two-loci system. In general, the data of Lewis and Crowe supported the assumption that pollen tube growth would not be restricted in $SC \times SI$ and $SI \times SI$ crosses but would be restricted in $SI \times SC$ crosses. However, several SC species were found which behaved as SI species when they were crossed to other SC species. These unexpected results were explained by assuming that such SC species had recently evolved from self-incompatible ones and that they represented an intermediate stage in the evolution of SI to SC.

The evolution of compatibility was thought to be as follows: $SI \rightarrow S_c \rightarrow S_{c'} \rightarrow SC$. Species or biotypes which were self-compatible but behaved in unilateral crosses as self-incompatible types were considered to be in the S_c phase. S_c mutants do not affect the stylar activity, because S_c-bearing styles will accept SI pollen but will reject SC pollen. Representatives of the $S_{c'}$ phase have not been located. Even if self-compatibility did not evolve from self-incompatibility in the manner suggested by Lewis and Crowe, such a change probably did not occur in a single step, because the SC mutant could not become established in the population, owing to unilateral incompatibility.

One of the first cases of unilateral incompatibility reported in the literature, although it was not referred to as such, was in the cross *Nicotiana alata* (SI) $\times$ *N. langsdorffii* (SC), described by Anderson and de Winton [1]. Such crosses were generally reciprocally cross-compatible; however, a biotype of *N. alata* was found which rejected pollen of *N. langsdorffii*. Self-incompatibility and rejection of S_F pollen were believed due to a gene, designated as S_F, an allele of the self-incompatibility locus. The self-incompatible F_1 hybrids obtained from the reciprocal cross ($SC \times SI$) were cross-incompatible as females with the *N. langsdorffii* parent. Anderson and de Winton explained their results by assuming that the *N. langsdorffii* parent contributed two recessive factors which when homozygous caused a self-incompatible hybrid to become pseudo-self-compatible.

Later Pandey [83] studied the incompatibility relations in the *Nicotiana langsdorffii* $\times$ *N. alata* cross and its hybrids. His findings supported those of Anderson and de Winton, except that S_F plants were found to be quite common rather than a rarity. However, only two S_F alleles, S_{F10} and S_{F11}, were found. Pandey also offered a different genetic interpretation from that of Anderson and de Winton. Pandey indicated that his theory on S gene structure would explain the results; namely that the S_F alleles produced two kinds of specificity in the style, one which rejected like- or self-pollen and one which rejected S_f pollen. The genetic units which controlled these reactions were considered to be closely linked. Evidence was presented which indicated that the S_F gene slowed down pollen tube growth. Such an effect was less on S_1 pollen than on S_F pollen; hence the possibility of sporophytic control.

Evidence of S gene polymorphism in interspecific crosses of *Nicotiana* spp. was reported by Pandey [86]. *N. alata* plants with S_{F10} and S_{F11} alleles rejected *N. langsdorffii* pollen, as expected, while the S_{F11} allele accepted and the S_{F10} allele rejected *N. noctiflora* pollen. At the same time the S_1 alleles (normal S alleles) or *N. alata* accepted *N. langsdorffii* pollen, as expected, but certain S_1 alleles accepted and others rejected *N. noctiflora* pollen. A similar situation existed for the acceptance or rejection of *N. bonariensis* pollen on *N. alata* styles.

Four types of UI (SI × SC, SI × SI, SC × SI and SC × SC) were found by Pandey [89] in interspecific crosses of *Nicotiana* spp. In general, the site of inhibition in intraspecific matings was in the style, but in the UI matings the site was in the stigma. Pandey concluded that SI, SC and UI were under the control of the S gene complex, 'a cluster of closely linked, physiologically integrated genetic elements'. He further stated that the polygenic background of progenies of certain crosses influenced the incompatibility reaction to such an extent that the S alleles lost their supremacy and that other major, non-allelic modifier genes then expressed themselves.

An exception to the usual incompatibility scheme (SI × SC), found in many UI studies, was observed between crosses of certain clones of *Solanum chacoense* (SI) and *S. soukupii* with *S. verrucosum* [31]. Later Grun and Aubertin [30], using the same *Solanum* material, studied the genic differences between clones which gave UI and those which did not when crossed to *S. verrucosum*. Unilateral incompatibility was conditioned by 2–4 dominant genes and these genes were considered to differ from the self-incompatibility alleles.

Interspecific crosses between *Solanum* spp. revealed that UI existed between SI as well as between SC and SI species [81]. One SI × SC cross, *S. vernei* × *S. verrucosum*, was fertile. In addition, all hybrids were SC and cross-compatible with both parents. One SC species, *S. polyadenium*, was cross-incompatible as both female and male, with all species tested. Pandey suggested that 'the failure of the principle of unilateral interspecific incompatibility in solanaceous species may be due to the action of alleles at the second incompatibility locus revealed in certain Mexican species'.

From unilateral incompatibility studies with SI genera of the

Cruciferae, Sampson [93] concluded that most intergeneric SI was not S allele specific. He suggested that a molecular code existed which permitted compatible pollen and stigma to recognise one another.

Recently Martin [66–69, 71, 72] has made intensive investigations into the SI and UI relationships in *Lycopersicon* spp. His first study [66] reported the occurrence of UI between SI and SC forms as well as between two SI forms of *L. hirsutum*. On the basis of the hypothesis by Lewis and Crowe [54], the former results were expected but the latter were not. Self-compatible *L. esculentum* was cross-incompatible as male and cross-compatible as female with all forms of *hirsutum*.

The same genetic system controlled self-incompatibility and unilateral incompatibility in the F_1, F_2, and backcross progenies of the cross *Lycopersicon esculentum* (SC) × *L. chilense* (SI) [67]. Self-incompatibility in *L. chilense* was of the oppositional S allele type. In the backcross progenies to *L. esculentum*, segregation of self-compatible and self-incompatible plants approximated a 3 SC:1 SI ratio. The data were interpreted as indicating that a major independent dominant gene from the *L. chilense* parent, in addition to the S allele, was required for the expression of self-incompatibility. Polygenes influenced the strength of the compatibility reaction. A second sterility barrier, resulting in poor pollen germination and embryo abortion, underlaid that due to the S allele system. Three genes on chromosome two and several quantitative characters were linked with SI and UI.

Martin's compatibility studies with *Lycopersicon hirsutum* f. *hirsutum* and *L. hirsutum* f. *glabratum* revealed that a wide range in degree of self-compatibility, self-incompatibility and unilateral incompatibility occurred in these lines [68]. The similarity between the stepwise pattern of the crossing reactions among the different unilateral incompatible groups and pollen tube growth led Martin to suggest that one physiological process controlled unilateral incompatibility and that the strength of the reaction was affected by polygenes. He further proposed that pollen tube growth and the resulting crossing patterns were due to 'a variable amount of pollen-tube, growth-promoting substance in pollen, complemented by an appropriate amount of an inhibiting substance in the style, which results in a controlled balance of growth substances'.

From backcross, F_1 and F_2 progeny data of crosses between *Lycopersicon hirsutum* f. *glabratum* and *L. hirsutum* f. *hirsutum* (SC × SI) Martin [69] reported that both self-incompatibility and unilateral incompatibility appeared to be controlled by two dominant genes from the SI parent, f. *hirsutum*. Similar data from crosses between SC forms of *glabratum* and *hirsutum* suggested that one dominant gene from the *hirsutum* parent controlled both compatibility reactions. Evidence of polygenic modifiers was found in progenies from both crosses.

Martin [71] reported that two major genes controlled self-incompatibility and unilateral incompatibility in progenies of crosses between *Lycopersicon esculentum* (SC) and *L. hirsutum* (SI form). In the segregating progenies of the cross *L. esculentum* × *L. hirsutum* (SC form), evidence was obtained which indicated that relic major incompatibility genes, as well as polygenes, conditioned self-incompatibility and unilateral incompatibility.

The self-incompatibility alleles from *Lycopersicon peruvianum* v. *dentatum* (SI) were transferred to *L. esculentum* (SC) through a series of six backcrosses [72]. Two major genes from the SI parent were believed to control unilateral incompatibility. Martin suggested that a switch gene regulated the activity and that the S alleles regulated the specificity of the incompatibility system. The normal S alleles in the hybrids possessed the usual stylar property of inhibiting pollen tubes with like S alleles as well as pollen tubes bearing the S_f allele. However, such S alleles in the style had lost the property of permitting unrestricted pollen tube growth of pollen grains possessing different S alleles. Consequently the likelihood of producing a SI form of *L. esculentum* seemed remote.

Martin [72] indicated that the SC types had a recessive, inactive allele at the switch gene locus and a non-specific (S_f) allele at the S locus. He further hypothesised that a one-step mutation at the switch gene locus resulted in self-compatibility and that this was followed by a loss of S gene properties. If this explanation is correct, the evolution of SI to SC can occur without a series of mutations at the S locus as was suggested by Lewis and Crowe [54].

In general, Martin's studies with *Lycopersicon* spp. have shown that: (*a*) unilateral incompatibility and self-incompatibility are controlled by the same genetic system; (*b*) two or more major genes from the SI parent of the SC × SI crosses control the compatibility reactions; (*c*) relic major incompatibility genes control the compatibility reactions in SC × SC crosses; and (*d*) the underlying strengths of the compatibility reactions are influenced by polygenes.

BIOCHEMISTRY

The biochemistry of incompatibility has received considerable attention in relatively recent years, and Linskens [56, 57] has reviewed the literature pertaining to this subject. Consequently only the highlights of these reviews will be discussed. The stigmas of some plants with sporophytic incompatibility have a cuticular layer which serves as an incompatibility barrier. Pollen of such plants possesses a cutinase enzyme system capable of 'destroying' the cuticular layer, thus overcoming the incompatibility barrier. Several unusual phenomena have been observed following incompatible matings of plants with the gametophytic system. The generative nucleus of many pollen grains failed to divide, while the vegetative nucleus disappeared shortly after pollen grain germination. Carbohydrate metabolism was influenced, as evidenced by increased fibril deposits, number of callose plugs, branching of the pollen tube tips, respiration rate, and activity of cytochrome oxidase and phosphatases. A change also occurred in the protein fraction, and differences were found among the unpollinated, incompatible pollinated and compatible pollinated styles for nucleic acid content.

Even though S alleles were suspected for many years of having individual antigenic properties, Lewis [47] was the first to demonstrate such properties. Antisera were produced by injecting pollen extracts from several S genotypes of *Oenothera organensis* into rabbits. Homologous pollen extracts gave stronger precipitation reactions than did heterologous pollen extracts. In related studies with stylar proteins of *Petunia*, Linskens [55] obtained similar results. Serological studies by Mäkinen and Lewis [65] confirmed the work of Lewis [47]. They considered the site of S protein action to be near the surface of the pollen tube because the S protein diffused equally well from intact and macerated pollen. Additional immunological studies by Lewis *et al.* [53] confirmed Mäkinen and Lewis' observation [65] that the S protein from pollen grains of *O. organensis* was diffusible. Amylase and invertase diffused from the pollen grains also, but the phosphatases did not. They also reported the separation of S proteins by electrophoresis on polyacrylamide gels.

Immunochemical studies with *Brassica oleracea* var. *capitata* by Nasrallah and Wallace [76a] demonstrated that antigens diffused from unmutilated stigmas. They also found an association of increased self-incompatibility with the appearance of certain antigens in flower development. Nasrallah and Wallace [76b] reported that the specific antigenic properties were due to specific S alleles.

Stanley and Linskens [94] also observed substance(s) which diffused from germinating pollen of *Petunia*. They indicated that the materials were protein or flavonoid compounds which might be involved in the incompatibility reaction. They stated 'that the freely diffusing proteins and constituents forming *in vivo* are the principal molecules that must be analysed if we are to understand incompatibility reactions at the biochemical level'.

Linskens [57] concluded that the immunological theory probably provided the best explanation of the incompatibility reaction, but he also considered the roles that the clonal selection theory and the concepts of enzyme induction, feedback and end-product-inhibition might play in the incompatibility reaction.

Linskens and Tupý [60] investigated the amino acid pools in the styles of SI strains of *Petunia* after self- and cross-pollination. They found evidence for two pools: 'a storage pool to serve as source of oxygen for respiration, and another small pool which provides amino acids for protein synthesis'. These findings were discussed in regard to the dimer hypothesis of Lewis [52].

Peroxidase isozymes were reported to be involved in the self-compatibility reactions of *Nicotiana alata*, *N. langsdorffii*, and their F_1 hybrids [85]. The enzyme components of the stylar extracts were separated on acrylamide gel by electrophoresis. The different S genotypes were characterised by definite electrophoretic patterns. The same S allele gave the same banding pattern even when in different genetic backgrounds. Pandey suggested 'that (1) peroxidase isozymes determine S gene specificity and (2) the basis of allelism of this gene lies in the particular combinations of the specific isozymes'.

Heslop-Harrison [36] discussed the possible relation among protein synthesis, pollen cytology and type of incompatibility system. The ribosome population of the pollen mother cell is drastically reduced in mid-prophase of meiosis. Consequently protein synthesis is probably lost or at a very low level until restoration of the ribosome population occurs at the conclusion of the meiotic mitoses. Since the generative nucleus of the binucleate pollen grain does not divide until pollen grain germination, there is competition among the protein-synthesising processes for ribosome sites. Inhibition of pollen tube growth for the gametophytic system does not occur until the tube has progressed some distance into the style. This is believed to be due to the delay in the synthesis of the inhibiting S protein. On the other hand, trinucleate pollen grains are characteristic of the sporophytic system. S protein synthesis may occur in the meiocyte before ribosome elimination or the tapetum may be involved in some manner. In any event, the S protein probably has been synthesised by the time of pollen grain maturation because the site of the inhibiting reaction is the stigma.

CONCLUSIONS

The self-incompatibility system has been attacked on various fronts. The effects of various physiological treatments on the self-incompatibility reaction have been investigated, but relatively high temperatures have been the most effective in breaking down or changing the incompatibility reaction in the species tested. Genetic investigations have continued to occupy much of the research effort devoted to self-incompatibility. While emphasis on induced mutation of the S locus has diminished somewhat, studies on the interaction between S genotype and environment have increased. Contributions to the genetics of unilateral incompatibility have been significant. The biochemical nature of incompatibility has received considerable impetus in relatively recent years. Increased activity in this very important area has paralleled the advent of more sophisticated instrumentation in the field of biochemistry.

Joint contribution of the Crops Research Division, Agricultural Research Service, U.S. Department of Agriculture, and Colorado Agricultural Experiment Station, Fort Collins, Colorado 80521. Scientific Series No. 1480.

REFERENCES

1 ANDERSON, E. and DE WINTON, D., 'The genetic analysis of an unusual relationship between self-sterility and self-fertility in *Nicotiana*', *Ann. Mo. bot. Gdn*, **18**, 97–116 (1931)

2 ANNERSTEDT, I. and LUNDQVIST, A., 'Genetics of self-incompatibility in *Tradescantia paludosa* (Commelinaceae)', *Hereditas*, **58**, 13–30 (1967)

3 ARASU, N. T., 'Self-incompatibility in angiosperms: a review', *Genetica*, **39**, 1–24 (1968)

4 ASCHER, P. D., 'A gene action model to explain gametophytic self-incompatibility', *Euphytica*, **15**, 179–83 (1966)

5 ASCHER, P. D. and PELOQUIN, S. J., 'Effect of floral aging on the growth of compatible and incompatible pollen tubes in *Lilium longiflorum*', *Am. J. Bot.*, **53**, 99–102 (1966)

6 ASCHER, P. D. and PELOQUIN, S. J., 'Influence of temperature on incompatible and compatible pollen tube growth in *Lilium longiflorum*', *Can. J. Genet. Cytol.*, **8**, 661–4 (1966)

7 ASCHER, P. D. and PELOQUIN, S. J., 'Pollen tube growth and incompatibility following intra- and inter-specific pollinations in *Lilium longiflorum*', *Am. J. Bot.*, **55**, 1230–4 (1968)

8 BALI, P. N., 'Some experimental studies on the self-incompatibility of *Oenothera rhombipetala* Nutt', *Phyton, B. Aires*, **20**, 97–103 (1963)

9 BALI, P. N. and HECHT, A., 'The genetics of self-incompatibility in *Oenothera rhombipetala*', *Genetica*, **36**, 159–71 (1965)

10 BRANDON, R. A. and LEFFEL, R. C., 'Pseudo-self-compatibilities of a diallel cross and sterility-allele genotypic ratios in red clover (*Trifolium pratense* L.)', *Crop Sci.*, **8**, 185–6 (1968)

11 BREWBAKER, J. L., 'Pollen cytology and self-incompatibility systems in plants', *J. Hered.*, **48**, 271–7 (1957)

12 BREWBAKER, J. L., 'Self-compatibility in tetraploid strains of *Trifolium hybridum*', *Hereditas*, **44**, 547–53 (1958)

13 BREWBAKER, J. L. and GORREZ, D. D., 'Genetics of self-incompatibility in the monocot genera *Ananas* (Pineapple) and *Gasteria*', *Am. J. Bot.*, **54**, 611–16 (1967)

14 BREWBAKER, J. L. and NATARAJAN, A. T., 'Centric fragments and pollen-part mutation of incompatibility alleles in *Petunia*', *Genetics, Princeton*, **45**, 699–704 (1960)

15 BREWBAKER, J. L. and SHAPIRO, N., 'Homozygosity and S gene mutation', *Nature, Lond.*, **183**, 1209–10 (1959)

16 COHEN, M. M. and LEFFEL, R. C., 'Cytology of pseudo-self-compatibility in Ladino white clover, *Trifolium repens* L.', *Crop Sci.*, **3**, 430–3 (1963)

17 COHEN, M. M. and LEFFEL, R. C., 'Pseudo-self-compatibility and segregation of gametophytic self-incompatibility alleles in white clover, *Trifolium repens* L.', *Crop Sci.*, **4**, 429–31 (1964)

18 DENWARD, T., 'The function of the incompatibility alleles in red clover (*Trifolium pratense* L.). I. The effect of grafting upon self-fertility', *Hereditas*, **49**, 189–202 (1963)

19 DENWARD, T., 'The function of the incompatibility alleles in red clover (*Trifolium pratense* L.). II. Results of crosses within inbred families', *Hereditas*, **49**, 203–36 (1963)

20 DENWARD, T., 'The function of the incompatibility alleles in red clover (*Trifolium pratense* L.). III. Changes in the S-specificity', *Hereditas*, **49**, 285–329 (1963)

21 DUVICK, D. N., 'Influence of morphology and sterility on breeding methodology'. In: K. J. FREY (Ed.), *Plant Breeding*, 85–138, The Iowa State University Press, Ames, Iowa (1966)

22 EAST, E. M., 'Self-sterility', *Bibliog. Genet.*, **5**, 331–70 (1929)

23 EAST, E. M. and MANGELSDORF, A. J., 'A new interpretation of the hereditary behavior of self-sterile plants', *Proc. natn. Acad. Sci. U.S.A.*, **11**, 166–71 (1925)

24 EL MURABAA, A. I. M., 'Effect of high temperature on incompatibility in radish', *Euphytica*, **6**, 268–70 (1957)

25 EMSWELLER, S. L. and UHRING, J., 'Interaction of temperature and growth regulator in overcoming self-incompatibility in *Lilium longiflorum* Thunb.', *Hereditas*, **52**, 295–306 (1965)

26 EVANS, A. M., 'Relationship between vegetative and sexual compatibility in *Trifolium*', *Rep. Welsh Pl. Breed. Stn Rept.*, 81–7 (1959)

27 EWENS, W. J., 'On the problem of self-sterility alleles', *Genetics, Princeton*, **50**, 1433–8 (1964)

28 EWENS, W. J. and EWENS, P. M., 'The maintenance of alleles by mutation—Monte Carlo results for normal and self-sterility populations', *Heredity, Lond.*, **21**, 371–8 (1966)

29 FISHER, R., 'A model for the generation of self-sterility alleles', *J. theor. Biol.*, **1**, 411–14 (1961)

30 GRUN, P. and AUBERTIN, M., 'The inheritance and expression of unilateral incompatibility in *Solanum*', *Heredity, Lond.*, **21**, 131–8 (1966)

31 GRUN, P. and RADLOW, A., 'Evolution of barriers to crossing of self-incompatible with self-compatible species of *Solanum*', *Heredity, Lond.*, **16**, 137–43 (1961)

32 HARRISON, B. J. and DARBY, L. A., 'Unilateral hybridization', *Nature, Lond.*, **176**, 982 (1955)

33 HECHT, A., 'Growth of pollen tubes of *Oenothera organensis* through otherwise incompatible styles', *Am. J. Bot.*, **47**, 32–6 (1960)

34 HECHT, A., 'Partial inactivation of an incompatibility substance in the stigmas and styles of *Oenothera*'. In: H. F. LINSKENS (Ed.), *Pollen Physiology and Fertilization*, 237–43, North-Holland, Amsterdam (1964)

35 HECHT, A., 'Inactivation of incompatibility', *Am. J. Bot.*, **53**, 615 (Abstr.) (1966)

36 HESLOP-HARRISON, J., 'Ribosome sites and S gene action', *Nature, Lond.*, **218**, 90–1 (1968)

37 HOPPER, J. E., ASCHER, P. D. and PELOQUIN, S. J., 'Inactivation of self-incompatibility following temperature pretreatments of styles in *Lilium longiflorum*', *Euphytica*, **16**, 215–20 (1967)

38 JOHNSTON, K., TAYLOR, N. L. and KENDALL, W. A., 'Occurrence of two homozygous self-incompatibility genotypes in I_1 segregates of red clover, *Trifolium pratense* L.', *Crop Sci.*, **8**, 611–14 (1968)

39 KENDALL, W. A., 'Growth of *Trifolium pratense* L. pollen tubes in compatible and incompatible styles of excised pistils', *Theoret. appl. Genet.*, **38**, 351–4 (1968)

40 KENDALL, W. A. and TAYLOR, N. L., 'Growth of red clover pollen', *Crop Sci.*, **5**, 241–3 (1965)

41 KENDALL, W. A. and TAYLOR, N. L., 'Effect of temperature on pseudo-self-compatibility in *Trifolium pratense* L.', *Theoret. appl. Genet.*, **39**, 123–6 (1969)

42 KUMAR, S. and HECHT, A., 'Inactivation of incompatibility in *Oenothera organensis* following ultraviolet irradiation', *Naturwissenschaften*, **52**, 398–9 (1965)

43 KUMAR, S. and HECHT, A., 'Growth of styles and pollen tubes of *Oenothera organensis* following emasculation', *J. Hered.*, **56**, 125–6 (1965)

44 KWACK, B. H., 'Stylar culture of pollen and physiological studies of self-incompatibility in *Oenothera organensis*', *Physiologia Pl.*, **18**, 297–305 (1965)

45 LEFFEL, R. C., 'Pseudo-self-compatibility and segregation of gametophytic self-incompatibility alleles in red clover, *Trifolium pratense* L.', *Crop Sci.*, **3**, 377–80 (1963)

46 LEWIS, D., 'The physiology of incompatibility in plants. I. The effect of temperature', *Proc. R. Soc. B.*, **131**, 13–26 (1942)

47 LEWIS, D., 'Serological reactions of pollen incompatibility substances', *Proc. R. Soc. B.*, **140**, 127–35 (1952)

48 LEWIS, D., 'Incompatibility and plant breeding', *Brookhaven Symp. Biol.*, **9**, 89–100 (1956)

49 LEWIS, D., 'Genetic control of specificity and activity of the S antigen in plants', *Proc. R. Soc. B.*, **151**, 468–77 (1960)

50 LEWIS, D., 'Chromosome fragments and mutation of the incompatibility gene', *Nature, Lond.*, **190**, 990–1 (1961)

51 LEWIS, D., 'The generation of self-incompatibility alleles', *J. theor. Biol.*, **2**, 69–71 (1962)

52 LEWIS, D., 'A protein dimer hypothesis on incompatibility'. In: S. J. GEERTS (Ed.), *Genetics Today* (Proc. XI Int. Congr. Genet., 1963), Vol. 3, 657–63, Pergamon, Oxford (1965)

53 LEWIS, D., BURRAGE, S. and WALLS, D., 'Immunological reactions of single pollen grains; electrophoresis and enzymology of pollen protein exudates', *J. exp. Bot.*, **18**, 371–8 (1967)

54 LEWIS, D. and CROWE, L. K., 'Unilateral interspecific incompatibility in flowering plants', *Heredity, Lond.*, **12**, 233–56 (1958)

55 LINSKENS, H. F., 'Zur Frage der Entstehung der Abwehrkörper bei der Inkompatibilitätsreaktion von *Petunia*. Mitteilung: Serologische Teste mit Leitgewels— und Pollen-Extrakten', *Z. Bot.*, **48**, 126–35 (1960)

56 LINSKENS, H. F., 'Biochemical aspects of incompatibility', *Recent Adv. Bot.*, **2**, 1500–3 (1961)

57 LINSKENS, H. F., 'Biochemistry of incompatibility'. In: S. J. GEERTS (Ed.), *Genetics Today* (Proc. XI Int. Congr. Genet., 1963), Vol. 3, 629–35, Pergamon, Oxford (1965)

58 LINSKENS, H. F., 'The influence of castration on pollen tube growth after self-pollination'. In: H. F. LINSKENS (Ed.), *Pollen Physiology and Fertilization*, 230–6, North-Holland, Amsterdam (1964)

59 LINSKENS, H. F., SCHRAUWEN, J. A. M. and VAN DEN DONK, M., 'Überwindung der Selbstinkompatibilität durch Röntgenbestrahlung des Griffels', *Naturwissenschaften*, **47**, 547 (1960)

60 LINSKENS, H. F. and TUPÝ, J., 'The amino acids pool in the style of self-incompatible strains of *Petunia* after self- and cross-pollination', *Züchter*, **36**, 151–8 (1966)

61 LUNDQVIST, A., 'The genetics of incompatibility'. In: S. J. GEERTS (Ed.), *Genetics Today* (Proc. XI Int. Congr. Genet., 1963), Vol. 3, 637–47, Pergamon, Oxford (1965)

62 LUNDQVIST, A., 'The nature of the two-loci incompatibility system in grasses. IV. Interaction between the loci in relation to pseudo-compatibility in *Festuca pratensis* Huds.', *Hereditas*, **52**, 221–34 (1965)

63 LUNDQVIST, A., 'The mode of origin of self-fertility in grasses', *Hereditas*, **59**, 413–26 (1968)

64 MAHESHWARI, P. and RANGASWAMY, N. S., 'Embryology in relation to physiology and genetics'. In: R. D. PRESTON (Ed.), *Advances in Botanical Research*, Vol. 2, 219–321, Academic Press, London (1965)

65 MÄKINEN, Y. L. A. and LEWIS, D., 'Immunological analysis of incompatibility (S) proteins and of cross-reacting material in a self-compatible mutant of *Oenothera organensis*', *Genet. Res.*, **3**, 352–63 (1962)

66 MARTIN, F. W., 'Complex unilateral hybridization in *Lycopersicon hirsutum*', *Proc. natn. Acad. Sci. U.S.A.*, **47**, 855–7 (1961)

67 MARTIN, F. W., 'The inheritance of self-incompatibility in hybrids of *Lycopersicon esculentum* Mill. × *L. chilense* Dun.', *Genetics, Princeton*, **46**, 1443–54 (1961)

68 MARTIN, F. W., 'Distribution and interrelationships of incompatibility barriers in the *Lycopersicon hirsutum* Humb. & Bonpl. complex', *Evolution, Lancaster, Pa.*, **17**, 519–28 (1963)

69 MARTIN, F. W., 'The inheritance of unilateral incompatibility in *Lycopersicon hirsutum*', *Genetics, Princeton*, **50**, 459–69 (1964)

70 MARTIN, F. W., 'Incompatibility in sweet potato, a review', *Econ. Bot.*, **19**, 406–15 (1965)

71 MARTIN, F. W., 'The genetic control of unilateral incompatibility between two tomato species', *Genetics, Princeton*, **56**, 391–8 (1967)

72 MARTIN, F. W., 'The behavior of *Lycopersicon* incompatibility alleles in an alien genetic milieu', *Genetics, Princeton*, **60**, 101–9 (1968)

73 MARTIN, F. W., 'The system of self-incompatibility in *Ipomoea*', *J. Hered.*, **59**, 262–7 (1968)

74 MATHER, K., 'Specific differences in *Petunia*. I. Incompatibility', *J. Genet.*, **45**, 215–35 (1943)

75 MAYO, O., 'On the problem of self-incompatibility alleles', *Biometrics*, **22**, 111–20 (1966)

76 NASRALLAH, M. E. and WALLACE, D. H., 'Immunochemical detection of incompatibility antigens in stigmas of self-incompatible and self-fertile inbreds of *Brassica oleracea* var. *capitata*', *Genetics, Princeton*, **54**, 351 (Abstr.) (1966)

76a NASRALLAH, M. E. and WALLACE, D. H., 'Immunochemical detection of antigens in self-incompatibility genotypes of cabbage', *Nature, Lond.*, **213**, 700–1 (1967)

76b NASRALLAH, M. E. and WALLACE, D. H., 'Immunogenetics of self-incompatibility in *Brassica oleracea* L.', *Heredity, Lond.*, **22**, 519–27 (1967)

76c NASRALLAH, M. E. and WALLACE, D. H., 'The influence of modifier genes on the intensity and stability of self-incompatibility in cabbage', *Euphytica*, **17**, 495–503 (1968)

77 PANDEY, K. K., 'Genetics of self-incompatibility in *Physalis ixocarpa* Brot.—A new system', *Am. J. Bot.*, **44**, 879–87 (1957)

78 PANDEY, K. K., 'Mutations of the self-incompatibility gene (S) and pseudo-compatibility in angiosperms', *Lloydia*, **22**, 222–34 (1959)

79 PANDEY, K. K., 'A theory of S gene structure', *Nature, Lond.*, **196**, 236–8 (1962)

80 PANDEY, K. K., 'Genetics of incompatibility behaviour in the Mexican *Solanum* species *S. pinnatisectum*', *Z. VererbLehre*, **93**, 378–88 (1962)

81 PANDEY, K. K., 'Interspecific incompatibility in *Solanum* species', *Am. J. Bot.*, **49**, 874–82 (1962)

82 PANDEY, K. K., 'Stigmatic secretion and bud-pollinations in self- and cross-incompatible plants', *Naturwissenschaften*, **50**, 408–9 (1963)

83 PANDEY, K. K., 'Elements of the S-gene complex. I. The S_{F1} alleles in *Nicotiana*', *Genet. Res.*, **5**, 397–409 (1964)

84 PANDEY, K. K., 'Centric chromosome fragments and pollen-part mutation of the incompatibility gene in *Nicotiana alata*', *Nature, Lond.*, **206**, 792–5 (1965)

85 PANDEY, K. K., 'Origin of genetic variability: Combinations of peroxidase isozymes determine multiple allelism of the S gene', *Nature, Lond.*, **213**, 669–72 (1967)

86 PANDEY, K. K., 'S gene polymorphism in *Nicotiana*', *Genet. Res.*, **10**, 251–9 (1967)

87 PANDEY, K. K., 'Elements of the S-gene complex. II. Mutation and complementation at the S_1 locus in *Nicotiana alata*', *Heredity, Lond.*, **22**, 255–83 (1967)

88 PANDEY, K. K., 'Colchicine-induced changes in the self-incompatibility behaviour of *Nicotiana*', *Genetica*, **39**, 257–71 (1968)

89 PANDEY, K. K., 'Compatibility relationships in flowering plants: Role of the S-gene complex', *Am. Nat.*, **102**, 475–89 (1968)

90 REIMANN-PHILIPP, R., 'The application of incompatibility in plant breeding'. In: S. J. GEERTS (Ed.), *Genetics Today* (Proc. IX Int. Congr. Genet., 1963), Vol. 3, 649–56, Pergamon, Oxford (1965)

91 ROWLANDS, D. G., 'Self-incompatibility in sexually propagated cultivated plants', *Euphytica*, **13**, 157–62 (1964)

92 SAMPSON, D. R., 'An hypothesis of gene interaction at the S locus in self-incompatibility systems of angiopserms', *Am. Nat.*, **94**, 283–92 (1960)

93 SAMPSON, D. R., 'Intergeneric pollen-stigma incompatibility in the Cruciferae', *Can. J. Genet. Cytol.*, **4**, 38–49 (1962)

94 STANLEY, R. G. and LINSKENS, H. F., 'Protein diffusion from germinating pollen', *Physiologia Pl.*, **18**, 47–53 (1965)

95 THOMPSON, K. F. and TAYLOR, J. P., 'Identical S alleles in different botanical varieties of *Brassica oleracea*', *Nature, Lond.*, **208**, 306–7 (1965)

96 THOMPSON, K. F. and TAYLOR, J. P., 'The breakdown of self-incompatibility in cultivars of *Brassica oleracea*', *Heredity, Lond.*, **21**, 637–48 (1966)

97 TOWNSEND, C. E., 'Self-compatibility studies with tetraploid alsike clover, *Trifolium hybridum* L.', *Crop Sci.*, **5**, 295–9 (1965)

98 TOWNSEND, C. E., 'Seasonal and temperature effects on self-compatibility in tetraploid alsike clover, *Trifolium hybridum* L.', *Crop Sci.*, **5**, 329–32 (1965)

99 TOWNSEND, C. E., 'Self-compatibility studies with diploid alsike clover, *Trifolium hybridum* L. I. Frequency of self-compatible plants in diverse populations and inheritance of a self-compatibility factor (S_f)', *Crop Sci.*, **5**, 358–60 (1965)

100 TOWNSEND, C. E., 'Self-compatibility response to temperature and the inheritance of the response in tetraploid alsike clover, *Trifolium hybridum* L.', *Crop Sci.*, **6**, 409–14 (1966)

101 TOWNSEND, C. E., 'Self-compatibility studies with diploid alsike clover, *Trifolium*

hybridum L. II. Inheritance of a self-compatibility factor with gametophytic and sporophytic characteristics', *Crop Sci.*, **6**, 415–19 (1966)

102 TOWNSEND, C. E., 'Self-compatibility studies with diploid alsike clover, *Trifolium hybridum* L. III. Response to temperature', *Crop Sci.*, **8**, 269–72 (1968)

103 TOWNSEND, C. E., 'Self-compatibility studies with diploid alsike clover, *Trifolium hybridum* L. IV. Inheritance of Type II self-compatibility in different genetic backgrounds', *Crop Sci.*, **9**, 443–6 (1969)

104 TOWNSEND, C. E. and DANIELSON, R. E., 'Non-translocation of temperature-induced self-compatibility substance(s) in alsike clover, *Trifolium hybridum* L.', *Crop Sci.*, **8**, 493–5 (1968)

104a WALLACE, D. H. and NASRALLAH, M. E., 'Pollination and serological procedures for isolating incompatibility genotypes in the Crucifers', *Mem. Cornell Univ. agric. Exp. Stn*, No. 406 (1968)

105 WILLIAMS, W., 'Genetics of incompatibility in alsike clover, *Trifolium hybridum*', *Heredity, Lond.*, **5**, 51–73 (1951)

106 WRIGHT, S., 'On the number of self-incompatibility alleles maintained in equilibrium by a given mutation rate in a population of given size: a re-examination', *Biometrics*, **16**, 61–85 (1960)

107 WRIGHT, S., 'The distribution of self-incompatibility alleles in populations', *Evolution, Lancaster, Pa.*, **18**, 609–19 (1964)

Incompatibility in *Trifolium pratense* L.*

W. A. Kendall and N. L. Taylor, *Crops Research Division, Agricultural Research Service, U.S.D.A., University of Kentucky, Lexington, Kentucky, U.S.A.*

Red clover (*Trifolium pratense* L.) is generally self- and inter-specifically incompatible. In both types of incompatibility there is a failure of pollen tubes to grow through the styles. Some early studies with red clover showed that a few seeds could be obtained when a large number of florets were self-pollinated. This phenomenon was named pseudo-compatibility. Pseudo-compatibility appeared to be influenced by the plant genotype, and it was particularly susceptible to the plant environment [5–7].

We initiated our *in vitro* studies of red clover pollen with a medium commonly used for many plant species. This medium contained sucrose, calcium nitrate, boric acid and agar. Red clover pollen germinated and formed pollen tubes less than 1 mm long when cultured on this medium. However, we were interested in pollen tubes 10–15 mm long; and in consideration of our needs we concluded that red clover pollen germinated, but did not elongate, on the general medium [1].

Pollen of red clover frequently germinated up to 90% and formed pollen tubes with the modal length of 2·5 mm, with the longest from 5 to 10 mm, while submerged in a medium that contained boric acid (50 p.p.m.), calcium nitrate (2000 p.p.m.), yeast extract (200 p.p.m.) and sucrose at 30%. Pollen tube elongation was inhibited by agar in the semi-solid and dispersed (unheated) condition. Various treatments to alter the oxygen content of the substrate did not influence elongation. Elongation was reduced by a few p.p.m. of potassium cyanide and [6]N-benzyladenine, which indicates that oxygen was used in some of the metabolic process of elongation [2].

We studied effects of various auxins, gibberellic acid and kinetin on red clover pollen tube elongation. The materials were used

* Abstract.

310

separately and in all combinations. High concentrations of auxin were inhibitory, and none of the treatments enhanced germination or tube elongation of red clover pollen.

The excised pistil or semi-*in-vitro* technique, which involved culturing only the stigma and style of the pistil on nutrient media, was used to study effects of chemicals on pollen growth in styles. Pollen tubes in matings that were compatible and incompatible did not grow through styles on the pollen medium. Raffinose was the most suitable carbohydrate, and boric acid was essential for pollen tube growth through excised pistils. None of the organic or inorganic nutrients commonly used for pollen *in vitro* had a specific effect on incompatibility. High temperatures during anthesis inactivated a part of the incompatibility mechanism located in the styles. The greatest pollen tube growth occurred at cool temperatures in both mating types [3].

The excised pistil technique was used to study the mechanism of incompatibility. Chemicals that might function as mutagens, auxins, anti-auxins and inhibitors for antigen–antibody-type reactions, protein synthesis and peroxidases were added to the medium. The chemicals were used with incompatible matings at several concentrations above and below the level that caused an inhibition in 50% of compatible matings. None of the chemicals had a specific effect on self-incompatibility.

A relatively high temperature treatment, applied during anthesis, enhanced self-seed production on excised stems. The stems were excised when petal colour was beginning to appear in the buds, and they were held in a solution of 2·5% sucrose. During anthesis the cultures were incubated with the flower heads at 40°C and the stems at 25°C. After pollination the cultures were held at 20°C during the period of pollen growth through the styles, and also during seed development. The quantities of seed obtained by this procedure have been satisfactory for plant breeding purposes [4].

Cooperative investigations of the Crops Research Division, Agricultural Research Service, U.S. Department of Agriculture and Kentucky Agriculture Experiment Station at Lexington, Kentucky.

REFERENCES

1 KENDALL, W. A. and TAYLOR, N. L., 'Growth of red clover pollen', *Crop Sci.*, **5**, 241–3 (1965)

2 KENDALL, W. A., 'Growth of red clover pollen. II. Elongation in vitro', *Crop Sci.*, **7**, 342–4 (1967)

3 KENDALL, W. A., 'Growth of *Trifolium pratense* L. pollen tubes in compatible and incompatible styles of excised pistils', *Theoret. appl. Genet*, **38**, 351–4 (1968)

4 KENDALL, W. A. and TAYLOR, N. L., 'Effect of temperatures on pseudo-self-compatibility in *Trifolium pratense* L.', *Theoret. appl. Genet.*, **39**, 123–6 (1969)

5 LEFFEL, R. C., 'Pseudo-self-compatibility and segregation of gametophytic self-incompatibility alleles in red clover, *Trifolium pratense* L.', *Crop Sci.*, **3**, 377–80 (1963)

6 MÜLLER, GERTRUD, 'Untersuchungen uber das pollen schlauchwachstum bei verschiedenen *Trifolium*-Artkeruzungen', *Züchter*, **33**, 11–17 (1960)

7 SILOW, R. A., 'A preliminary report on pollen tube growth in red clover, *T. pratense* L.', *Bull Welsh Pl. Breed. Stn, Ser. S*, **12**, 228–40 (1931)

Genetic and Environmental Variation of Pseudo-Self-Compatibility*

R. C. Leffel, *Crops Research Division, Agricultural Research Service, U.S.D.A., Beltsville, Maryland, U.S.A.*

Self-incompatibility is defined by Brewbaker [2] as the inability of a plant, producing functional male and female gametes, to set seed upon self-pollination. The gametophytic, one-locus incompatibility system is recorded for more than 60 Angiosperm families, including the important *Trifolium* species, red, white and alsike clovers (*T. pratense* L., *T. repens* L., and *T. hybridum* L.). In this system of incompatibility the mating type of the pollen grains is determined gametophytically; the numerous alleles at the locus designated as S have individual action in the style. Haploid pollen cannot effect fertilisation in a diploid pistil bearing the same allele; inhibition of incompatible pollen tubes is in the style. Thus, theoretically, a diploid plant with this system is always self-incompatible. The infrequent exception (i.e. self-seed set by a normally self-incompatible plant) was termed pseudo-self-compatibility (PSC).

It is now apparent that species with gametophytic one-locus type of incompatibility vary in the intensity of the incompatibility reaction, and that PSC may be caused by many factors, classified as environmental, artificial and genetic by Pandey [7]. The upper range of PSC may approximate that of self-compatibility; the inability to distinguish between PSC and self-compatibility prevents interpretation of the genetics of PSC. Denward [5] concluded that PSC is influenced by: (1) environment, especially in species with a 'weak' incompatibility system, (2) modifier genes randomly segregating, (3) variance among S alleles in ability to allow the pollen tubes to grow in incompatible pistils, (4) mutations, spontaneous or induced, (5) polyploidy, and (6) interspecific hybridisation and

* Abstract.

313

disturbance of the polygenic basis of the incompatibility reaction.

A review of the literature reveals that PSC may be caused by the external environmental factors of temperature, soil moisture, reduced vigour, length of light and dark periods, and by internal factors of bud pollination, age of flower or plant, and end-season and seasonal self-compatibilities. Artificial causes of PSC include the manipulation of the washed stigma; wounding, shaving or removal of stigma; grafting of plant parts; pollination density, whether sparsely or abundantly applied; emasculation; radiation of style with X-rays or ultra-violet light; and the transfer of stigmatic substances upon germination of pollen on foreign stigma. Chemicals affecting PSC include boric acid, naphthalene acetamide and colchicine. Genetic explanations of PSC include variation among the S alleles in their control of pollen tube growth; major genes and polygenes other than the S factors; polyploidy, the mutual weakening effect in heterogenic pollen; revertible and permanent mutations at the S locus; centric chromosome fragments containing the second S allele within the pollen grain; and ecotype or genotype of the species concerned. Additional suggested causes of PSC include contamination, virus, chimera and cytoplasmic factors.

We began our studies on PSC at the University of Maryland in 1960; most of this work was published by Leffel [6], Cohen and Leffel [3, 4] and Brandon and Leffel [1]. We concluded that PSC was a clonal characteristic, transmitted from parent clones to F_1 progenies, and affected greatly by temperature. The S allele genotypes did not appear in expected ratios in some I_1 and F_1 progenies; we encountered higher levels of PSC within I_1 progenies and within F_1 crosses between I_0 clones high in PSC than we found among I_0 clones. We thought the explanation of PSC in red and white clover by a theory of mutation at the S locus (admittedly always possible to some degree) to be unrealistic. We questioned the role of the pistil as a 'perfect sieve for mutations', and considered selective fertilisation, modifier genes and the disturbance upon enforced inbreeding of a polygenic system normally supporting outcrossing as more reasonable explanations of PSC with these insect-cross-pollinated species.

In addition to academic interests, we have very practical interests in genetic incompatibility and PSC in *Trifolium* species. As suggested many years ago, it may be possible to use the S allele system and variance therein to produce F_1 single-cross and F_1 double-cross hybrids. We need answers to many questions before we can establish the merits of F_1 hybrids in clovers. We do not know that such a hybrid will be sufficiently superior in performance to synthetic varieties, or how practical it will be to obtain and maintain inbred lines for use in hybrids. Furthermore, we do not know whether the self-incompatible, cross-compatible features of the S allele system can be maintained in highly inbred lines. However, the role of PSC, a 'magic wand' giving PSC or self-incompatibility as desired by the plant breeder, seems quite important in the pursuit of breeding plans; thus we have continued studies on PSC of red clover at Beltsville since 1962.

We assumed that head rolling was more practical than the tripping method for self-pollinating large populations of plants. We envision self-seed set achieved via PSC, and we desire to convert from PSC to self-incompatibility at our whim. Since the literature includes 30 possible explanations of PSC, we investigated the effect of some of these variables upon PSC of intact red clover plants. In most of these simple and exploratory studies we sought gross differences. Choice of variables was made on the basis of facility and applicability within our laboratory.

All of our studies used clonally propagated red clover plants, affording replication of genotype, or random samples of plants of specific populations. We grew propagules or plants in 6 in pots as good rosettes, and then subjected materials to long days for floral induction and development. A well-developed plant will produce 100 or more flowering heads; for non-critical comparisons we can assume 100 florets per flowering head. We continued to find that PSC was a definite clonal characteristic and that the clone by temperature interaction was especially significant. We found no evidence that PSC was affected by plant infection with Alfalfa Mosaic Virus or with Bean Yellow Mosaic Virus, or by daily or weekly applications of naphthalene acetamide or gibberellic acid as floral and foliage sprays. Ecotypes of red clover from various parts of the world did not vary greatly for degree of PSC.

Currently, we are investigating the effect of soil moisture stress on PSC and the practicality of operating a small greenhouse at temperatures optimal for PSC. Eventually we anticipate a concentration of research on manipulation of PSC, self-incompatibility, and cross-compatibility by pollinating insects; it is difficult to envision practical seed production in red clover by other means. Perhaps sufficient variation exists in the behaviour of S alleles and associated polygenic systems in red clover to allow us to select only those S alleles and plant materials that we find satisfactory for a breeding plan.

REFERENCES

1 BRANDON, R. A. and LEFFEL, R. C., 'Pseudo-self-compatibilities of a diallel cross and sterility-allele genotypic ratios in red clover (*Trifolium pratense* L.)', *Crop Sci.*, **8**, 185–7 (1968)

2 BREWBAKER, J. L., 'Pollen cytology and self-incompatibility systems in plants', *J. Hered.*, **48**, 271–7 (1957)

3 COHEN, MAIMON M. and LEFFEL, R. C., 'Cytology of pseudo-self-compatibility in Ladino white clover, *Trifolium repens* L.', *Crop Sci.*, **3**, 430–3 (1963)

4 COHEN, MAIMON M. and LEFFEL, R. C., 'Pseudo-self-compatibility and segregation of gametophytic self-incompatibility alleles in white clover, *Trifolium repens* L.', *Crop Sci.*, **4**, 429–31 (1964)

5 DENWARD, THORE, 'The function of the incompatibility alleles in red clover (*Trifolium pratense* L.). I. The effect of grafting upon self-fertility. II. Results of crosses within inbred families. III. Changes in the S-specificity. IV. Resume', *Hereditas*, **49**, 189–202, 203–36, 285–329, 330–4 (1963)

6 LEFFEL, R. C., 'Pseudo-self-compatibility and segregation of gametophytic self-incompatibility alleles in red clover, *Trifolium pratense* L.', *Crop Sci.*, **3**, 377–80 (1963)

7 PANDEY, KAMLA KANT, 'Mutations of the self-incompatibility gene (S) and pseudo-compatibility in Angiosperms', *Lloydia*, **22**, 222–34 (1959)

Pollen Size and Incompatibility in *Nicotiana**

K. K. Pandey, *Grasslands Division, D.S.I.R., Palmerston North, New Zealand*

In a study of interspecific compatibility relationships in the genus *Nicotiana* a correlation has been detected between pollen grain size and incompatibility. Seeds of all Australian species were obtained from the Commonwealth Scientific and Industrial Research Organisation, Canberra, Australia, and seeds of South American species from U.S.D.A., and from Professor D. R. Cameron, University of California, Berkeley, U.S.A.

For pollen-size measurements a small, approximately constant amount of freshly mature pollen was stained in acetic-orcein. Thirty pollen grains were measured (diameter in μm) in each species, and an average for the species was calculated. The technique used for the study of pollen tube growth has been reported earlier [14].

Interspecific incompatibility was studied in 27 species of *Nicotiana*, including 21 self-compatible (SC) and all six self-incompatible (SI) species occurring in this genus [4]. Of the SC species, 10 are indigenous to South America and 11 to Australia [16].

Pollen size was measured in 34 species of *Nicotiana*, including 17 South American and 17 Australian species. The pollen size measurements along with the chromosome number, breeding behaviour and country of origin of each of the species are given in Table 1. The South American species can be classified into two distinct groups according to their pollen size, one having the smaller size range of 26 μm–37 μm and the other, with only one species, *N. repanda*, having the high measurement of 47 μm. The Australian species, however, form almost a continuous series from small to large pollen size ranging from 31 μm to 49 μm.

All six SI species fell in the lower range of pollen size, varying from 29 μm to 35 μm (the SI strain GM8 of otherwise SC species *N. glauca* has a pollen size of 26 μm, the lowest). In contrast, the

* Abstract.

317

Table 1. POLLEN SIZE, CHROMOSOME NUMBER AND BREEDING BEHAVIOUR OF 34 SPECIES OF *Nicotiana*

Species	Chromo-some No.	Pollen size (mean diam. in μm)	Breeding behaviour
South American:			
N. tomentosa	12	29·0	
N. bonariensis	9	30·0	
N. noctiflora	12	31·0	
N. petunioides	12	31·6	Self-incompatible
N. forgetiana	9	32·3	
N. alata	9	35·4	
N. glauca GM8	12	26·2	
N. glauca N29	12	28·8	
N. glauca C	12	31·2	
N. langsdorffii	9	29·3	
N. sylvestris	12	30·8	
N. tabacum	24	32·2	
N. paniculata	12	32·5	
N. glutinosa	12	33·4	Self-compatible
N. plumbaginifolia	10	33·7	
N. rustica	24	34·7	
N. longiflora	20	34·8	
N. corymbosa	12	37·6	
N. repanda*	24	47·3	
Australian:			
N. benthamiana	19	31·6	
N. amplexicaulis	18	32·9	
N. debneyi	24	33·0	
N. occidentalis	21	34·6	
N. velutina	16	36·7	
N. exigua	16	37·0	
N. umbratica	23	37·2	
N. maritima	16	37·9	
N. simulans	20	38·1	Self-compatible
N. cavicola	23	38·5	
N. rotundifolia	22	39·6	
N. goodspeedii	20	41·3	
N. megalosiphon	20	41·7	
N. ingulba	20	43·0	
N. rosulata	20	43·0	
N. hesperis*	21	47·4	
N. suaveolens*	16	49·1	

* These three species belong to the higher pollen size range (measurements underlined).

pollen size of SC species varies greatly, from 29 μm in *N. glauca* to 49 μm in *N. suaveolens*. Species belonging to both upper and lower ranges of pollen size occurred in both American and Australian SC species. The two Australian species with the largest pollen size, *N. hesperis* (47 μm, variable between plants of different stocks, this being the largest size) and *N. suaveolens* (49 μm), compare favourably with *N. repanda*, which has the largest pollen size (47 μm) among the South American species.

There is an association between the pollen grain size and the pollen tube width in the style, the larger pollen grains having thicker pollen tubes, and vice versa.

Pollen size is not directly related to chromosome number. Although all diploid species, whether SI or SC, fell into the lower range of pollen size, aneuploid and polyploid SC species fell into the upper or lower range. Thus, of the three tetraploid American species, two, *N. rustica* and *N. tabacum*, have a pollen size in the lower range while one, *N. repanda*, has a pollen size in the upper range. Pollen size is also not directly related to the general vigour, or size of organs. *N. longiflora*, with the longest flowers in the genus, measuring over 15 cm, has pollen grains in the lower size range. Nevertheless, the relationship between pollen size and stylar length in certain other genera is too obvious to be ignored. For example, maize, in which pollen may have to grow more than 30 cm to reach the egg, has the largest grass pollen known, about 100 μm [19].

In spite of the lack of clear-cut relationships, the following points may be emphasised regarding incompatibility and pollen size in *Nicotiana*:

1. All SI species belong to the lower pollen size range. In this range also belong the SC species, *N. langsdorffii* and *N. glauca*, which have rare SI races (polymorphic) and are believed to have become SC comparatively recently. This observation may be related to the fact that smaller pollen size [3] and self-incompatibility are both believed to be primitive characters [2, 6–9, 11, 13, 17].

2. *N. tomentosa*, *N. glauca* race GM8 and *N. langsdorffii*, the three species having the smallest pollen size, are also the species which show the most extensive interspecific incompatibility as pollen parents among the SI or polymorphically SI species. *N. repanda*, *N. hesperis* and *N. suaveolens*, the three species having the largest pollen size, are also the species which show the most extensive interspecific incompatibility as pollen parents among the SC species.

The question arises: Is the relationship between the pollen size and pollen interspecific incompatibility real? If it is, what accounts for it? Certain general observations regarding pollen size and pollen incompatibility may be made. First, the range of pollen size difference between the SC species of *Nicotiana*, which is 20 μm, is larger than that occurring in the two forms of many of heterostylous species of other genera with pronounced dimorphism in the pollen size. The latter is often less than 15 μm and rarely over 20 μm [1, 12, 18]. In Vuilleumier's recent review of heterostyly [18] no definite case is cited where pronounced pollen size dimorphism is not associated with self-incompatibility, although there are three genera, *Fagopyrum*, *Primula* and *Nymphoides*, in which pollen size dimorphism occurs with variable self-incompatibility. Therefore the suggestion that the large difference in pollen size between different species of *Nicotiana* may, in some way, be related to the interspecific compatibility behaviour is not unreasonable.

However, a difference of the same order as occurs between the SC species in *Nicotiana* may also occur between the SI species in other genera, as has been shown by Levin and Kerster [5] in the genus *Phlox*. *P. pilosa* and *P. glaberrima* are both self-incompatible. The former has a pollen diameter of 30 μm and the latter of 55 μm. Interestingly, as is to be expected on the basis of the similar large pollen size difference and the associated interspecific cross-incompatibility behaviour occurring in the SC *Nicotiana*, pollen of the larger pollen size species *P. glaberrima* is cross-incompatible on styles of the smaller pollen size species *P. pilosa*.

Another noteworthy relationship of pollen size with breeding behaviour is found in connection with the pollen/ovule index in intraspecific pollination. Pollen/ovule index is the ratio of the total number of pollen grains and the total number of ovules per flower. SC species or strains, as compared to related SI species, generally have a smaller pollen/ovule index [10]; i.e. SC species seem to need fewer pollen grains per ovule than SI species. The association of larger pollen size in SC species with lower pollen/ovule index and conversely, of smaller pollen size in SI species with higher pollen/ovule index may perhaps be understood in terms of reproductive economy. With the comparative assurance of fertilisation in SC species, the need for a high pollen/ovule index is diminished; and this may, in turn, make available more nutrients per pollen grain in the anther, thereby leading to an increase in the pollen size.

Genetically, the relationship between pollen size and interspecific incompatibility discovered in *Nicotiana* is in agreement with the general hypothesis of the S gene evolution put forward to explain the interspecific incompatibility [16]. It has been proposed that the evolution of the S complex involved the development of two forms of specificity, each governed by a separate but closely linked set of genetic elements: (1) the *primary specificity*, variability in which controls interspecific incompatibility, and (2) the *secondary specificity*, which is superimposed upon the primary specificity, and variability in which controls intraspecific incompatibility. The primary specificity developed early during the evolution of pre-angiosperms, as a means of avoiding indiscriminate fertilisation by anemophilous pollen; the secondary specificity developed early during the evolution of angiosperms, as a means of excluding self-fertilisation.

Normally, in intraspecific pollination the secondary specificity is dominant over the primary specificity in the pollen. In interspecific pollination, however, where the two parental species are sufficiently distinct polygenically, the dominance of the secondary specificity over the primary specificity is lost, with the resultant independent expression of both specificities. In an interspecific pollination, if the primary specificities expressed in the pollen are evenly matched by those present in the style, the pollination is incompatible. Compatibility occurs when pollen has an additional primary specificity not present in the style. As a result of past isolation of species, a limited variability of the primary specificity occurs in the SI species, thereby giving rise to polymorphism of self-incompatibility alleles in relation to interspecific incompatibility. Once a mutation involving a loss of the secondary specificity occurs, leading to self-compatibility, and there is a subsequent adaptation for self-pollination, the erosion of the primary specificity is accelerated. The interspecific cross-incompatibility of the pollen is thus related to the extent of the erosion of the primary specificity in the pollen [14–16].

N. repanda, *N. hesperis* and *N. suaveolens* belonging to the SC class of species, and having the largest pollen on the one hand, and *N. tomentosa*, *N. glauca* race GM8 and *N. langsdorffii* belonging to the SI or polymorphically SI class of species, and having the smallest pollen on the other, are similar in one respect. They both, in their respective classes, have supposedly suffered maximum erosion of their primary specificity in the pollen and are therefore most cross-incompatible as pollen parents.

REFERENCES

1 BAKER, H. G., 'Studies in the reproductive biology of West African Rubiaceae', *Jl W. Afr. Sci. Ass.*, **4**, 9–24 (1958)
2 BAKER H. G., 'The evolution, functioning, and breakdown of heteromorphic incompatibility systems. I. The Plumbaginaceae', *Evolution, Lancaster, Pa.*, **20**, 349–68 (1966)

3 ERDTMAN, G., *Pollen Morphology and Plant Taxonomy*, Almqvist and Wicksells, Uppsala (1952)

4 GOODSPEED, T. H., *The Genus* Nicotiana, Chronica Botanica Company, Walthum, Mass. (1954)

5 LEVIN, D. A. and KERSTER, H. W., 'Natural selection for reproductive isolation in *Phlox*', *Evolution, Lancaster, Pa.*, **21**, 679–87 (1967)

6 LEWIS, D., 'Incompatibility in relation to physiology and genetics', *Proc. 8th Int. bot. Congr.*, **9**, 124–32 (1954)

7 LEWIS, D., 'Comparative incompatibility in angiosperms and fungi', *Adv. Genet.*, **6**, 235–85 (1954)

8 LEWIS, D., 'Incompatibility and plant breeding', *Brookhaven Symp. Biol.*, **9**, 89–100 (1956)

9 LEWIS, D. and CROWE, L. K., 'Unilateral incompatibility in flowering plants', *Heredity, Lond.*, **12**, 233–56 (1958)

10 LLOYD, D. G., 'Evolution of self-compatibility and racial differentiation in *Leavenworthia* (Cruciferae)', *Contr. Gray Herb. Harv.*, **195**, 3–134 (1965)

11 LUNDQVIST, A., 'The genetics of incompatibility'. In: S. J. GEERTS (Ed.), *Genetics Today* (Proc. XI Int. Congr. Genet., 1963), 637–47, Pergamon, London (1965)

12 ORNDUFF, R., 'The origin of dioecism from heterostyly in *Nymphoides* (Menyanthaeceae)', *Evolution, Lancaster, Pa.*, **20**, 309–14 (1966)

13 PANDEY, K. K., 'Evolution of gametophytic and sporophytic systems of self-incompatibility in angiosperms', *Evolution, Lancaster, Pa.*, **14**, 98–115 (1960)

14 PANDEY, K. K., 'Elements of the S-gene Complex. I. The S_{F1} alleles in *Nicotiana*', *Genet. Res.*, **5**, 397–409 (1964)

15 PANDEY, K. K., 'Elements of the *S*-gene Complex. IV. *S*-allele polymorphism in *Nicotiana* species', *Heredity, Lond.*, **24**, 601–19 (1969)

16 PANDEY, K. K., 'Elements of the *S*-gene complex. V. Interspecific cross-compatibility relationships and theory of the evolution of the *S* complex', *Genetica*, **40**, 447–74 (1969)

17 STEBBINS, G. L., 'Self-fertilization and population variability in the higher plants', *Am. Nat.*, **91**, 337–54 (1957)

18 VUILLEUMIER, B. S., 'The origin and evolutionary development of heterostyly in the angiosperms', *Evolution, Lancaster, Pa.*, **21**, 210–26 (1967)

19 WODEHOUSE, R. P., *Pollen Grains: Their Structure, Identification and Significance in Science and Medicine*, McGraw-Hill, New York (1967)

Indexes

Author Index

325

Genera Index

331

Kigelia, 266

Lathyrus, 20, 29
Leavenworthia, 322
Lilium, 4, 6, 8–15, 18, 21–22, 24, 25, 29,
30, 32, 33, 49–51, 56, 58, 61, 81, 82,
85, 88, 89, 91, 92, 96–98, 101–106,
108–111, 119, 120, 155, 157–161,
168, 172, 177–185, 190–200, 204,
205, 209, 210, 215, 217, 221, 223,
229, 239–254, 256, 260, 267–278,
284, 285, 304, 305
Lobelia, 203, 213, 219
Lotus, 203
Lychnis, 155, 178, 179, 182, 184
Lycopersicum, 5, 159, 186, 300, 301, 307
Lycopodium, 78, 79, 96, 100, 101, 104–
106, 110, 111, 153

Malus, 143, 154, 158, 159, 161, 168
Malvaviscus, 172
Mirabilis, 87, 180
Mucor, 106, 110, 111, 219
Mus, 14, 35

Narcissus, 57, 153, 203
Nelumbo, 121, 122
Nicotiana, 28, 172, 203, 204, 209, 220,
221, 234, 236, 254, 283, 289, 299,
303, 308, 318, 319, 321
Nuphar, 57, 58, 122
Nymphoides, 320, 322

Oenothera, 87, 159, 161, 162, 174, 248,
255–261, 282–284, 287–290, 302,
304, 306, 307
Orchis, 30, 148
Ornithogalum, 154
Oxalis, 49, 50, 51, 61, 91, 92, 98

Paeonia, 6, 8, 10, 13, 14, 29, 33, 34, 36,
37, 51, 61, 159, 171, 217, 218
Parkinsonia, 83, 85, 86, 97
Pediastrum, 106, 110, 111
Pelargonium, 180
Petunia, 146, 154, 155, 157–159, 168, 172,
179, 182, 185, 210, 220, 221, 232,
233, 245, 252, 253, 265, 266, 282,
289, 290, 302, 303, 305, 307

Pharbitis, 52
Phleum, 101
Phlox, 320
Phoenix, 100, 133, 136, 148, 155
Physalis, 290, 291, 308
Picea, 100, 143
Pinus, 95, 98, 100, 101, 105, 132, 133,
136–143, 145–148, 152–154, 172,
174–176, 209, 219, 221, 222
Pisum, 20, 29
Plantago, 174
Poa, 79, 91, 96, 98
Podocarpus, 70–74
Populus, 91, 98, 101, 119, 144, 158, 168
Prasinocladus, 110
Primula, 172, 320
Prunus, 134, 283, 289
Pyrus, 135, 149, 174, 175, 176

Ranunculus, 108, 109
Raphanus, 184, 305
Rhoeo, 8, 10, 12, 17, 29
Ribes, 12
Riccardia, 97
Rosa, 136, 153

Saccharomyces, 5, 12
Salix, 98, 119
Salsola, 174
Sambucus, 101
Saponaria, 83
Selaginella, 87, 97, 106, 110, 111
Silene, 83, 85, 90–92, 96, 123, 172
Solanum, 291, 297, 299, 305, 308
Sparganium, 86
Spinacia, 193, 221
Strelitzia, 263, 264, 266
Streptocarpus, 203

Taraxacum, 154
Tasmanites, 106, 111, 113
Taxus, 100, 209, 221
Tephrosia, 159
Tetrahymena, 4, 14
Torreya, 209, 221
Toxoplasma, 69
Tradescantia, 6, 10, 14, 21, 23, 25, 30, 43,
57–61, 69, 97, 168, 170, 202, 203,
205, 209–213, 215, 217–220, 222,
230, 231, 291, 304

Subject Index